The Geoarchaeology of Lake Michigan Coastal Dunes

Environmental Research Series, Vol. 2
James A. Robertson, Series Editor

The Geoarchaeology of Lake Michigan Coastal Dunes

William A. Lovis, Alan F. Arbogast,
and G. William Monaghan

Michigan State University Press
East Lansing

Copyright © 2012 by the Michigan Department of Transportation

This research was funded through federal Enhancement Funds provided by the Federal Highway Administration and a match provided by the Michigan Department of Transportation.

∞ The paper used in this publication meets the minimum requirements of ANSI/NISO Z39.48-1992 (R 1997) (Permanence of Paper).

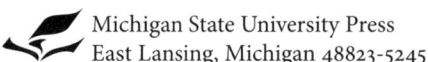

Printed and bound in the United States of America.

18 17 16 15 14 13 12   1 2 3 4 5 6 7 8 9 10

LIBRARY OF CONGRESS CATALOGING-IN-PUBLICATION DATA
Lovis, William A.
The geoarchaeology of Lake Michigan coastal dunes / William A. Lovis, Alan F. Arbogast, and G. William Monaghan.
p. cm. — (Environmental research series ; vol. 2)
Includes bibliographical references and index.
ISBN 978-1-61186-051-1 (pbk. : alk. paper) 1. Indians of North America—Michigan—Antiquities. 2. Indians of North America—Michigan, Lake—Antiquities. 3. Archaeological geology—Michigan, Lake. 4. Coastal archaeology—Michigan, Lake. 5. Sand dunes—Michigan, Lake. 6. Michigan, Lake—Antiquities. I. Arbogast, Alan F. II. Monaghan, G. William. III. Title.
E78.M6L78 2012
977.4'01—dc23
2011050507

ISBN 978-1-61186-051-1 (paper) / ⓔ ISBN 978-1-60917-348-7 (e-book)

Environmental Research Series, Vol. 2
James A. Robertson, Series Editor

Cover design by Erin Kirk New
Layout and typography by Charlie Sharp, Sharp Des!gns, Lansing, MI
Cover photo: Camp Miniwanca dunes and archaeological site.

g green press INITIATIVE Michigan State University Press is a member of the Green Press Initiative and is committed to developing and encouraging ecologically responsible publishing practices. For more information about the Green Press Initiative and the use of recycled paper in book publishing, please visit *www.greenpressinitiative.org*.

Visit Michigan State University Press at *www.msupress.org*

Contents

LIST OF FIGURES	*vii*
LIST OF TABLES	*xi*
ACKNOWLEDGMENTS	*xiii*
1. Introduction: The Geoarchaeology of Lake Michigan Coastal Dunes	*1*
2. The Archaeology of the Lake Michigan Coastal Zone: A Perspective from the Coastal Dunes	*15*
3. Coastal Dunes, Eolian Processes, and Activation-Stabilization Cycles	*37*
4. Middle and Late Holocene Lake-Level Variation, Isostatic Rebound, and Environmental Changes in the Upper Great Lakes	*59*
5. Archaeological Sites in Dune Contexts around Lake Michigan	*83*
6. Discussion and Synthesis of the Processes and Timing of Dune Formation and Archaeological Site Burial in Coastal Settings of Lake Michigan	*111*
7. The U.S. 31 Case Study: Torch Bay to South Point	*137*
APPENDIX A. Description of Methods Employed	*149*
APPENDIX B. Tables of Radiocarbon and OSL Dates	*157*
APPENDIX C. Descriptions of Sample Locales	*161*
REFERENCES	*207*
INDEX	*221*

Figures

1-1. Map of major dunefields in Michigan showing the location of the study area — 2

1-2. A small Late Woodland hearth south of Ludington (sample locale Camp Miniwanca) that has been buried beneath several meters of eolian sand within a coastal foredune complex — 3

1-3. Buried soils in a small back dune near Ludington State Park — 5

2-1. Locations of important buried or stratified archaeological sites within coastal zone eolian contexts — 18

3-1. Major dunes and dunefields in Michigan — 38

3-2. Near-surface velocity gradient of wind — 39

3-3. Method of sediment transport by the wind — 40

3-4. Cross-section of a typical sand dune — 41

3-5. Classification of dune morphological types within typical Lake Michigan shoreline setting — 44

3-6. Locations of dunes and dunefields sampled and analyzed during this study — 53

3-7. Temporal variation in dune construction in the northern half of Lake Michigan — 54

4-1. Maps of the upper Great Lakes showing isostatic rebound and major outlets — 60

4-2. Time stratigraphy for geological and cultural events during the late Wisconsin and Holocene in the upper Great Lakes — 61

4-3.	Lake-level curves for Lakes Michigan and Huron	67
4-4.	Isostatic uplift and subsidence in the upper Great Lakes	70
4-5.	Geomorphology of area between Brevort and St. Ignace	75
4-6.	Lake hydrographs for Lake Michigan adjusted for isostatic rebound in areas of northern Lake Michigan	76
4-7.	The effects of differential uplift and lake-level changes around Lakes Michigan and Huron	79
4-8.	The effects of differential uplift and lake-level changes around Lakes Michigan and Huron	80
5-1.	The Winter site location	84
5-2.	Generalized, composite stratigraphic column of the Winter site from Richner's archaeological units and as noted from the GeoProbe core	86
5-3.	The Ekdahl-Goudreau site location	89
5-4.	Generalized, composite stratigraphic column of the Ekdahl-Goudreau site from UMMA field notes and as noted from the GeoProbe core	90
5-5.	The Scott Point site location	93
5-6.	Generalized, composite stratigraphic column of the Scott Point site	94
5-7.	The Mt. McSauba site location	96
5-8.	The Mt. McSauba site sample location	97
5-9.	Generalized, composite stratigraphic column of the Mt. McSauba site	99
5-10.	Fisherman's Island State Park archaeological sites and dune sample locations	100
5-11.	Generalized composite stratigraphic column of the Solomon Seal site, Fisherman's Island State Park	102
5-12.	The Camp Miniwanca site location	103
5-13.	The Camp Miniwanca exposure showing buried hearth feature	104
5-14.	Generalized, composite stratigraphic column of the Camp Miniwanca exposure	105
6-1.	Locations of archaeological sites and dune-sampling locales around Lake Michigan	112
6-2.	Bathymetric map of Lake Michigan in northwestern Lower Michigan	118

FIGURES

6-3. Dune ages and magnitude of lake-level changes with dune activation during the middle and late Holocene — *121*

6-4. Eastport and Torch Lake areas — *126*

6-5. Zones on the Lake Michigan shoreline where groups of processes dominate to create specific types of beaches, dunes, and archaeological sites — *129*

6-6. Descriptive model of middle and late Holocene shoreline and dune development in areas of subsidence or stability along southern and southeastern shore of Lake Michigan — *131*

6-7. Descriptive model of middle and late Holocene shoreline and dune development in areas of rapid uplift on northern and northeastern shore of Lake Michigan — *132*

6-8. Diagrammatic model of the formation, burial, and stratification of archaeological sites in areas of rapid uplift on the northern northeastern shore of Lake Michigan — *134*

7-1. Locations of the four U.S. 31 case study segments — *139*

7-2. Geological and cultural features and locales of interest within U.S. 31 case study Segment 1 — *141*

7-3. Geological and cultural features and locales of interest within U.S. 31 case study Segment 2 — *143*

7-4. Geological and cultural features and locales of interest within U.S. 31 case study Segment 3 — *145*

7-5. Geological and cultural features and locales of interest within U.S. 31 case study Segment 4 — *147*

C-1. The Summer Island site location — *162*

C-2. Generalized, composite stratigraphic column of the Summer Island site — *163*

C-3. The Manistique Sand Pit location — *167*

C-4. The two sample locales at the Manistique Sand Pit locale — *167*

C-5. Generalized, composite stratigraphic column of the Manistique Sand Pit — *167*

C-6. The Moran and Round Lake dunefield sample locales — *169*

C-7. Detailed maps of the Moran Sand Pit and Round Lake dune sample locales — *169*

C-8. The Moran dunefield sites showing the Moran Sand Pit and Round Lake sample locales — *170*

C-9. Generalized, composite stratigraphic column of the Moran Sand Pit and Round Lake dune samples locales — *170*

FIGURES

C-10.	Wilderness State Park sample locations	*171*
C-11.	Wilderness State Park sample exposures	*172*
C-12.	Generalized, composite stratigraphic columns of the Wilderness State Park exposures	*172*
C-13.	The Wycamp Creek site location	*175*
C-14.	Generalized, composite stratigraphic column of the Wycamp Creek site	*177*
C-15.	The Portage site location	*181*
C-16.	Generalized, composite stratigraphic column of the Portage site	*181*
C-17.	Fisherman's Island State Park archaeological sites and dune sample locations	*185*
C-18.	Fisherman's Island transect across increasingly younger dune ridges	*185*
C-19.	Generalized, composite stratigraphic column of the O'Neil site, Fisherman's Island State Park showing probable correlation of archaeological and sedimentary units	*186*
C-20.	The Antrim Creek Natural Area sample location	*192*
C-21.	A dune swale that includes an ephemeral buried cultural horizon at the Antrim Creek Natural Area	*192*
C-22.	Sample transect through dune ridges between the Nipissing wave terrace and the Lake Michigan beach	*193*
C-23.	Sample locations near Torch Bay and Eastport	*194*
C-24.	GeoProbe sampling a dune in the Torch Bay Nature Preserve	*194*
C-25.	Sample transect through the Torch Bay Nature Preserve between the Nipissing shoreline and Lake Michigan	*195*
C-26.	The Porter Creek site location	*198*
C-27.	Generalized, composite stratigraphic column of the Porter Creek site	*199*
C-28.	Ludington State Park sample locations, Ludington Sand Quarry Site 1 and Ludington Sand Quarry Site 2	*201*
C-29.	Ludington State Park exposures in Ludington Sand Quarry Site 1 and Ludington Sand Quarry Site 2	*202*
C-30.	Generalized, composite stratigraphic column of Ludington Sand Quarry Site 1	*203*
C-31.	Stratigraphy of Ludington Sand Quarry Site 2	*203*

Tables

2-1.	Michigan Prehistory: Time Periods, Ages, and Key Events	*19*
5-1.	Correlation of 1972 Site Stratigraphy with the Current Study	*85*
5-2.	Camp Miniwanca artifact assemblage, Locales A and B, 26 October and 16 November 2007	*106*
A-1.	Conventions for Reporting the Ages of Samples from ^{14}C and Optical (OSL) Dating	*151*
B-1.	^{14}C Ages	*157*
B-2.	OSL Ages	*159*

Acknowledgments

As is normally the case with a multiyear project of some magnitude, a large number of individuals and institutions contributed in various degrees and fashion to its successful completion. The Dune Activation, Cycling, and Site Taphonomy Project was made possible by funding from the Michigan Department of Transportation (MDOT) and Federal Highway Administration (FHWA) through the Safe, Accountable, Flexible, Efficient Transportation Equity Act: A Legacy for Users (SAFETEA-LU) . SAFETEA-LU funding was largely brought about through the efforts of Dr. David Ruggles, then MDOT Staff Archaeologist in the Environmental Section at MDOT, and MDOT Enhancement Grant staff. With the departure of Dr. Ruggles from MDOT in 2008 the project came under the sequential management of Mr. Paul McAlister and Ms. Christine Stephenson. Ultimately the project came under the oversight of MDOT Staff Archaeologist, Dr. James A. Robertson, who assisted in bringing the project to completion. Given the delays incurred by the project for a variety of reasons, we appreciate the patience of MDOT in providing us with appropriate extensions to complete what we believe is a high-level and practical piece of research.

Our access to pertinent records and collections central to the conduct of our research was facilitated by a variety of individuals and institutions. Dr. Marla Buckmaster, Northern Michigan University, graciously provided unpublished information on the Scott Point site, including photographs, field notes, stratigraphic sections, and collections access, prior to transfer of these materials to the Office of the State Archaeologist. Drs. Claire McHale Milner, Terrance Martin, and Michael Hambacher likewise provided unpublished information and data from the site. Our understanding of the Winter site was assisted by access to the ceramic collections at Western Michigan University, provided by Dr. William Cremin and Mr. Michael Fournier. Mr. Jeffrey Richner, Midwest Archaeological Center, National Park Service, and excavator of the Winter site, provided us with copies of pertinent reports and commentary on their interpretations, as well as assisting us with information on National Park Service activities on the Lake Michigan coastal zone. The records housed at the Great Lakes Range, Museum of Anthropology, University of Michigan were kindly made available to us by Dr. John O'Shea. This included survey reports of early work on the Lake Michigan coastal zone, as well as documents on the Summer Island site, and field notes, photographs, and maps from the Ekdahl-Goudreau site. Requests for information from the Office of the State Archaeologist, Michigan Department of Arts, Libraries and History, were always accommodated by State Archaeologist Dr. John Halsey and Assistant State Archaeologist Barbara Mead. We thank all of these individuals for their assistance, without which we could not have brought this project to successful completion.

The research for this project took place at two institutions. At Michigan State University this included the Consortium for Archaeological Research, especially the Department of Anthropology and its chair, Dr. Robert Hitchcock, and the MSU Museum and its Director Emeritus, Dr. C. Kurt Dewhurst. Likewise, the Department of Geography and department chair, Dr. Richard Groop, provided facilities particularly for graphics and Optically Stimulated Luminescence (OSL) sample processing. At Indiana University space and facilities were provided by the Glenn A. Black Laboratory of Archaeology and then Director Dr. Christopher S. Peebles. To all of these individuals and both institutions we extend our heartfelt gratitude.

Our ability to access various field locations both archaeological and geological was the product of many hours of effort by a very large number of people and institutions interested in our work. Permission to work at the Winter site was provided by the current property owners, Frank and Sandra Sakowski. The Portfleets of Goudreau's Harbor, particular Dr. Diane Portfleet, kindly allowed access to the Ekdhal-Goudreau site. The site at Camp Miniwanca was brought to our attention by Dr. Edward Hansen of Hope College, and permission to access the site on two occasions was provided through the camp Director, Mr. Thomas Moore. The Moran dunefield is located within the Hiawatha National Forest. Mr. John Franzen and Mr. Eric Drake facilitated our negotiation of the USDA permitting system, for which we cannot thank them sufficiently. St. Ignace District Ranger Steve Christiansen, Special Use Permit Coordinator Mr. James Phillips, EPA Coordinator Ms. Lynne Hyslop, and Realty Specialist Ms. Susan Alexander reviewed and approved our requests. Permission to undertake OSL sampling at the Torch Bay Township Hall dune and the Torch Bay Nature Preserve was obtained through the good offices of Township Supervisor Mr. Robert Spencer and the Torch Bay Township Board of Directors. Likewise, our ability to perform geomorphological work at the Antrim Creek Natural Area was made possible by permission of the Antrim County Parks Commission.

Properties under management by the State of Michigan, particularly the Department of Natural Resources, required a variety of permit clearances. At the Michigan Department of Transportation Messrs. David W. Schuen, Endangered Species Specialist, and Jeremie Wilson, Environmental Clearance Coordinator, assisted with the permit process. Critical Dunes Environmental Quality Analyst Mr. Matthew Warner at the Department of Environmental Quality reviewed and approved our plans to work on properties designated as part of the Critical Dunes Inventory. Umbrella permits were reviewed and approved through the good offices of Mr. Glenn Palmgren and Ms. Kris Tunney of the Parks and Recreation

Division. Monitoring of our attenuated work at Scott Point was conducted by Ms. Erynn Call, Naturalist at the Sault Ste. Marie Office, and we also received assistance from Conservation Officer Kellie Nightlinger. The extensive work conducted at Fisherman's Island State Park was approved and facilitated by Young and Fisherman's Island State Park Manager Ms. Sue Rose, and Fisherman's Island State Park Supervisor Ms. Jeanne Kokx. Without the coordinated efforts of all of these individuals it is safe to say that we could never have brought our fieldwork to fruition.

Dating of our OSL samples was undertaken by two laboratories. The initial set of field samples was sent to Sheffield University, UK, where they were processed at the Sheffield Centre for International Drylands Research laboratories by Dr. Mark D. Bateman. Notably, a number of our samples did not survive international security inspections intact. The bulk of our OSL samples were processed by Professor Steven L. Forman and the Luminescence Dating Research Laboratory, Department of Earth and Environmental Sciences, at University of Illinois–Chicago, who has also contributed to this book in a substantive fashion. We have all benefited greatly from our conversations. Our attempts to apply focused geophysical prospection to our work were coordinated by Dr. Remke Van Dam, Department of Geological Sciences, Michigan State University. Dr. Van Dam conducted a series of transects at the Antrim Creek Natural Area, and joined us during our brief foray at the Scott Point site. His assistance is tremendously appreciated. Carbonized plant remains from the Camp Miniwanca locale were kindly identified by Dr. Frank Telewski, Department of Biology and Beal Botanical Garden, and Dr. Catherine Yansa, Department of Geography, Michigan State University. We most certainly would not have been able to bring this project to its current state without the labor of our research assistants, Ms. Marieka Brouwer and Ms. Jennifer Holmstadt. Ms. Brouwer contributed in a major fashion to our summary of the Summer Island site, and along with Mr. Brad Blumer joined us in fieldwork at the Camp Miniwanca locale. Ms. Holmstadt's MA research forms the core of our discussion of the Antrim Creek Natural Area sample locale. Lastly, the presentation of our research was brought to a higher level by our external reviewer, Dr. Paul Hanson, whose substantive and editorial insights enhanced the final product in a variety of ways, and which we hope are evident to him. To all of the above we offer our heartfelt thanks.

Lastly, the people who deserve our warmest thanks and appreciation are our families, who suffered our periodic absences and anxieties for, in this instance, several years while we brought the project to fruition. Without their forbearance and support we could not have successfully accomplished what we have produced.

Given the large number of people and institutions who have assisted us in our research over the past three years we certainly hope we haven't inadvertently omitted anyone. Despite the abundant assistance provided to us, we take full responsibility for any errors of substance or interpretation that might be present.

CHAPTER ONE

Introduction

The Geoarchaeology of Lake Michigan Coastal Dunes

Coastal sand dunes are common along the eastern shore of Lake Michigan (figure 1-1) and likely represent the largest body of freshwater dunes in the world (Peterson and Dersch 1981). They are certainly among the most prominent natural features of the modern Michigan coastal landscape, and contain a unique flora and fauna. Compared to other landscapes the dunes are highly dynamic, changing their shape and configuration over both the short and the long run. Moreover, coastal dunes are attractive to contemporary populations, who use them regularly for a variety of recreational activities. Government agencies from the municipal to the federal level have placed large tracts in the public trust, make them publicly accessible, interpret them for the larger community, and responsibly manage them for future generations. The inherent sensitivity of dunes to small-scale human and natural perturbances, however, renders them particularly susceptible to rapid changes—some desirable, and some not. This sensitivity and susceptibility to change requires a clear knowledge of the processes involved in coastal dune formation, their long-term dynamics, and the factors that influence their activation and stabilization. Our multifaceted study of the evolution of Lake Michigan coastal dunes and the processes responsible for their dynamism, although primarily directed at the way in which archaeological sites are buried and preserved along Lake Michigan, also provides a wealth of information for the understanding and managing of dunes throughout the Great Lakes.

Use of the coastal dunes has long precedent and predates European contact. The Lake Michigan coastal dunes have been used regularly over several millennia by the native people indigenous to the Great Lakes region. Our bibliography reveals that precontact use of dune environments is well documented by many individual archaeological sites along the coastal zone, especially in the northern part of the Lake Michigan basin (figure 1-2). This cumulative research documents the timing and types of human use at specific locations, often in buried, stratified, or layered soil deposits that also allow interpretations of stability and change over

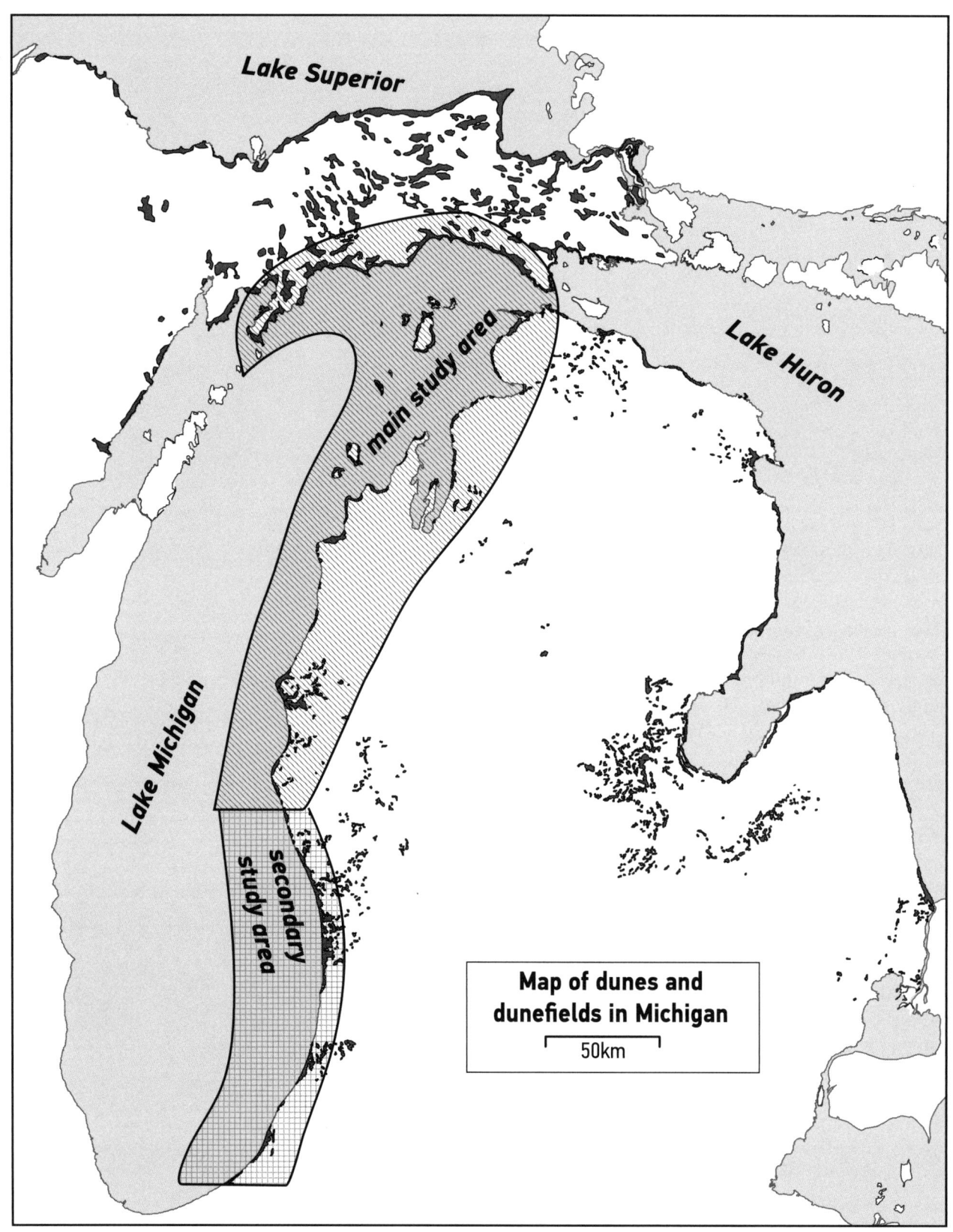

1-1. Map of major dunefields in Michigan showing the location of the study area.

extended periods of time. The chronology of some, but not all, archaeological sites is known through the use of both conventional and small-sample radiocarbon dating techniques. Part of this research has been reported in detail in journals, chapters, and monographs, while other sites have neither been formally analyzed nor published.

However, despite the half century or more during which intensive research has been undertaken by multiple institutions, no attempt has been made to synthesize the abundant data from the many buried, stratified, archaeological sites found within dunes along the coastal zone of Lake Michigan. The problem is one of scale. As data related to individual sites, and to sets of sites within subregions or time periods, have accumulated, the focus of archaeological research and related earth science disciplines has shifted to more inclusive perspectives. Among the several goals of the research presented here is to integrate pertinent components of the related archaeological, geological, and dune geomorphological research into a systematic framework for use by archaeologists, earth scientists, interpreters, land managers, the general public, and especially MDOT environmental scientists and transportation planners.

Intimately linked to many of the individual archaeological studies, which have often been undertaken without the participation of geologists, geoarchaeologists, or physical geographers, is analysis of the ongoing processes by which local dunes formed and changed over time, sometimes remaining stable and at other times

1-2. A small Late Woodland hearth (^{14}C age ca. 730 cal BP; appendix B) south of Ludington (sample locale Camp Miniwanca; see chapter 5) that has been buried beneath several meters of eolian sand within a coastal foredune complex.

eroding or building. While not completely understood, the processes of local-level change are often related to long-term regional variation in climate (e.g., precipitation and temperature), as well as to changing levels of the postglacial Great Lakes basins over the past 12,000 or more years. These factors, precipitation and temperature, can affect local shoreline conditions by altering other environmental conditions, such as macro-level changes in groundwater availability, vegetation destruction, and bluff or shoreline erosion.

Furthermore, any of these individual potential catalysts or agents of change may be cyclically exacerbated or ameliorated by changes in the physical elevation of coastal zones related to postglacial isostatic rebound (isostatic rebound is the process by which the earth's crust, depressed by the weight of glacial ice, readjusts by steadily rising after the ice load is removed; see chapter 4). Because rebound varies across the Great Lakes region, the northernmost areas continue to be subject to the relatively high rates of postglacial uplift compared to more southern areas.

In the Lake Michigan basin the area in which such uplift is rapid and greatest occurs north of a line extending from south of Muskegon to the Door Peninsula in Wisconsin. The shoreline south of this line is rising only minimally, or may even be undergoing subsidence. In general, isostatically induced rises in elevation are greater to the north than to the south, mainly because southern areas became free of glacial ice earlier than northern areas. As a result of isostatic rebound, former coastal areas are not only many tens of meters higher today than they were millennia ago, but are also commonly removed far inland of modern coastal zones and environments. As a result of greater uplift rates, early (older) beaches formed on former, now uplifted, coastlines in the northern parts of the Great Lakes basins are higher and further away from the modern coastline. In the southern parts of the basin, on the other hand, where uplift is minimal, beaches have generally not been significantly raised and removed inland, and remain on or near the modern shoreline. As a result, these former shorelines have been continually inundated and eroded throughout the middle and late Holocene to such an extent that they are often absent. The spatial differences in preservation of coastal shorelines and related dunes are accentuated by temporal variations in climate and lake levels.

Just the few regional processes noted above show that understanding the formation processes of dunes and their periodicity or activity cycling over extended periods of time and space, not to mention how or when archaeological sites are incorporated into the dunes (figure 1-3), requires close integration of data and theory from both archaeologists and earth scientists. Thus, the second major goal of this volume is to synthesize the available information on the timing and nature of change in coastal dunes over extended periods of time, and to relate evident patterns of change in dunes to larger-scale paleoenvironmental changes. We accomplish this by synthesizing data from existing geological studies of Lake Michigan dunes, incorporating information from well-dated and stratified dune archaeological sites, and supplementing deficiencies in dated contexts through exploration of selected new locales and datable materials.

The third major goal of this study is to test and refine current models of dune formation, including the chronology and periodicity of such formation. Standing geological models of dune activation, which are presented in great detail in chapter 3, have neither sufficiently availed themselves of, nor appropriately incorporated the data from, buried and stratified dune archaeological sites. Existing geological/

1-3. Buried soils in a small back dune near Ludington State Park. Stacked, thin, discontinuous paleosols are labeled on diagram. Note that beach gravel forms an erosion-resistant base for the dunes.

geomorphological explanations have presented models of varying periodicity and amplitude through which to better understand cyclic dune activation. These models suggest a relationship between lake-level fluctuations and cycles of dune activity. In addition, these studies imply that there are, potentially, century or multicentury "lags" in dune activation following increases in temperature or decreases in precipitation, or changes in the elevation of the lake basins. In other words, dunes do not necessarily respond rapidly to other changes in the environment—while clearly coupled to environmental conditions, they may have an extended "response" time.

The validity and utility of applying a single overarching or omnibus model of dune activation and cycling to the Lake Michigan coastal zone are also questionable. Again, based largely on the geological information, current models of dune activation suggest that the dunes on the north shore of Lake Michigan, the northeast shoreline, and the southeast shoreline may all behave somewhat differently over extended periods of time. This would suggest that a single model of dune activation may not readily explain the observable long-term changes in coastal dunes. Our research demonstrates that significant temporal and spatial variations do exist across the Lake Michigan basin, and are related to many different local and regional variables. By implication, these observations also imply that archaeological site burial (or erosion), stratification, and location may also vary in a similar regional or temporal fashion, which has important implications for developing site location and subsistence models for various cultural periods. We address these multiple issues directly, and show how our work has import for understanding issues of differential site preservation and interpretation of past human adaptations, in addition to contemporary management practice.

As we hope is evident by this juncture, the several goals of our research cannot be properly achieved through the perspective of a single discipline. Rather, as we have both articulated and demonstrated in past research (e.g., Lovis et al. 2005; Monaghan and Lovis 2005; Monaghan et al. 2006), we take the position that the combined views of geology, physical geography, geoarchaeology, and archaeology are necessary to properly understand the complexities of the past natural environment, how humans chose to use that environment, and how the evidence of human use is preserved, altered, or eradicated from the archaeological record. This is a fundamentally multidisciplinary task that necessitates bridging the natural and social sciences, focusing on a discrete set of common goals and questions, and developing the common research design that will allow our questions to be answered.

The Dunefield Activation and Evolution Model

Our research into site formation and preservation was initially conceived around and conditioned by the void between what the earth sciences knew about dune formation processes, and the scale and direction from which archaeologists approached similar problems. Current models of dune activation and evolution along Lake Michigan have been undertaken almost exclusively by geologists and physical geographers, but largely in the absence of the types of detailed information available from archaeological sites. Archaeological sites provide detailed information on the activation and stabilization cycles from both absolute and relative dating of cultural contexts. For example, both the age of a buried horizon, obtained from ^{14}C dating, and its duration, represented by the various cultural periods found within it, indicate the length of time a dune has been stable. Thus, social scientists and earth scientists, who have largely worked in isolation from one another, focus on different aspects of related problems of coastal landform evolution and the human use of eolian features. Our research was specifically designed to redress this interdisciplinary isolation by coalescing the information available to both schools of researchers into an integrated model of dune taphonomy and a more refined understanding of archaeological site preservation and human use.

Geologists, geomorphologists, and geoarchaeologists often distinguish between coastal and interior dunes. Although this distinction is primarily to the location of such dune types, it is also related to subsequent forces that might act on their evolution. Interior dunes are often the result of deposition during earlier stages of deglaciation, related to sediment outwash during glacial recession that often results in extensive sand sheets mantling large areas. These same areas, when associated with later processes of lake basin evolution including higher-water stands in the peri- and postglacial Great Lakes, might be subsequently mantled by additional lacustrine sand or further modified by coastal formation processes. Coastal dunes, on the other hand, most often derive from the activation of lacustrine or littoral sand deposits associated with higher-water stands of the conjoined Michigan/Huron basin.

Recent research on dune formation and activation cycles has resulted in a general model that links dune activation to climate change and precipitation, as well as to the partially coupled hydrological regime of the Great Lakes basins (e.g., Arbogast et al. 2002; Loope and Arbogast 2000). This research suggests that eolian activation in the past has been cyclic, pulsating, or periodic rather than a continuous process, and is related to regional- or subregional-scale secular changes in the environment. Thus, even though different classes of dunes can form through multiple and sometimes quite different processes, they are nevertheless linked by larger-scale regional cycles of environmental change. Dunes within the interior of Michigan, for example, are directly affected by climate change and were commonly activated during periods of increased drought and vegetation loss (Arbogast et al. 2002). Loope and Arbogast (2000), who specifically directed their research at the evolution of coastal dunes, show that many dunes in these landscapes activate during periods of high water (transgressions) and stabilize during low-water intervals (regressions). These observations suggest that coastal dunes respond to regional climate fluctuations indirectly through their impact on hydrological cycles. This relationship is contrary to the dominant, previously existing, qualitative model (e.g., Dorr and Eschman 1970) that linked the growth of most coastal dunes to processes associated with the Lake Nipissing high-water phase of the ancestral Great Lakes (ca. 5,000 to 4,000 years ago). Whether coastal or interior, however, dunes may not necessarily respond to such environmental changes quickly, which may result in century-scale, or even greater, lags in the eolian activation. In this research, we take the position that buried and stratified archaeological occupations represent stable surfaces within coastal dunes. Therefore, the analysis of these occupations can assist in understanding the age and duration of lag intervals present within dunes.

The chronology and activation cycling of both coastal and interior dunes have been addressed in several recent works for different subregions across the Great Lakes region. We will embed our present research on archaeological site formation processes into this larger body of research. Our current work will also provide data to refine, or at least partially test, standing models. For example, interior dune activation cycles have been extensively studied in low lake-plain areas near the margins of Saginaw Bay, which resulted in a preliminary understanding of interior dune dynamics (Arbogast et al. 2002). Chronometric dating of basal peats buried by eolian sand deposited when Lake Grassmere dropped in elevation to that associated with Lake Algonquin (Monaghan and Lovis 2005) suggests dune formation in this area began circa 12,000 years ago. This age and stratigraphic relationship also suggests that dunes probably formed when the broad areas of the Saginaw lake plain were exposed, although it might also relate to cool/dry environmental conditions that are commonly associated with the Younger Dryas climatic interval (Alley 2000; Alley et al. 1993; Taylor et al. 1997). Based on Optically Stimulated Luminescence (OSL) dating from various sampled exposures in the region, sand mobilization apparently continued until circa 10,000 years ago (i.e., for ca. 2,000 years), which suggests a relatively long interval in which dry and windy conditions prevailed (Arbogast et al. 2002). Sand mobilization may also have been driven by groundwater declines related to the drop in the Michigan/Huron basin elevation known as Lake Stanley (Larson and Schaetzl 2001; Monaghan and Lovis 2005). This study of interior dunes, particularly the relationship between Lake Stanley and the interior activation cycle, reveals that a significant geomorphic link may exist between the postglacial dynamics of the Michigan/Huron lake basin and the interior landscape, beyond that of fluvial system impacts resulting from base-level fluctuations noted by Monaghan and Lovis (2005).

The landscape in the Saginaw region essentially stabilized after this late Wisconsin and early Holocene interval of initial and intensive dune formation. Combined OSL and ^{14}C dating programs (Arbogast et al. 2002), however, reveals a subsequent, localized reactivation of dunes in the area. Key cycles of subsequent sand movement occurred at circa 8,000, 6,000, 1,100, and 200 years ago, revealing that the landscape is much more sensitive to such change than previously believed. To a degree, the cycles at circa 8,000 and 6,000 years ago are expectable given their association with cyclic episodes of Hypsithermal warming during this period. Of particular interest, however, is the activation cycle circa 1,100 years ago, which seems to occur coincident with the Medieval Warm Period (ca. 1,000–700 years ago); and also shortly followed by recorded rises in the elevation of the Michigan/Huron basin (Monaghan and Lovis 2005). While the association with the Medieval Warm Period is tantalizing, this late activation cycle may well reflect multiple causes.

In contrast, coastal dunes are treated as a somewhat separate, but certainly linked, phenomenon because they are subject to fundamentally different processes than interior dunes. Bluff destabilization and changes in sand supply resulting from macro-, meso-, and microscale fluctuations of lake levels, and consequent sand deposition, may result in more dynamic and more variable coastal physiographical and depositional settings. This is well reflected by dated episodes of dune activation along the east-central coastline of Lake Michigan between Manistee and Grand Haven. A clear north-to-south trend in the timing of dune activation can be discerned: at the Nordhouse Dunes between 4,900 and 4,500 years ago, between 4,300 and 3,900 years ago at the Jackson and Nugent Quarry sites, and between 3,300 and 2,900 years ago at the Rosy Mound site (Arbogast and Loope 1999). The time of Nordhouse Dune activation appears to be related to the high-water Nipissing lake stage, whereas data from the latter three sites suggest that the later stages of dune growth may have occurred from a combination of microscale lake-level fluctuation and lake terrace/bluff destabilization.

Microscale high-level fluctuations of the Michigan basin on the order of about 150 years become particularly important in explaining the formation and activation of perched coastal dunes during the past 1,500 years. This phenomenon has been attributed to low sand supply by Loope and Arbogast (2000), who argue that the supply of eolian sand decreases during lower-level cyclic basin fluctuations, which results in stability and soil formation on dune surfaces. These soils are subsequently buried when sand supply increases during lake-level peaks. The weakly developed soils suggest to Loope and Arbogast (2000) that the cycles lasted approximately 150 years. Large coastal dunes along the southeastern shore of Lake Michigan near Holland (Arbogast et al. 2002) have also been studied with the goal of reconstructing the cycling of activation processes. Much like the other data presented, evidence from this area demonstrates eolian sand deposition as early as circa 5,500 years ago, during the Lake Nipissing high-water stage, but also correlates with the waning periods of Hypsithermal warming. Episodic dune growth continued until about 2,500 years ago and resulted in the formation of several weakly developed, buried soils stacked within the eolian sand sequence. The dunes subsequently stabilized until about 1,000 years ago. During this relatively long period of stability a weakly developed Spodosol formed in most dunes. This activity increased in intensity during the past 1,000–500 years, presumably as a result of lakeshore recession and wave erosion of the modern bluff base (Arbogast et al. 2002).

Unless each of these necessary factors occurs at the same time, then regardless of sand volume or wind energy, dunes will not form or at best will only occur as localized sand sheets. Although this research is mainly focused on the age and construction of the more prominent coastal dune, sand-sheet deposition may also have import for understanding stratification of cultural deposits as well as some of the more fine-scale formation processes within archaeological sites found in dune contexts.

Ultimately, a sand dune can only form when three major factors occur simultaneously: a supply of sand that is erodible, enough replenishment of the sand to maintain dune construction, and enough wind to move sand into a dune. Previous research in the Great Lakes region indicates that at the local, most basic level, the activation or stabilization of coastal dunes, the category that forms the focus of this research, was affected by three primary variables: climate change, including warm and dry intervals; fluctuations of lake level at the macro-, meso-, and microscales; and available sand supply. In addition to factors that control the formation of individual dunes and dunefields, a broader, subregional variation overlies local processes in the coastal zone. This variation is most apparent in north-south differences in the timing, preservation potential, and primacy of different processes in dune activation cycling. Some of this variability is plainly related to the effects of isostatic rebound, but some may also be linked to local, idiosyncratic differences in sand supply, shoreline configuration, or other geomorphology variations. From the standpoint of archaeological sites and their burial, we are particularly interested in the fashion by which subregional variation might affect both the formation and the destruction of buried archaeological sites along different segments of the coastal zone.

The Dune Activation Research Design

Our research on dune activation, cycling, and human use required the coordination of several categories of information drawn from multiple disciplines. Certain categories of information, for example, already existed in published and unpublished documents. Many of these documents and reports, however, had not been synthesized in a fashion that allowed their direct application to our problem. In addition, other categories of information had to be newly generated and incorporated into the existing data in a fashion that allowed their focused application. Thus, to achieve these ends the larger project was partitioned into several archival, field, and analytic stages. Our field strategy was developed almost from whole cloth by the information gleaned from our archival research.

Dune-Based Stratified Archaeological Sites

The preceding historical overview is based on better than a half century of published and unpublished information on dune-based archaeological sites. The published data are available in a variety of sources, including journals such as *Michigan Archaeologist, Wisconsin Archeologist, Papers of the Michigan Academy of Science, Arts, and Letters, Historical Archaeology*, and *Fieldiana: Anthropology*, monograph series such as the Anthropological Papers of the University of Michigan Museum of Anthropology, and doctoral dissertations and master's theses from multiple institutions available through University Microfilms International. Books containing synthetic information such as *Retrieving Michigan's Buried Past* (Halsey 1999), *The Archaeology of Michigan* (Fitting 1970, 1975), and *Great Lakes Archaeology* (Mason 1981) were significant sources of compendium data and bibliographic citations to both published and unpublished works. Of particular initial value was the early coastal zone study of Peebles and Black (1976).

Perhaps more important to our work, however, was the unpublished information available in the so-called "gray literature": contract- or compliance-based studies either by private firms or undertaken internally for federal and state agencies. To properly mine this information, it was necessary to contact the National Park Service Midwest Archaeological Center, the United States Department of Agriculture Hiawatha, Ottawa, and Huron-Manistee National Forests, and the State of Michigan Office of the State Archaeologist, as well as personnel from several private archaeological contract firms. In addition, educational institutions including Northern Michigan University, Western Michigan University, and the University of Michigan were approached about unpublished data that might be available.

Our research revealed that the majority of stratified dune-based archaeological sites have some kind of compendium report, albeit containing greater or lesser detail. The Ekdahl-Goudreau site, for example, was test excavated by the University of Michigan Museum of Anthropology (UMMA) in the 1960s, but no complete summary report is available. UMMA through the good offices of Curator John O'Shea made available the pertinent information on this site. The Scott Point site was partially excavated by Northern Michigan University in the 1970s, and analyses of the data are ongoing. Emeritus Professor Marla Buckmaster at Northern Michigan University made unpublished data, field drawings, and photographs available prior to transferring the collections and field data to the Office of the State Archaeologist. This access proved particularly important given our inability to perform fieldwork at the site. Likewise, several sites excavated by Michigan State University did not have complete reports available, including the Wycamp Creek site and the Mt. McSauba site. Field notes and draft reports were accessed through the MSU Museum to compile site summaries.

The resulting inventory of stratified dune-based archaeological sites in the Lake Michigan coastal zone of Michigan and adjacent parts of Wisconsin is impressive, a total of 15 sites. Summary data on each of these sites is presented in a partially standardized format in chapter 5 and appendix C of this book. These data include site locations; descriptions of the stratified dune deposits; associated absolute dates either OSL, standard ^{14}C, or Accelerator Mass Spectrometer (AMS), the latter largely relying on carbonized ceramic sherd residues; and a summary of pertinent occupation information, with topographic and landform position including elevations and, as available, assemblage composition, season of occupation, and site functions. These data will now be readily available to future researchers.

As mentioned, a central focus of our research into dune activation, cycling, and human use was assessing the timing of discrete formation and occupation events. A substantial inventory of absolute dates was available from the numerous stratified archaeological sites in the region of interest (appendix B). However, many important sites lacked absolute dates, or the available dating was insufficient to our larger purposes. It was necessary to formally evaluate the quality of this data relative to our focus. Moreover, no prior research had systematically calibrated this entire suite of dates according to contemporary standards, which was essential to the comparative outcome of our work. Calibration was also necessary to directly compare the calendar ages derived from OSL methods with the calibrated radiocarbon years derived from ^{14}C dating methods.

Thus, among the earliest steps in our research was to compile a complete list of radiocarbon (^{14}C) dates available from stratified dune archaeological sites, keyed to discrete strata and cultural units within each site. All of the ^{14}C data were uniformly calibrated using CALIB v. 5.0.2 (Stuiver et al. 2006), which at times resulted in rather substantial alterations or changes between the radiocarbon age as reported, and the most likely calibrated age range for the sample. This synthesis formed the basis for our evaluation of the absolute dating available, and thereby allowed decisions to be made about which sites or strata within sites required enhanced dating to make them susceptible to analysis. We then systematically requested appropriate samples of charcoal or carbonized sherd residues from the appropriate institutions housing the archived collections, prepared the samples, and sent them to Beta Analytic for either conventional radiocarbon dating or AMS assay. These new dates are incorporated into chronological discussions for each site as appropriate.

Clearly, the recorded stratigraphy of each of these sites figures prominently in our research. Among the more salient aspects of stratigraphic contexts are the tandem variables of the altitudes of specific stable and unstable surfaces, the specific topographic or landform contexts of the sites, and their subregional position along the northwest, north, northeast, or southeast shorelines of the Lake Michigan basin. In combination with the calculated ages of discrete strata, the initial phases of research sought to assess regularities in these variables. This compiled information provided substantial background to both our further field assessment of individual archaeological and geological sites, and evaluation and alteration of standing models of dune activation. Results of this work are presented in detail elsewhere.

Was there seasonal or functional regularity to the use of Lake Michigan coastal dunes by precontact populations in Michigan? Faunal, floral, and assemblage data contribute to our research into this significant behavioral question. Our initial compilations of dune archaeological site data incorporated existing information on site function and seasonality of occupation in the individual site summaries. Given the substantial temporal span of the site sample, and its distribution across many different environmental zones, however, we did not expect complete uniformity. Rather, we expected some regional variation both in time and in space, as local economies and social groups underwent transformation between circa 2,500 years ago and the period of European contact (see Halsey 1999; Monaghan and Lovis 2005 for summaries).

To address this issue directly required that we select a sample of buried and/or stratified archaeological sites that allowed us to effectively fill the evident information voids. In this process of selection we weighted how well reported sites were (Summer Island, O'Neil, and Portage, for example, have been described in the published literature) and whether they posed logistic and economic problems in the transport of heavy equipment such as the tractor-mounted GeoProbe (locations such as South Manitou Island raised difficulties). We also tried to limit accessing sites experiencing high visitor usage, such as Porter Creek South. Of particular interest were several locales in the northern and northeastern part of the basin, the parts subject to the greatest isostatic rebound, which had not been well dated or for which the stratigraphy required clarification. This suite of sites included Winter, Ekdahl-Goudreau, and Scott Point in the north, and Wycamp Creek, Mt. McSauba, Solomon Seal, Eastport, and the sites in the Antrim Creek Natural Area. Ultimately we were not able to access all of these locales.

Geological Dune Sites

This study is the first to fully integrate the evolution of coastal dunes along the entire length of the Lake Michigan coast in Michigan. Although a great deal of research (e.g., Olson 1958a, 1958b, 1958c; Dorr and Eschman 1970; Loope and Arbogast 2000; Arbogast and Loope 1999; Arbogast et al. 2002) has been conducted on the geomorphology of coastal dunes along the eastern shore of Lake Michigan, the vast majority of it focuses on dunes along the southeastern shore of the lake. Thus, these studies describe the evolution of dunes south of the isostatic hinge line and therefore contain no record of dune response to postglacial uplift. In contrast, only a few studies have been conducted on dunes in the northern and northeastern parts of the basin, where uplift was significant. Most of these studies, such as Dow's (1937) work, are qualitative in nature and simply describe the basic processes of sand transport for dunes relative to high bluffs. Exceptions to this research design are the work at Sleeping Bear Dunes National Lakeshore (Snyder 1985) and at Petoskey State Park (Cordoba-Lepczyk and Arbogast 2005). These studies are highly localized, however, and do not fully integrate the effects of isostatic uplift in their conclusions.

Because one of the primary goals of this study is to create a more holistic model for coastal dune formation along the eastern coast of Lake Michigan, and because previous research focused on Lake Michigan's southeastern shore, fieldwork in the current study centered almost exclusively on dunes in the northern and northeastern parts of the basin. Results from this study therefore fill a major void in our understanding of coastal dune evolution along the entire eastern and northern shore of Lake Michigan. These new results are integrated with the extensive research that was conducted previously along the southeastern shore to assess the geographical variations that occur in coastal dune geology along the entire coast of Lake Michigan in Michigan.

Field Data Collection

In addition to the general systematic synthesis of information on both dune-based archaeological sites and geological dune sites, our data search provided us with substantial background information through which to assess the need for supplementary field information on both categories of site. As discussed earlier, we employed our compiled data in a similar fashion when determining where we should supplement existing radiocarbon chronologies with additional dates from specific contexts. Here, however, our multiple goals could only be achieved through field visits to specific locales.

Some of the information that we evaluated related quite specifically to physical features of the environment, the nature of site stratigraphy, and sites' spatial distribution within the area of interest. Did we have precise enough information, for example, on landform types and elevation? Were the existing descriptions of locale stratigraphy sufficient to extract detailed information on formation processes? Was it necessary to supplement existing radiocarbon date sequences with additional and complementary data derived from dune sand deposits requiring the generation of new field samples and employing OSL dating? Were there spatial gaps in our data points that we could fill with judicious exploration for, and coring of, additional locations, and where did these gaps exist? Of particular import in our selection of nonarchaeological dunes for sampling was to fill in spatial gaps left by prior geological and geomorphological research, and second to place high priority on dunes presumed to be of mid-Holocene age based on their size and position relative to other landform features.

Based on these considerations, site selection focused on the coastal zone north of Muskegon, the least-investigated part of the Lake Michigan coastal dunefields. We placed a priority on selecting the larger dunes within this coastal zone in a concerted effort to test the hypothesis that the largest dunes are of mid-Holocene or Nipissing age. Selected large dunes were deep sampled so that multiple age assessments could be generated, allowing the formulation of models of dune growth rates. In the areas subjected to rapid rates of isostatic rebound, we selected well-developed coastal dune series, displaying dune sequences extending lakeward from known mid-Holocene features. This subset was approached using transects that sampled three or more dune series, specifically to assess the effects of isostatic rebound on coastal dunefield development. We could not easily overcome the problem of sample point clustering, however. Dunes do not occur ubiquitously along the Lake Michigan coastal zone, not all are on accessible properties, and not all can be negotiated by the sampling equipment.

To address the data issues we engaged in fieldwork during the summers of 2006, 2007, and 2008 with the goal of filling information gaps. Because each locale in some respects posed an idiosyncratic set of issues, no omnibus design could be employed; we collected the necessary and appropriate data at each location. Results of this additional effort are found in our descriptions of individual locales. Uniformly, however, when additional deep stratigraphic information was required we employed a GeoProbe coring apparatus from the Glenn A. Black Laboratory of Archaeology at Indiana University. The GeoProbe, importantly, has the ability to generate continuous, largely undistorted solid-earth cores that retain both matrix integrity and relative stratigraphic position, thereby allowing the extraction of datable organic and inorganic soil samples from known contexts. We employed the GeoProbe at several key archaeological sites, including the Winter site, the Ekdahl-Goudreau site, and the Solomon Seal site, as well as dune locations lacking cultural materials, such as those at the Torch Bay Nature Preserve and Fisherman's Island/Young State Park.

Of particular importance in the use of the coring device was extraction of samples for OSL dating, both as an independent method for dune or site dating, and for comparison with new and existing radiocarbon dates. OSL dates were processed at laboratories initially at Sheffield, UK, but subsequently at the University of Illinois–Chicago. A comprehensive listing of cores, their locations, and dating results is presented in appendix C.

CHAPTER ONE

Summary

This book is organized in a fashion aligned with the major topical areas of the larger research problem addressed by our multidisciplinary work. Subsequent sections develop in greater detail the three major arenas that, when coupled, underpin the research theme. Chapter 2 provides a historical and chronological overview of coastal dune archaeological research, first at a general level, and subsequently in sufficient detail to articulate what has been observed, what can be confidently stated, and what is unknown about the formation of buried and stratified archaeological sites in coastal dunes. This chapter attempts to move the taphonomy of such sites to a level more synthetic than idiosyncratic. Chapter 3 presents a summary of coastal dunes, their formation processes, and in particular the current models associated with coastal dune activation, stabilization, and cycling, as well as broadly outlining the results of our research. The dune activation models interface closely with broader questions attached to lake-level fluctuation and alterations in sand supply. Our goal was to coalesce models of dune geomorphology with the taphonomy and chronology, as well as to incorporate regional histories of archaeological site burial that provided the catalyst for our work. Chapter 4 summarizes those basin-wide issues of long-term lake level fluctuation related to postglacial stabilization of the Michigan(-Huron) lake basin, drawing heavily on our recent related research into this topic. The relationship between lake-level fluctuation and postglacial isostatic readjustment in the basin is summarized in an attempt to understand the coupling of the two phenomena. Chapter 5 presents detailed information on the preserved, buried, and stratified archaeological sites visited or enhanced as part of the current study. Much of the information presented here is new, and provides significant insights into the modeling of site formation processes and their timing among the Lake Michigan coastal dunes.

With the three major underlying disciplinary perspectives in hand, we move our presentation to analysis and synthesis. Our synthesis of field data on site stratigraphy, deposition, formation processes, and absolute dating is presented in chapter 6. Here we modify several standing conceptions of coastal dune age and formation processes, develop a general model for archaeological site burial, and then develop a subregional model of archaeological site formation and burial that has predictive value. Much of the supporting data for this discussion is to be found in chapter 5 and the several appendices, which detail suites of OSL, ^{14}C, and AMS dates from our sample locales, the methods we employed in the field and subsequently, and summary information on all of the locales newly sampled, revisited, or reassessed as part of our research. We conclude our presentation in chapter 7 with a summary of the larger dune cycling/stabilization and archaeological site burial model, and then offer a case application of the model to one coastal region of northwest Michigan as a demonstration of its utility for planning purposes. Discussion of sampling issues attached to archaeological site exploration in dunes is an integral component of the more practical perspective.

CHAPTER TWO

The Archaeology of the Lake Michigan Coastal Zone

A Perspective from the Coastal Dunes

The research presented in this book is specifically directed at delineating the conditions under which archaeological sites are formed, buried, and stratified within dunes of the coastal zone of Lake Michigan. Understanding such taphonomic processes, and their variability or consistency in time and space, underpins our ability to assess whether coastal dune archaeological site populations are representative of past site populations. Are we missing some sites because they are deeply buried within dunes, or because they have been selectively destroyed by dune erosion or building episodes? Whether or not we ultimately find that the remnants of past archaeological site populations in such contexts are truly representative, an understanding of where, why, and how such populations may have been altered or destroyed provides significant corrective information, both for archaeological interpretation and for heritage management and planning.

As described in chapter 1, the history of research into archaeological site formation processes, also known as site taphonomy, in the Lake Michigan coastal zone goes back half a century. Despite substantial historical interests in coastal zone archaeological site formation processes, until relatively recently detailed geoarchaeological analysis has largely been applied at the scale of the individual archaeological site. Because of the disparate geomorphological and geological training of researchers undertaking such investigations, the quality and detail of this research has also been quite variable. Given that buried and stratified deposits, which because of their potential to yield chronological information are of particular interest to archaeologists, are common in coastal dune settings, these sites have garnered the most attention by archaeological researchers. Consequently, considerable detailed information is available on the depositional histories and formation processes of a variety of dune-based archaeological sites in an arc around Lake Michigan. In Michigan, these sites stretch from Summer Island off the southern tip of the Garden Peninsula in the west, around the south shore of the Upper Peninsula and then south along the west shore of the Lower Peninsula to the vicinity of Muskegon.

Such idiosyncratic site-level approaches to research, however, do not provide the broad regional framework for understanding why the current population of archaeological sites is preserved, where it is preserved, and under what conditions it can best be discovered. Such a research goal requires a larger synthetic context, as well as more detailed and comparable information, and is the key issue that drives this research. That said, the cumulative historical record of information from past work is central to understanding what is known, what is not known, and what can be learned about archaeological site populations in the Lake Michigan coastal dunes. With this caveat in mind, subsequent parts of this presentation will place archaeological coastal and dune site research in a broader context, provide key summary information on coastal archaeological research, summarize the results of fieldwork conducted by this project, and conclude with some basic summary information to be synthesized with other dimensions of the research in chapters 5 and 6.

Coastal Archaeology in the Context of Coastal Dunes

The importance of coastal zones to the precontact inhabitants of Michigan cannot be overstated, as witnessed by the abundance of archaeological sites spanning several millennia recorded in coastal contexts (e.g., Peebles and Black 1976). Coastal zones in general, but the Great Lakes coastal zones in particular, are highly visible, important, and dynamic locations on the broader landscape. The interface of land with substantial expanses of water, aside from being dramatic, also signals substantial changes in the locations and distribution of economically important and procurable food resources. At the same time this interface allows for more efficient modes of coastal transportation than does terrestrial travel. No doubt these were all significant issues to the indigenous peoples of precontact Michigan. However, the ever-changing dynamics of coastal zones also posed both solvable and unsolvable problems to these same people over both the short and long run, that is, at different scales of time. From the perspective of the current study, these same coastal zone dynamics pose multiple archaeological problems as well. This study is directed toward understanding the impact of these processes on the formation, burial, stratification, and preservation of Michigan's population of archeological sites.

As synopsized in chapter 1 and presented in greater detail in chapters 3 and 4, the recorded dynamism of coastal zones is in large part a product of the following factors. First, the position of the coastline in geographic space has changed dramatically since the retreat of the late Wisconsin, Lake Michigan lobe glaciers. It took several millennia of geomorphological evolution to develop the modern configuration of Lake Michigan, the geographic focus of this study. These coastline changes largely reflect the changing altitudes and outlets of the glacial and postglacial lakes as they adjusted to Holocene conditions, as well as other processes such as isostatic rebound (see chapter 4 for a discussion of isostatic processes). Importantly, isostatic readjustment continues to this day, and is still a factor across the Great Lakes. In the Lake Michigan basin, the area north of about Muskegon and Ludington is particularly subject to this process, rebounding on the order of circa 10 cm/century (Larsen 1985b). Given that Michigan is a land of peninsulas, this coastline change had a major impact on the amount and type of landforms available to its occupants prior to the relatively recent stabilization of the system (Lovis et al. 2005; Lovis 2009).

Second, whatever the position of the coastline in space or in elevation/altitude, the coastal zone is subject to alterations from natural processes such as wind, wave action, and ice. All of these processes modify the coastal environment in a variety of ways, either individually or in concert. However, these processes can also be amplified or muted by changes in the water plane of Lake Michigan, which can fluctuate on the order of more than a meter over short periods of time, such as a decade, or even annually, a process abundantly evident in the recent past. Although the effects of certain of these fluctuations on coastal archaeological sites have been documented (e.g., Monaghan and Lovis 2005), the broader impacts of millennia of such processes on archaeological sites both individually and collectively are not well understood.

Finally, because the sequence of ancestral phases of the Great Lakes deposited large amounts of sediment across Michigan, a mantle of sand was common across much of the state. The veneer of sand is particularly evident along the northern and eastern coastal zones of Lake Michigan, where it has been modified by waves, wind, ice, fluctuations of the lake, and other factors such as vegetation change. However, regardless of these processes, eolian landforms are among the more dynamic landforms. This is overtly due to the abundant supply of sand, which has been formed into a variety of different coastal sand dunes. Dependent on local formation processes, these features range from large, prominent, and complex coastal landscapes (such as Sleeping Bear Dunes) to smaller foredune systems common in most beach settings. Sand dunes are also not stable landforms and are sensitive to small changes in their local environment, such as vegetation, temperature, and precipitation. From an archaeological standpoint, this not only can promote site stratification but also poses significant problems of understanding site formation or taphonomy, post-depositional change, and long-term preservation. Such environments also create more practical management issues, including predicting preserved site locations, managing them in an appropriate framework, and ultimately directing site discovery methods. Until the present research, none of these questions had been systematically addressed by archaeologists despite the fact that the coastal zone figures into virtually every model of precontact settlement for almost every time period. These issues are discussed in greater detail in chapters 4 and 6.

The remainder of this chapter places the archaeology of the Lake Michigan coastal zone into historical and interpretive perspective. Our goal is not to produce yet another overview of the prehistory of Michigan, since an abundance of such works are already available (Halsey 1999; Lovis 2009; Monaghan and Lovis 2005, 69–96). It is also decidedly not our intent to produce or update a site inventory of the Lake Michigan coastal zone, such as that of Peebles and Black (1976). Rather, our goal is to summarize key pieces of research that shed light on the nature of coastal dune archaeology in the circum–Lake Michigan basin and provide chronological control for prehistoric usage of coastal environments. Throughout the ensuing discussion we will employ standard nomenclature for cultural time periods as employed in the previously cited works and as presented in table 2-1.

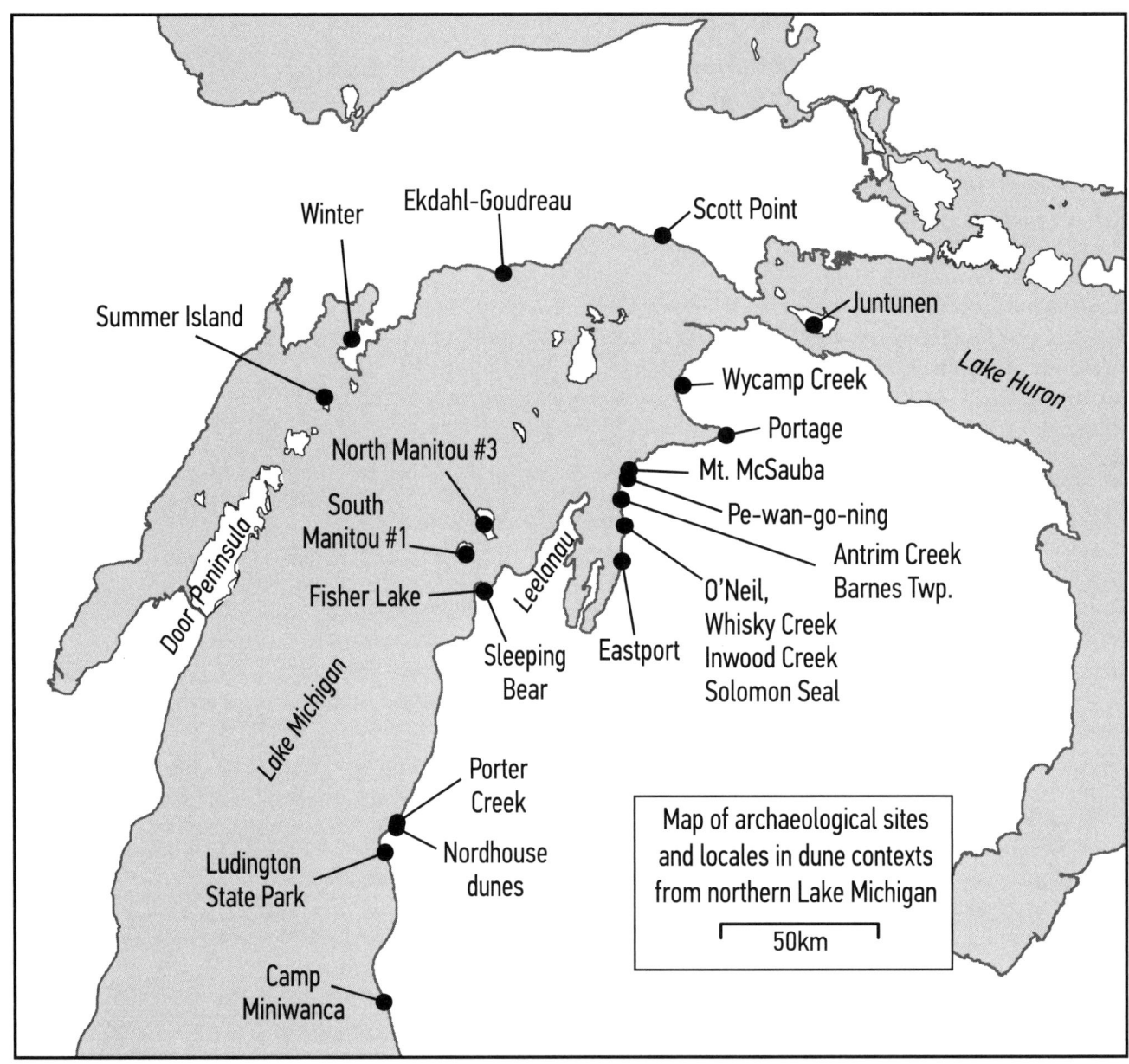

2-1. Locations of important buried or stratified archaeological sites within coastal zone eolian contexts.

The locations of key archaeological sites discussed in the text are presented in figure 2-1.

Coastal Zone Archaeology in the Michigan Basin to the 1960s

Archaeological interest in, and directed professional survey of, the Lake Michigan coastal zone had its inception in the first quarter of the twentieth century, and its first summary compilation as a well-integrated component of Wilbur B. Hinsdale's omnibus work *An Archaeological Atlas of Michigan* (1931), which incorporated

TABLE 2-1. Michigan Prehistory: Time Periods, Ages, and Key Events

CULTURAL PERIODS		INCEPTION OF CULTURAL PHASES		KEY CULTURAL EVENTS	KEY GEOLOGICAL EVENTS AND TIME STRATIGRAPY	
		CALENDAR	¹⁴C AGE			
EUROPEAN CONTACT		A.D. 1640	0.36 kyBP	Europeans arrive in Michigan	High levels of Great Lakes; "premodern" flood phase (<0.8 kyBP: Little Ice Age)	H O L O C E N E
UPPER MISSISSIPPIAN/ LATE WOODLAND		A.D. 1200	0.8 kyBP	Upper Mississippian in southwest Michigan		
					Low levels of Great Lakes; few floods (1.0–0.8 kyBP: Medieval Warm Period)	
WOODLAND	LATE	A.D. 500	1.5 kyBP	Complex egalitarian social systems		
	MIDDLE	0 A.D./B.C.	2.0 kyBP	Complex mound burial; use of tropical and indigenous cultigens	High levels of Great Lakes; post-Algoma flood phase (2–1.5 kyBP)	
	EARLY	550 B.C.	2.55 kyBP	Initial mound/earthwork construction; first ceramics introduced	Low levels of Great Lakes; "low" flood interval (3.0–2.0 kyBP)	
ARCHAIC	LATE	3000 B.C.	5.0 kyBP	First cultigens (squash)	Nipissing-Algoma high-water phase; flooding common (3.5–5.0 kyBP)	
	MIDDLE	5000–6000 B.C.	7–8 kyBP	First extensive regional exchange	Nipissing transgression; Chippewa-Stanley Phase initiated; Michigan-Huron basins ice-free	
	EARLY	8000 B.C.	10 kyBP	Early and Middle Archaic occupations now submerged under Great Lakes		
PALEO-INDIAN		9000–10000 B.C.	11–12 kyBP	Earliest Paleo-Indian penetration in Michigan	Algonquin Phase initiated; Greatlakean advance; high-level lake in Lake Michigan basin; Early Algonquin in Lake Huron basin	LATE WISCONSIN
GLACIATED		11000 B.C.	13 kyBP	Port Huron advance; much of southern lower Michigan ice-free; high-level glacial lakes in all basins (low-level lakes between 13.5–13.0 kyBP)		

NOTE: Time designations given in thousand years before present (kyBP)
SOURCE: Modified from Monaghan and Lovis (2005, table 3-1).

material from the same author's *Primitive Man in Michigan* (1925). Given that the only archaeological program, faculty, and facility in the state of Michigan at the time were at the University of Michigan, it is no surprise that much of this early history of Michigan archaeology resides in repositories or collections at Ann Arbor.

In large part, early knowledge of the archaeology of the coastal zone derived from published histories of various better-developed and populated regions of the state, in particular areas such as Berrien County (e.g., histories by Champion 1926; Coolidge 1906; Cowles 1871; Winslow 1876) or the Grand Traverse Bay region (Leach 1903; Winchell 1866), in turn supplemented by various maps such as the well-known Farmer map of 1926, which annotated the locations of Indian sites within its coverage. By 1910 a sufficient amount of such information had accumulated from various sources that a nonarchaeologist, Harlan I. Smith, was able to develop a comprehensive list of so-called aboriginal sites in Michigan, of course including sites arrayed along the Lake Michigan coast (1910).

Archaeological surveys began in earnest during the 1920s, by both professionals and interested avocationalists, and included some of the earliest systematic work along the Lake Michigan coast (e.g., Stevens 1924). In some respects these early surveys began to fill in some evident spatial voids. For example, apart from his abundant research in the southern part of Lower Michigan, Emerson F. Greenman also performed survey to the west of St. Ignace at the Straits of Mackinac, where he described the Fort Hill site (1926). Based on his work along the Lake Michigan coast in Emmet County south of the Straits of Mackinac, he recorded the prehistoric and historic sites at Wycamp Creek (a site later excavated by Michigan State University), as well as describing burials and surface finds in the vicinity of Cross Village (1927). Later in his career Greenman returned to the region to visit the Eastport site in Antrim County (1941), an Archaic site subsequently reported upon in some detail by Binford and Papworth (1963). Both the Eastport and Wycamp Creek sites are components of our current study and shed light on dune chronology in the middle and late Holocene. Less systematic recording is evident in Vreeland's early (1924) logbook of sites from northern Michigan (but used to good effect by Peebles and Black 1976). In the southeastern part of the basin, other University of Michigan personnel, such as Vernon Kinietz, performed survey in the Oceana-Mecosta-Newaygo County vicinity (Kinietz 1929). He may have relied on earlier documents and maps, including those penned by recorders such as Southwick in Oceana County (1922). Significantly, though, almost all of these early records were incorporated into Hinsdale's 1931 *Archaeological Atlas of Michigan,* although with somewhat inaccurate locational information intentionally designed to foil potential site looters.

As may be evident, this history of dune-based archaeological site exploration in Michigan to a degree follows the evolutionary history of archaeological institutions in the state. From before World War II until the 1960s, the only institution engaged in sustained archaeology in Michigan was the University of Michigan, in particular the University of Michigan Museum of Anthropology (UMMA). Consequently, much of the earlier research on dune-based archaeological sites derives from the work of faculty and students at the UMMA (e.g., Spaulding 1948). Their work, in turn, was often supplemented by the efforts of informed avocational archaeologists, a group with whom they regularly interacted. With expansion of higher-education programs and the state university system in Michigan during the 1960s the archaeology program at the University of Michigan was joined by programs at several other colleges and universities, including Michigan State University, Grand Valley State University, Western Michigan University, and Wayne State University. From the 1970s through the present these schools was joined by Northern Michigan University, Alma College, and Michigan Technological University, among others.

During the 1970s, with the rise of federal compliance-based legislation related to the National Environmental Policy Act (1969), the National Historic Preservation Act (1966), and the Archaeological and Historic Preservation Act (1974), the number of project-related, impact-based projects in Michigan (and elsewhere), increased significantly. This work was initially undertaken by educational institutions, but subsequently a growing number of private-sector environmental, archaeological, and historical preservation firms entered the compliance research arena. As a further consequence of the provisions of this federal legislation various government land management agencies, including the USDA Forest Service and the National Park Service expanded their environmental, historic preservation, and recreational staffs to include archaeologists. In Michigan this was most notable in the National Forest system. Increasing acquisition of properties in Michigan by the National Park Service, specifically the Pictured Rocks and Sleeping Bear Dunes National Lakeshores, and in the adjacent Indiana Dunes National Recreation Area, brought enhanced work on these properties by the National Park Service's Midwest Archaeological Center, based in Lincoln, Nebraska. The cumulative, and cascading, effect of increased institutional engagement in Michigan resulted in a substantial increase in the number of coastal dune archaeological sites recorded, and in the exploration of both coastal dunes and these sites.

To a degree, developing a history of dune-based archaeological site research is a difficult proposition. That a site is present in dune contexts, has been systematically recorded, and is present in institutional site files does not necessarily translate well to "research." An excellent example of this disjunction is the Juntunen site, which is situated in coastal dunes on Bois Blanc Island in the Straits of Mackinac. It may also be the first major stratified dune site excavated in the upper Great Lakes. Alan McPherron (1967, 2), who directed excavation at the site, reveals that the location of Juntunen was first recorded in 1932 by Robert Braidwood, then a student at the University of Michigan. Braidwood's report on the site remained generally untouched on file at the UMMA for almost 30 years. It was not until 1960, however, that any formal excavation took place at Juntunen, and this was a direct consequence of the discovery of burials at the site during construction on the property.

Coastal Zone Archaeology in the Michigan Basin from the 1960s

The decades of the 1960s and the 1970s were the most significant periods for dune site research in the Lake Michigan basin in Michigan and Wisconsin, by multiple institutions including the Neville Public Museum, the University of Michigan, and Michigan State University. Although this work was often directed at large regional problems of subsistence and settlement, the magnet that catalyzed reconnaissance for and investigation of dune sites was their stratified deposits, which allowed study of change over time and the establishment of regional chronological sequences tied to ceramics, lithics, copper, and other formal artifact categories and styles. Of particular interest in this regard is the fact that, to date, no stratified dune-based sites predating circa 2,500 years ago have been located or investigated along the Lake Michigan coastal zone. Or, at least, none producing ^{14}C dates or diagnostic artifacts prior to this date have yet been recognized. Given that this observation is based on half a century of concerted archaeological research, it poses a potentially significant issue of both archaeological site and dune formation and taphonomy that warranted investigation by our research.

Discussion of the more recent work in the Lake Michigan basin in Michigan will be presented in terms of geographic subregions. These include the north shore of Lake Michigan including the Garden Peninsula across to the St. Ignace at the Straits of Mackinac, the northeast shoreline from Mackinac City south to the Leelanau Peninsula and Sleeping Bear Dunes, the east shoreline from Sleeping Bear Dunes to about Muskegon, and finally the extreme southeastern part of the basin from Muskegon to the Michigan-Indiana border. Notably, there are few sites of import to this study within the segment from Sleeping Bear Dunes to Muskegon. Within each subregion surveys and sites will be described sequentially, from west to east across the northern part of the basin, and north to south along the east coast of the basin.

North Shore of Lake Michigan, from Garden Peninsula to St. Ignace

Research on the western and northern coasts of the Michigan lake basin took place concurrently with the work at the Juntunen site, and is included in this prefatory discussion because it contributes substantively to our broader regional perspective. Ronald J. Mason, then a student at University of Michigan, conducted excavation at the stratified Heins Creek and Mero sites on Wisconsin's Door Peninsula in 1960 and 1961 under the auspices of the Neville Public Museum (Mason 1966). Mason supplemented this work by excavation at the Porte des Morts site, also in Door County, during 1961 under auspices of the same institution (Mason 1967). This combined work not only established clear chronological sequences for the area, but also contributed to knowledge of site formation processes and lake levels on the west side of the lake basin. In fact, it was Mason who first recognized that there might have been a period of higher than modern water plane altitudes in the Lake Michigan basin during the period circa 2,000 to 1,500 years ago, which he attributed to storm or wind seiche activity. We now know that this was a basin-wide phenomenon that had effects much more broadly based than just the Door Peninsula, and which in fact was a reflection of basin-wide processes of lake altitude oscillation (chapter 4; Branstner and Cleland 1994; Monaghan and Lovis 2005). Not directly related to our current research, but warranting mention, is the UMMA work by James Fitting in 1963 and 1965 at the Spider Shelter on the Garden Peninsula (Cleland and Peske 1968).

The 1970s saw the entry of other institutions into coastal zone research on stratified dune archaeological sites, including the Western Michigan University (WMU) Department of Anthropology at the Winter site near Valentine Creek on Big Bay de Noc (Richner 1973; Martin 1980; Bianchi 1974), and Northern Michigan University (NMU) through its research at the Scott Point site (Buckmaster personal communication; Martin 1982).

The 1980s witnessed further research by Ronald Mason along the northwestern shoreline of the Michigan basin at the Rock Island site, on Washington Island between the Door and Garden peninsulas. Although the primary goal was to investigate the historic seventeenth-century Huron occupation, much like the sites on the Door Peninsula to the south and the Garden Peninsula to the north, Rock Island produced stratified dune deposits that included occupation from the Middle Woodland through the seventeenth century (Mason 1988).

University of Michigan 1962 and 1963 Survey

From the standpoint of site discovery, one of the more productive coastal surveys was that carried out along the north shore of Lake Michigan in 1962 and 1963 by the UMMA (Binford and Quimby 1963; Peske and Kent 1963). This program of survey and test excavation recorded numerous sites across the coastal zone, including several that were subsequently excavated by University of Michigan or other institutions, such as the Summer Island site, Spider Shelter, Ekdahl-Goudreau (also known as Seul Choix), and Scott Point (or Point Scott, or incorrectly Point Patterson depending on which document one is reading). From the perspective of the current study's focus on buried and stratified sites in coastal dunes, the Summer Island site (Brose 1970a, 1970b, 1970c), the Ekdahl-Goudreau site, and the

Scott Point site are of particular importance. The former is near the southernmost point of uplift on the northwest side of Lake Michigan, just south of the Garden peninsula. The site displays minimal eolian deposition, but sufficient separation of occupations to provide clear burial, preservation, and stratification. Ekdahl-Goudreau was eventually excavated by the University of Michigan, displayed clear stratification, and produced two radiocarbon dates. Scott Point is an abundantly stratified deposit excavated by NMU in two consecutive field episodes, and displays regular and abundant deposition sufficient for multiple occupation layers to be preserved and buried intact.

Summer Island Site

Overlapping somewhat with Mason's research on the Door Peninsula were investigations at the Summer Island site off the southern tip of Michigan's Garden Peninsula, spanning the period from 1959, when George I. Quimby first visited the locale, to 1967, when the site was formally excavated by David S. Brose. The Summer Island site is situated on the east coast of the island, fronting Lake Michigan. This site had apparently been known and recorded long before Quimby's visit, although largely on the basis of its historic Indian settlement. The UMMA revisited the site in 1963, performed test excavations, and established that the site was stratified, a fact verified by a revisit to the site in 1965, and Brose's preliminary and subsequent work in 1967 revealing Laurel Middle Woodland through historic period stratigraphy and occupation (Brose 1970a, 1970b, 1970c). The results of the major work at this largely stratified, multicomponent occupation has been the topic of several publications (Brose 1970a, 1970b, 1970c). The site has a well-dated occupation sequence (appendix B) ranging from circa 2,000 to 300 years ago. Readers are referred to the original reports for detailed assemblage descriptions.

Of particular interest to our work on site taphonomy in coastal dune settings is the fact that given the position of Summer Island just north of the "hinge line," or point of zero uplift, in the Lake Michigan basin, it has only been subject to uplift or rebound rates of about 0 to 5 cm/century. Thus, the marginally separated and superimposed, but nonetheless stratified occupation sequence may be expectable given its position in the basin. We were also particularly interested in the age of the earliest deposits relative to the characteristics and activation cycles of coastal dunes. Given the abundance of published information, our current research did not revisit the Summer Island site, but more detailed information on the site stratigraphy and age is presented in appendix C.

The Winter Site

Further clarification of both Middle (or Initial) Woodland on the north shore of Lake Michigan, as well as the potential for stratification, came with excavations by WMU at the Winter site. The Winter site is situated on the west side of the Garden Peninsula adjacent to the outlet of Valentine Creek on Big Bay de Noc, a northern arm of Green Bay that, in turn, is connected with the larger Lake Michigan basin to the east through an outlet between the Door Peninsula and the Garden Peninsula. Thus, the coastline dynamics of this part of the basin are more than likely controlled by basin-wide processes. Nonsystematic survey of the coastline by Thomas Bianchi resulted in discovery of the Winter site in 1967; it became the subject of a WMU field school in 1972, and formed the core of a master's thesis by Jeffrey Richner (1973).

Extensive excavation at the site revealed stratified Middle (or Initial) Woodland deposits, as evidenced by an assemblage with diagnostic Laurel and North Bay ceramics and lithics, with as many as three separate, discontinuous, or merged cultural strata buried and arrayed along the face of an uplifted post-Nipissing beach and including evidence that the site had once been inundated. Of particular interest is the fact that the site is situated almost 400 meters from the current Lake Michigan coastline (Bianchi 1974; Martin 1980; Richner 1973). This is consistent with the uplift of former coastlines to higher elevations and consequently more interior locations that are today removed from modern shorelines. This is the westernmost buried or stratified coastal site we could document in our literature survey, although unfortunately the site had not been absolute dated by radiocarbon or other means, but rather had been cross-dated by comparisons of diagnostic ceramics to as early as circa 2,000 years ago (Bianchi 1974; Richner 1973), an age estimate not disputed by anyone familiar with the assemblage. Given, however, that the Winter site was coastal, stratified and buried, undated by absolute dating, and at the westernmost position of our region of interest, it became one focus of our fieldwork in 2007. We also intended to access the ceramic assemblage in an attempt to provide a suite of dates on the occupation that could be associated with the site stratigraphy and with OSL dates we planned to obtain from other sites.

The Ekdahl-Goudreau (aka Seul Choix) Site

The site at Seul Choix point on the north shore of Lake Michigan had been known from as early as the 1950s, and as noted earlier was surface collected in the early 1960s (Binford and Quimby 1963). The site has been given multiple names associated with either the landform or property owners, and from records at the Office of the State Archaeologist had also been given several geographical locations, all but one of which was erroneous. Regardless of these confounding issues, construction of a house foundation in 1968 revealed a deeply stratified series of deposits in low-lying dunes perched as part of a thin sand mantle on the surface exposures of limestone bedrock in the area.

The Ekdahl-Goudreau site was excavated by a field crew from the UMMA in 1968, when the deep, stratified nature of the site was revealed by house foundation construction. The test excavations at the stratified and as yet unpublished site revealed both Laurel Middle Woodland and Late Woodland occupations (Fitting 1975, 136; Prahl, unpublished field notes). Although no formal report has ever been published on these excavations, field notes are available at the UMMA Great Lakes Range. Additionally, the assemblage has been referred to and even partially described in several major compendium summaries (Binford and Quimby 1963; Fitting 1970, 1975; Brose and Hambacher 1999).

Two ^{14}C dates (appendix B) suggest a late Middle/Initial Woodland occupation circa 1,400 to 1,300 years ago, and a Late Woodland occupation circa 1,000 years ago. The date on the basal Initial Woodland zone may be overly recent, although a suite of late Middle/Initial Woodland Laurel dates is present in the region (Lovis and Holman 1976). As with other stratified sites mentioned in this section, little in the way of soil development is present in several of the buried organic horizons containing occupation material, raising the question of anthropogenesis for these strata. This site is situated quite far north in our study area, and has been subjected to rebound effects on the order of 10 cm or more per century. This may in turn, we believe, account for the depth of the deposits at the site. Although nominally the stratification at Ekdahl-Goudreau consisted of two major zones, inspection of the photographs and field drawings suggests far more complicated formation processes, with cyclic episodes of stabilization and variable amounts of intervening eolian sand. The site description reveals this complexity in detail.

Because of its deep and stratified deposits spanning what appeared to be as much as two millennia, its location along the north shore of Lake Michigan, and its early occupation consistent with that at the Winter site, Ekdahl-Goudreau was included in our 2007 fieldwork.

The Scott Point (aka Point Scott and Point Patterson) Site

The Scott Point site was recorded during the University of Michigan Peske and Kent (1963) survey of the north shore of Lake Michigan, but other than limited surface collection was not investigated further. The site is situated in a series of unstable dune deposits along the west side of Scott Point, made increasingly unstable by regular all-terrain vehicle traffic. Much of the site is deflated, with pavements of natural pebbles, firecracked rock at times *in situ* as remnant hearth features, and other artifacts. Hints of stratification can be observed in vertical erosional faces in the dunes.

Scott Point site was the subject of a series of NMU archaeological field schools, most prominently in 1979. No report is yet available on this work, and much like the Winter site the Scott Point site had not been absolute dated. The NMU excavations at Scott Point revealed complex buried stratigraphy at least 1.5 meters in depth, although excavations were terminated because of slumping of excavation unit walls rather than as a consequence of encountering sterile deposits. The ceramic assemblage suggested that the earliest of the excavated occupations was early Late Woodland Mackinac Phase (ca. 1,400 to 1,000 years ago), and the most recent were a late Late Woodland Juntunen Phase (ca. 800 years ago to the time of European contact ca. 400 years ago) and an Oneota occupation (ca. 900/800 years ago to European contact 400 years ago) presumably associated with areas of Wisconsin to the west. Historic materials dating through the 1800s and early 1900s are prominent on the site surface and speak to continued use of the locale after European contact.

Field photographs of the stratigraphy made available to us by Dr. Marla Buckmaster suggest that at this site, much like the Winter, Ekdahl-Goudreau, and O'Neil sites, there was little natural soil development in the buried and apparently stable organic occupation horizons. Once again, this raised the question of anthropogenic processes either contributing to or being primarily responsible for the stabilization and burial, and perhaps reactivation, of the sequence. Much like Ekdahl-Goudreau, Scott Point is quite far north in our study area, and is subject to substantial rebound effects, again raising the question of the relationship between isostatic rebound and depth of deposition.

The Scott Point site was important to our research, and we undertook a program of dating on carbonized ceramic residues from the assemblage. Attempts to visit the Scott Point site and perform systematic coring in both 2007 and 2008 were blocked by difficulties in obtaining a formal permit and in gaining safe access.

Northeast Shore of Lake Michigan, from Mackinac City to Sleeping Bear Dunes

In 1965 Charles Cleland of the Michigan State University Museum (MSUM) initiated the first year of a multidecade program of research along the coastal zone of northeast Lake Michigan between Traverse City and the Straits of Mackinac. This program resulted in the investigation of several significant stratified coastal dune sites through the early 1970s. Among the most significant of these was the Wycamp Creek site, which was excavated in 1967 and produced partially stratified Laurel Middle Woodland, Late Woodland, and historic occupations (MSUM field notes; see also Lovis 1973). Although our current research employed information from the Wycamp Creek site excavations, we were unable to return to the site and perform additional stratigraphic coring. A summary of the archaeology and geoarchaeology of the Wycamp Creek site is presented in appendix C.

The MSUM 1967 survey of Grand Traverse, Kalkaska, and Antrim counties, although primarily focused on interior lakes and various rivers and streams in the region, incorporated large parts of the Lake Michigan coastal zone as well. Several coastal sites, some important to the current study, were relocated or newly recorded during this work (MSUM field notes, 1967). In particular, sites in Antrim County including the Barnes Township Park site and the Antrim Creek site were recorded, and limited test excavations were conducted at the Antrim Creek site. The Barnes Township Park displayed a substantial stable A horizon with debitage and ceramics; it is not a coastal dune site despite the fact that it is a coastal sand deposit. As part of the same 1967 survey parts of the coastal zone in Charlevoix County were also covered. This reconnaissance also recorded the location of the Inwood Creek site (20CX34), but no test excavations were conducted. The Inwood Creek site was observed to be on an uplifted eolian beach terrace, and therefore was technically not a dune site despite the sand deposits.

The 1968 MSUM survey recorded several additional coastal sites in Charlevoix County including Neff's A.H. Map site, the O'Neil Site, the New Toilet Area site, and the Whisky Creek site. This information was later supplemented with survey data from Lake Charlevoix resulting in the discovery of the Lake Michigan coastal Pine River Channel site (Cleland 1970), subsequently excavated in 1973 (Holman 1978). Test excavations were conducted on the O'Neil, New Toilet Area, Whisky Creek, and Inwood Creek sites and produced highly variable outcomes. The work at the Whisky Creek and the O'Neil sites is particularly important from the perspective of the current research. Located on the south bank of Whisky Creek in what is now Fisherman's Island/Young State Park, test excavations revealed a very deep, aceramic artifact deposit in the post-Nipissing age foredune adjacent to the creek outlet into Lake Michigan. No intact soil or occupation horizons were encountered, no clear stratification of occupation was observed, and no organics were recovered. Whisky Creek mimics many other coastal dune sites in these characteristics and has been heavily reworked or redeposited by eolian activity.

The test excavations at the O'Neil site, situated at post-Algoma elevations inland of the foredunes on the north bank of Inwood Creek, revealed a deep, buried, and stratified Late Woodland and historic period site almost unique to the northeast Lake Michigan basin. Continued MSUM research resulted in excavation of the O'Neil site in 1969 and 1971 (Lovis 1973), revealing a complex dune depositional sequence (Lovis 1990a). This research clearly demonstrated that there was certain potential for such sites to be present and discoverable in the coastal zone if the systematics of their preservation could be understood. Our research is an attempt to accomplish this goal four decades later. Consequently, the O'Neil site is the subject of more detailed discussion elsewhere in this book.

Portage Site at L'Arbre Croche

By 1973 MSUM had partially shifted its research to the inland lake chain between Petoskey and Cheboygan in Michigan's northern Lower Peninsula. Although the focus of the MSUM Inland Waterway Project of 1974 and 1975 was explicitly *not* the coastal zones of Lakes Michigan or Huron (Lovis 1976b), the survey strategy nonetheless incorporated some areas of the Lake Michigan coastline north of Petoskey on the eastern part of Little Traverse Bay. Specifically, these included the prominent sand dunes on the north edge of Petoskey State Park, and in privately owned properties that abut the north park boundary. Most of the recorded archaeological sites in this vicinity expectably appeared as lag artifact "pavements" on deflated dune surfaces (including some recent nineteenth-century historical occupations). It was during this work that the partially stratified Middle and Late Woodland Portage site was discovered. This site was tested in 1974, and then excavated in 1975 (Lovis et al. 1998). The Portage site is wholly situated within the L'Arbre Croche development atop the dune deposits fronting the Lake Michigan coastline of Little Traverse Bay (Lovis et al. 1998). Excavations at the site demonstrated that intact early, buried, and stratified archaeological deposits could occur in swales or depressions behind the high foredunes. The earliest buried surfaces are radiocarbon-dated to two millennia or more in age.

Mt. McSauba Site

During fieldwork at the O'Neil site, stratified Late Woodland dune occupations were recorded north of Charlevoix at Mt. McSauba City Park (MSUM field notes). In the summer of 1972 a small MSUM field party performed a long-ranging and peripatetic survey of parts of the Lake Michigan coastal zone from Grand Traverse Bay to the western Upper Peninsula with the intent of assessing the condition of a number of previously reported archaeological sites. Their work included survey of several Leelanau County coastal locations, revisits to sites in Antrim and Charlevoix counties (as well as performing limited surface collections on several sites along the north shore of Lake Michigan including the Scott Point site). As other observers had recorded, the dune sites they revisited in both Leelanau and Antrim counties were largely artifact scatters on deflated eolian surfaces with no intact organic horizons visible.

Of importance to the current work was their revisit to the Mt. McSauba locale north of Charlevoix. The Mt. McSauba site had previously been visited by MSUM field crews on a periodic basis. They undertook test excavations, recorded the by now unsurprising presence of artifacts on deflated surfaces, and performed unsystematic surface collection. More important, however, was the recognition that these pavements had derived from a significant and visible paleosol or buried soil surface containing cultural material. This meant that at least one intact cultural horizon was present at variable depths below the dune crest. The current project was particularly interested in further investigating the age and dune geomorphic context of this site. Mt. McSauba had not been absolute dated, although the ceramics suggested a date after 1,000 years ago.

Michigan State University 1971 and 1972 Survey and Excavation

During the summer of 1971 the MSUM and Department of Anthropology continued their research program along the coastal zone of Lake Michigan in Emmet, Antrim, and Charlevoix counties. One prong of the research was to conduct intensive excavation at the O'Neil site, which resulted in several publications on the site (Lovis 1973, 1990c). Multiple buried occupation strata were found in the swale behind the foredune at the site. The earliest occupation is Mackinac/Skegemog phase early Late Woodland, and occupations range from about 1300 to 300 BP. Data from multiple ^{14}C assays date the occupation strata (appendix B). Of particular interest, and as has been observed at the Scott Point, Ekdahl-Goudreau, and Winter sites, among others, the occupied buried organic strata do not reveal much in the way of soil development, but do reveal episodes of stabilization and occupation. This observation in turn raises the question of whether or not anthropogenic processes may be responsible for stabilization, burial, and stratification.

Also significant to the current study was the coastline survey of Nipissing age features in the region (Brashler 1972). Some of these features included active and stabilized dunes. A number of additional sites were discovered as a result of this work, including one potentially significant to the current study; the Solomon Seal site. Although the majority of the other recorded sites were situated on well-defined uplifted beach terraces or in swales behind coastal and inland dunes, the Solomon Seal site was located at high elevations on the peak of a very substantial, and presumably of Nipissing-age, dune crest. Moreover, Brashler recorded artifacts eroding down slope from what appeared to be a deeply buried and organic-rich feature below the dune crest. This was a very different and intriguing geographic or landscape position for such a site. Although Solomon Seal and other sites begged for exploratory excavation, none was conducted in 1971.

Fisherman's Island State Park Survey, 1976

Much like the Sleeping Bear Dunes survey, the Fisherman's Island State Park survey was precipitated by development of the local area into the system of Michigan State Parks. The 1976 survey of the affected area, in concert with limited test excavations at certain newly discovered as well as previously recorded sites, was conducted by the MSUM under contract with the Michigan Division of History, and the results summarized in a major report (Lovis 1976a). Although the work augmented the inventory of known coastal archaeological sites, certain of the results are of particular importance to our current research into dune taphonomy and site burial and preservation. First, most of the site inventory revealed some form of post-depositional disturbances from natural processes in the form of deflation or erosion. It appeared that for the most part previously active dune contexts were not good situations for site burial and preservation. This was consistent with the information gleaned from the earlier MSUM test excavations at the Whisky Creek site. Second, however, was the discovery of partially stratified, and probably Archaic age, deposits at post-Nipissing elevations on an inland terrace south of Inwood Creek. Thus, two stratified and buried archaeological sites, O'Neil and Inwood Creek, occurred on the south side of Inwood Creek. This observation once again raised the question about timing and conditions that might result in such preservation. Finally, limited test excavations at the Solomon Seal site, initially recorded by Brashler in her 1971 survey, had revealed partially preserved stratified and buried cultural deposits in intact organic horizons over one meter in depth. The site was situated at elevations greater than 630 feet/192 meters above mean sea level. At the time, and actually up through our current research, such elevations were considered to be consistent with the Nipissing stage of the Michigan-Huron basin, suggesting that the Solomon Seal site was formed early in the eolian activation of post-Nipissing sand deposits.

Solomon Seal Site

As part of the 1976 survey of Fisherman's Island State Park near Charlevoix, the MSUM performed limited test excavations at the site initially reported by Janet Brashler in 1971, the Solomon Seal site (Brashler 1972). The site revealed multiple partially disturbed paleosols with occupation material extending at least several meters below the crest of the dune surface. Among the most interesting aspects of this partially stratified dune site was its position at elevations well above modern lake level and consistent with the Algoma or earlier lake stages in the Michigan basin (Lovis 1976a). This site may actually date as early as the Early Woodland or Late Archaic period, and may be the oldest dune-based archaeological site on the coastal zone. However, no absolute dates were available for any of the stratigraphy, let alone the occupation strata. Given its unique position, we decided that Solomon Seal would be a focus of site-specific fieldwork as part of our project.

Antrim Creek Site

The Antrim Creek site is situated in coastal dunes that were at one time the focus of a dune buggy tour of the Michigan coastline, but which are now managed by Antrim County as the Antrim Creek Natural Area. As with many other such locations on the coastal zone, the site was heavily reworked by both all-terrain vehicles and eolian activity, resulting in no visibly intact occupation horizons, and only occasional lag deposits in the dune blowouts. Some intact A horizons were observed in remnant stable "pedestals" (a term also used by Greenman in his 1927 description of Wycamp Creek site) across the southernmost part of the site area. The site was subsequently the subject of extensive survey and test excavation as part of its incorporation into the Antrim Creek Natural Area (Cleland 2002). A revisit to the site by an MSU Department of Anthropology field school in 2004 revealed small, deeply buried but discontinuous spatial remnants of an occupied paleosol (Raviele 2006). The paleosol occurred 1.3 m deep behind foredunes north of the creek. These remnants were sufficiently discontinuous that concerted attempts to relocate the buried surface through an intensive coring program as part of dune geomorphological and geoarchaeological research in 2007 were unsuccessful (Holmstadt 2008).

Due to the concentration of significant sites in this area of the coastal zone between South Point and Antrim Creek, and their pertinence to the current research, systematic revisits were conducted in both archaeological and nonarchaeological contexts in the zone between Antrim Creek and Charlevoix along the Lake Michigan coastline.

Eastport Site

The Eastport site, in Antrim County, is a well-known Archaic site initially reported on by avocational archaeologists who had performed surface collection and test excavation at the site (Davis and Gillis 1959). The Eastport site subsequently became the subject of test excavations by an excavation group including E. Gillis and G. Davis, and a UMMA field crew consisting of J. B. Griffin and M. L. Papworth. A report of their fieldwork, and a detailed description and analysis of the lithic assemblage, was then published by Papworth with L. R. Binford (Binford and Papworth 1963). The Eastport site has figured prominently in various interpretations of the Michigan Archaic since its publication. The following synopsis is drawn from the Binford and Papworth report.

Three test pits were excavated at the Eastport site, revealing a well-developed spodic sequence with a substantial artifact-bearing zone embedded within it. The assemblage was exclusively Eastport chert, now known as Norwood chert after its source location, from a primary source in close proximity to the north, and primarily comprised of chipping debris from the early stages of chipped stone tool manufacture. The artifact zone, at least as can be discerned from such limited excavation, varied in depth/elevation and density. A current view of the report suggests that this variability may relate to preservation of dune slopes by eolian activity and burial. Based on geochronological interpretations related to the elevation of postglacial Lake Nipissing, the timing of its maximum altitude and subsequent drop to lower altitudes as this was known in 1960 largely through the work of Hough (1958), and reconstruction of the altitude of the artifact bearing zone at the site, it was estimated that Eastport site had to date after (be more recent than) circa 4,200 years ago. Associating it with similar sites in the Saginaw basin of Michigan, Binford and Papworth (1963) concluded that the Eastport site dated between circa 3,300 and 2,500 years ago.

Due to a lack of datable organics, however, no absolute dates have been available to assess the accuracy of this estimate. Given that the Eastport site appeared to be in a post-Nipissing dune context that had subsequently stabilized and become buried, we were interested in clarifying the age and formation processes of the site using contemporary methods and perspectives.

Leelanau Peninsula/Sleeping Bear Dunes/South Manitou 1 Site

During the 1967 field season, the MSUM sent a field crew to North Manitou Island, then owned by the Shakespeare Fishing Reel Company, under sponsorship of the William R. Angell Foundation to conduct a combined survey and test excavation program (Cleland 1967). Although the survey recorded seven sites ranging in age from Late Archaic through Late Woodland, and performed test excavations at the substantial and ^{14}C-dated Late Archaic North Manitou 3 Site, none of the recorded sites were in dune deposits. All were situated on either modern or fossil beach features. Importantly, North Manitou Island was to come under federal ownership in 1975 as part of the Sleeping Bear Dunes National Lakeshore, putting it under some protective umbrella from development.

Subsequently, during 1974 and 1975 the MSUM, under contract with the National Park Service, conducted a substantial survey of areas to be included within the Sleeping Bear Dunes National Lakeshore. This survey was not limited to the coastal zone, although substantial coastal survey was undertaken particularly in Leelanau and Benzie counties and on South Manitou Island. The dune areas adjacent to the Sleeping Bear Dune were of particular interest in this regard. This prominent landform is a classic perched dune, and due to its size is often presumed to be of great age—a persistent misconception true of archaeological sites found on similar landforms. As revealed elsewhere in this volume, this is not the case, and the deposit actually dates between 3,000 and 2,000 years ago based on absolute dates from ^{14}C (Snyder 1985). A major report on this research was produced, detailing results of survey and test excavation at selected locales (Lovis et al. 1976).

Two observations pertinent to the current work derived from the initial Sleeping Bear Dunes survey. Of particular interest to the ongoing postdepositional processes operating in coastal dune environments was the observation that one of the earliest sites recorded in Leelanau County and noted by Hinsdale (1931), Site 20LU2, has probably been covered by eolian activity under the inland or slip face of a dune, a precursor to some of our field observations during the current research. No surface indications remain (Lovis et al. 1976, 49). Second, several archaeological sites were located by the 1974/1975 Sleeping Bear Dunes survey, but only one assists us in understanding the potential distribution of intact and dune-buried sites in the coastal zone, South Manitou 1 Site (20LU24). The ensuing description is drawn from Lovis et al. (1976, 51–52).

South Manitou 1 Site was exposed by eolian deflation of an active dune surface. The site is located adjacent to and southwest of the Old South Manitou Island Lighthouse and is situated on the protected or southeast side of the island. This deflation resulted both in the creation of a pavement of artifacts on a lag surface as well as in sweeping clean any buried organic material. The deflated surface contained lag artifacts, including prehistoric cultural material and cordmarked Late Woodland ceramics. In areas where it had not been scoured clean the surface lay at depths more than *10 meters* below the crest of the current dune. Test pit excavation revealed what can currently be best interpreted as an A-E-C stratigraphic sequence, suggesting stabilization and the development of a soil. South Manitou 1 Site is highly reminiscent of the situation observed at the Mt. McSauba site north of Charlevoix.

Continued archaeological research in the Sleeping Bear Dunes National Lakeshore has not discovered any further buried or stratified sites in the coastal dunes of Lake Michigan. Although of potential interest to our project, the logistic difficulties of GeoProbe transport to South Manitou 1 Site precluded direct observation of the deposits.

Fisher Lake Site

The Fisher Lake site was excavated under the direction of David S. Brose of Case Western Reserve University in 1971 (Brose 1975). Although only marginally stratified, the deposits occur on an interior sand dune ridge adjacent to Fisher Lake. Excavations at the Fisher Lake site revealed both a northern tier Middle Woodland occupation, and a superimposed Late Woodland occupation. Fisher Lake, as an interior site, was not directly pertinent to our current research and was not revisited.

Eastern Lake Michigan, from Sleeping Bear Dunes to Muskegon

There has been little in the way of systematic coastal dune research on archaeological sites in this segment of the Lake Michigan basin despite the fact that there are numerous sites listed in the State of Michigan archaeological files. Of the many sites in the inventory, those in dune contexts are not reported as having buried or stratified deposits, with few exceptions: the Porter Creek and Camp Miniwanca sites. Camp Miniwanca was recorded as part of the current research project. Porter Creek clearly demonstrated that stratified dune deposits could potentially be preserved high on the crests of perched dunes. Both the Camp Miniwanca and Porter Creek sites are described in chapter 5 and appendix C.

Southeastern Lake Michigan, from Muskegon to the Michigan-Indiana Border

Despite the presence of several small and dispersed parcels of state-owned land in the coastal area south of Muskegon and Ludington, and the presence of an impressive inventory of coastal archaeological sites, we had difficulty identifying major research efforts, either larger-scale systematic surveys or site excavations, in the coastal region. To be sure, a number of small, project-specific reports available provide idiosyncratic spatial information (for example, see Fitting 1967 on the assemblage from the Hamlin Lake site, one of many deflated sites near the Nordhouse Dunes and Ludington State Park), and even some major research reports on significant archaeological sites such as the Dumaw Creek site, which is actually an interior and not a coastal dune site (Quimby 1966). Most systematic survey in the region, however, has been along the interior parts of major drainages that ultimately empty into Lake Michigan and along proposed inland highway corridors. Thus, unlike more northern areas, no synthetic body of systematically collected site data exists to guide our research.

As other parts of this book reveal, however, the converse holds for dune geomorphological research. Importantly, the coastal dunes in this region are a form of perched system that mantles topographically low, pro-glacial lake plains. In some cases, they are found on higher bluffs composed of glacial sediment. These dunes are also largely stabilized, with occasional eolian exposures. They are topographically and evolutionarily different from most of the areas to the north where intensive archaeological survey has taken place. In addition, as noted in chapters 2 and 6, shorelines in southern Lake Michigan have also undergone considerable erosion during the past 5,000 years and may have receded several hundred meters from their former position.

As noted earlier in this discussion, the work of federal agencies in assessing archaeological and historical properties on federally owned land in and adjacent to Michigan has increased, and provides an abundant body of systematically collected data from which to draw. To that end, we employed the cumulative information from the Indiana Dunes National Lakeshore as a proxy signature for other parts of southern Lake Michigan with more poorly resolved information. Although the history of research within what is now the National Lakeshore goes back to the 1800s, more recent work is abundant, some being project specific (e.g., Lynott et al. 1998; Sturdevant and Bringelson 2007a), and some synthetic (Sturdevant and Bringelson 2007b). It is the latter that is particularly useful in the current context.

Given our interest in site burial and stratification in dune contexts, it was instructive to read the following summary statement relative to vertical artifact concentrations within the many sites recorded in the dunes; "Obvious cultural layers, with buried paleosols and associated artifacts, have not been identified in association with these concentrations" (Sturdevant and Bringelson 2007b, 73–74). This observation suggests that site burial processes have been insufficient, for as yet unknown reasons, to result in the association of cultural material with preserved and buried organic soil surfaces. However, much like the observations at the Whisky Creek site mentioned earlier, artifacts have been recovered as deep as 60 cm, perhaps due to dune formation processes, or bioturbation, or other unknown factors (Sturdevant and Bringelson 2007b, 74). Sturdevant and Bringelson make the case, however, that Archaic age artifacts occur largely within the B horizon, while Woodland artifacts are primarily found in the A horizon (2007b, 74). This would suggest that earlier, Archaic age, materials may have been deposited earlier than the remaining and presumably initial stabilization episodes. Or they were initially in stable soils that were reactivated. This reactivation resulted in deflated surfaces that were subsequently buried and restabilized during the Woodland period beginning circa 2,500 years ago.

Of further interest to us were the reported positions of sites within the stabilized dune system. For example, "Elevation on individual dunes vary; many sites are found on or near dune crests . . . , often on benches or saddles sheltered by hilltops, but sometimes on the hilltops themselves. Some sites are located on benches midway between a dune's top and its base" (Sturdevant and Bringelson 2007b, 78). Moreover, the entire cultural chronological sequence is represented from late Paleoindian to modern. Notably, sites are often found in the lee or on the slip face of the dunes, or on downslope locations within dune sequences.

Summary

Numerous archaeological sites are located on the Lake Michigan coastal zone in Michigan and the adjacent states of Wisconsin and Indiana. Although some are situated in eolian dunes, most are not. However, many of the sites situated in these coastal dunes are evidently not well preserved due to activation cycles resulting in deflation and erosion. This often leads to the deposition of cultural materials as secondary lag pavements rather than as intact deposits in clearly defined stratigraphic contexts. Buried sites in eolian contexts with intact occupation horizons, either single occupation zones, or multiple and stratified occupation zones, are rare. To date, based on almost a century of coastal survey data, very few, if any, stratified and preserved sites are present in dunes south of the hinge line, which demarks relatively rapid uplift to the north from limited uplift or subsidence to the south (see discussion in chapter 4) in the Michigan basin, unless they occur at undiscoverable depths. Even in intensively surveyed areas such as the Indiana Dunes National Lakeshore at the southern end of the basin, which is under federal management and has an impressive cumulative site inventory, none of the sites are in buried or stratified contexts, although they may be in surficially preserved soil horizons suggesting primary deposition (Sturdevant and Bringelson 2007b). Detailed literature searches revealed no buried or stratified dune sites of interest in this subregion of the Lake Michigan basin, despite the fact that recorded sites are quite numerous. Why is this the case? Is the lack of sites from such contexts related to incomplete survey and site discovery methods or related to differential preservation of site populations?

The age of such sites is also of interest. As we have articulated elsewhere (Monaghan and Lovis 2005) there is a long association between archaeologists using fictively, marginally, or speculatively dated landforms as a basis for making chronological estimates of site age. So, for example, the Solomon Seal site is situated at the crest of a large dune in Charlevoix County, with an altitude greater than that apparently achieved by the Nipissing stage maximum. According to common wisdom this means, ipso facto, that it is a landform formed during a post-Nipissing event. The occupation must be Archaic in age as a consequence. However, neither the Solomon Seal site nor the dune within which it is situated had been absolute dated by ^{14}C or other means. At present, the oldest buried and stratified cultural deposits in the Lake Michigan basin date to circa 2000 BP based on dates from Summer Island site and Portage site (as well as other dates obtained as part of this project). From the luxury of retrospection, the Eastport site in Antrim County may be best interpreted to be a lag pavement in eolian context. There are, however, far more buried and stratified sites in coastal dunes that postdate circa 1,500 to 1,400 years ago based on both radiocarbon dating and ceramic cross-dating, such as the O'Neil, Ekdahl-Goudreau, Scott Point, Summer Island, and South Manitou Island sites. Is this greater abundance of middle to late Late Woodland sites an accurate reflection of site formation processes within dunes, and if so are these processes predictable?

Finally, one needs to ask where stratified and buried archaeological sites are situated on landforms north of the hinge line. As will become evident, there are multiple associations with different variables, such that the same site may have multiple overlapping locational characteristics. With few exceptions, west-facing exposures expectably prevail given that primary wind directions are from the southwest and northwest. These include the west side of points projecting into Lake Michigan (Winter, Scott Point, Mt. McSauba sites), as well as at creek mouths and embayments (Winter, Wycamp Creek, O'Neil, Antrim Creek sites). Almost all are situated in the swale behind the foredune, which seems to be the case south of approximately Charlevoix (e.g., Mt. McSauba, O'Neil, Antrim Creek sites, although the Portage site is in a high dune swale north of Petoskey). Exceptions to foredune swale locations occur in situations where prevailing southwest winds can sweep across broad surfaces, such as the areas north of Charlevoix including Wycamp Creek, Scott Point, and Winter sites (and perhaps Ekdahl-Goudreau). Other prominent, and currently unexplained, local exceptions to the above include

South Manitou 1 Site, with its southeast exposure. Notably, too, there are fewer sites with multiple occupation strata than buried sites with single strata. These latter sites are confined to the shoreline from Charlevoix north and include the O'Neil, Scott Point, Ekdahl-Goudreau, Winter, and Summer Island sites. South Manitou 1 Site is also the most southern of the buried sites of any type unearthed by the literature survey. Clearly the spatial systematics of the different formation processes across the Lake Michigan basin need to be better defined.

Implications for the Dune Activation and Cycling and the Archaeological Site Taphonomy Project

At the outset of this project, our proposal did not stipulate specific locales that would be investigated, but rather articulated the need to assess which sites might best assist us to answer the questions we had posed about the relationships between archaeological site formation processes and the cycling of dune activation and stabilization. The information gained from our survey of the literature on coastal dune archaeology in the Lake Michigan basin affected our weighting of our field and lab research for the archaeological component of the taphonomy project.

For example, we believed it was necessary to place high priority on visits to several known and only partially reported archaeological sites within the project area, including the Winter, Ekdahl-Goudreau, Scott Point, Wycamp Creek, Mt. McSauba, and Antrim Creek sites, and possibly the Eastport site. The goal was to perform additional coring and/or OSL or AMS sampling at these sites to assist in the development of chronologies for the region. Our visits were only partially successful, and in some cases alternative field and lab methods were implemented. As noted above, we encountered access difficulties at Scott Point site, and the property owners of the Wycamp Creek site were not responsive; we could not core either site. Given the nature and depth of the deposits at the Mt. McSauba site, coring was not feasible, but OSL sampling was nonetheless conducted using standard techniques on the face of a parabolic dune exposing the archaeological horizon. We could not directly core and sample the disturbed surfaces of the Eastport site, but we were able to sample a small dune deposit nearby and place a *terminus ante quem* on the age of the archaeological deposits. Serendipity likewise played a role in our fieldwork. During a field visit to a parabolic dune near Muskegon that displayed stabilization episodes in the form of exposed paleosols, we encountered archaeological deposits including a hearth buried under 20 meters of sand.

For certain sites lacking absolute dates for various reasons, including a lack of datable materials with good context, we decided to augment the AMS or ^{14}C record in an attempt to tie down points of site stabilization. This was one component of our work at the Winter site, and particularly important at the Scott Point site when it was clear that we would not be able to core the deposits. At both sites we employed carbonized food residues extracted from ceramic sherds of known type and vertical provenance to extract AMS dates from various segments of the stratified sequence. This is a proven technique and yielded excellent results (Hart and Lovis 2007a, 2007b; Lovis 1990b, 1990c). Details of methods, individual site sequences, and our results are found in other parts of this book (appendix A, chapters 5 and appendix C, respectively).

CHAPTER THREE

Coastal Dunes, Eolian Processes, and Activation-Stabilization Cycles

This chapter presents an overview of research on the environmental variables associated with the evolution of Michigan's coastal dunes. The first section centers on eolian processes, specifically the way that wind moves and deposits sediment. The second portion of the chapter explores the early research on coastal dunes along Lake Michigan, including how vegetation affects the evolution of foredunes and the development of large dunes in various topographic positions (figure 3-1). The third section describes modern research on the geomorphology and evolution of Lake Michigan coastal dunes.

Dune Formation

Eolian Processes

Within the context of this research it is important to understand how flowing air modifies sandy landscapes. This discussion falls within the general framework of eolian processes, which are those that erode, transport, and deposit sediment by flowing air. At a fundamental level, flowing air is very similar to flowing water because wind is also a fluid that behaves according to specific physical laws (Bagnold 1941). For example, flowing air and water both exhibit a predictable velocity gradient in the vertical dimension in which flow speed close to the surface (or boundary layer) is zero with a rapid increase as elevation increases. The ability of wind to erode sediment is also consistent with water in that sediment particles begin moving once a critical threshold velocity is passed. This threshold velocity varies according to the size of any given particle, with small grains naturally more easily moved than larger ones (figure 3-2).

In this context, the most significant difference between flowing air and water is fluid density, with water being much denser than air. This difference is most important with respect to sediment transport because flowing air must move nearly

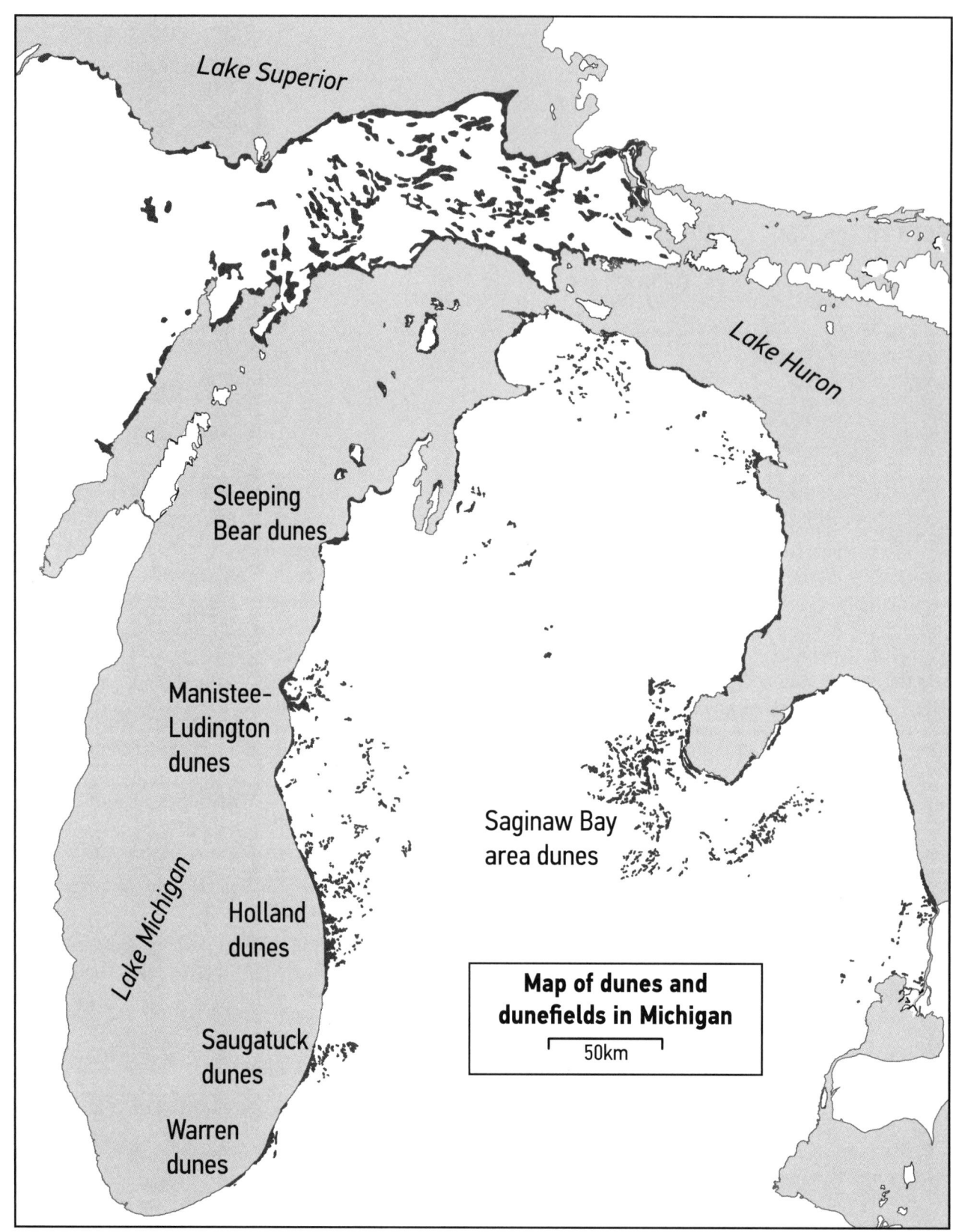

3-1. Major dunes and dunefields in Michigan; after Schaetzl et al. (2009).

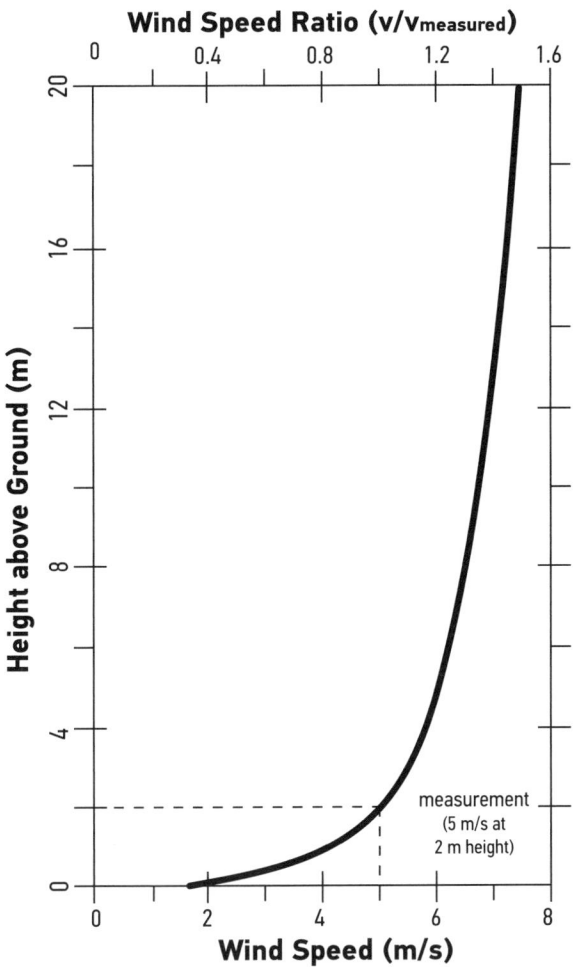

3-2. Near-surface velocity gradient of wind. Wind speed at ground level is very low due to boundary layer conditions. Above the level of friction, wind speed increases rapidly.

30 times faster than flowing water to move a particle of the same size. As far as wind is concerned, the average threshold velocity ranges from 10^{-6} m/sec for clay-sized (< 0.002 mm) particles to about 10 m/sec for the coarsest (2 mm) sands (Bagnold 1941). Within this continuum relatively small clays and silts (< 0.002–0.05 mm in size) can be transported hundreds of kilometers in suspension due to their low fluid thresholds. Sands (0.05–2.0 mm in size), in contrast, are typically transported only short distances because of their relatively large size.

The mobilization of eolian sand follows a very predictable pattern that was first recognized in wind-tunnel experiments by Bagnold (1941). He demonstrated that a general relationship exists between flow velocity and the way sand grains move (figure 3-2). As the threshold velocity is approached, sand grains begin to oscillate and roll downwind in a process called *creep* (figure 3-3). If wind velocity continues to increase, individual sand grains will begin to bounce downwind in a process called *saltation*. Saltating grains typically move within about 3 cm of the surface and have individual trajectories that are based on their density, shape, and mineralogy (Bagnold 1941; Gerety and Slingerland 1983; Greeley et al. 1983). As these grains fall to the surface, they often promote additional saltation through ballistic impacts that lift new sand particles into the wind stream.

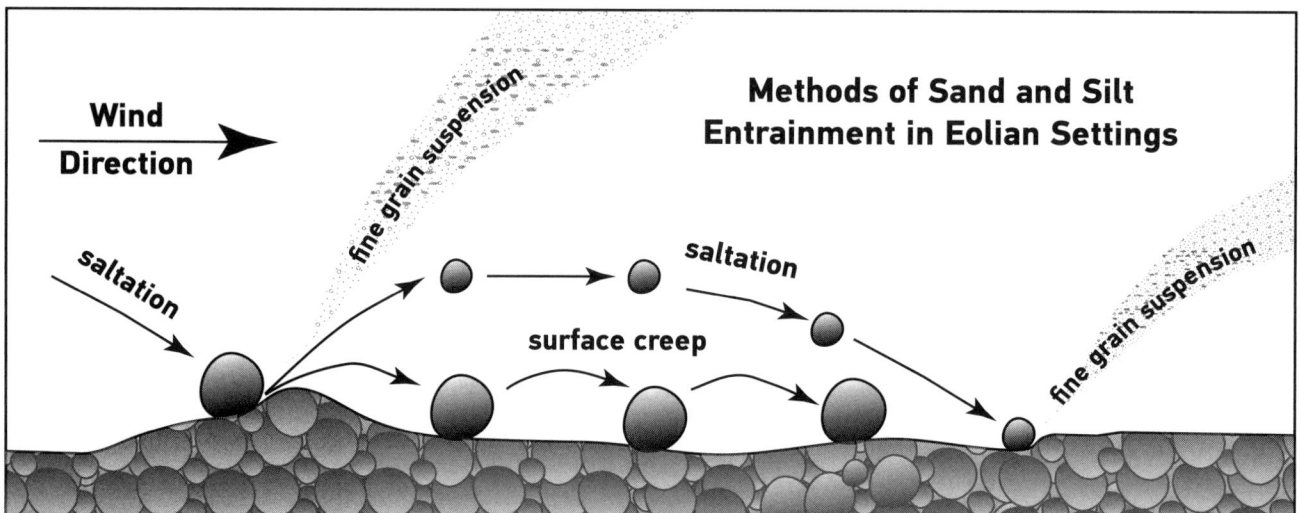

3-3. Method of sediment transport by the wind. Small pebbles and large sand grains move on the surface by creep. Somewhat smaller sand grains bounce along by saltation. The smallest grains, such as silts and clays, move through the air by suspension; after Blumer (2008).

Formation of Sand Dunes

Once sand grains begin moving, they will continue doing so as long as wind speed remains above a critical velocity. If wind speed drops below that threshold, sand grains will begin to drop out of the wind stream and be deposited upon the surface. Such deposition usually begins locally when obstacles such as vegetation or a man-made structure cause wind speeds to decrease (Olson 1958a). During the early stages of deposition, a small mound of sand will begin to form.

As additional sand is deposited the mound begins to acquire a characteristic cross-sectional profile that is associated with sand dunes, with a distinct *backslope*, *crest*, and *slip face* (figure 3-4). The backslope is a shallow (~ 8° to 13°) windward slope where erosion dominates (Livingstone and Warren 1996). At the peak of the dune is the crest, where erosion is in equilibrium with deposition. Leeward of the crest is the slip face, which is a depositional surface formed because sands arcing over the crest fall out of the wind stream and settle to the ground. The slip face is usually at or near the angle of repose, which for eolian sand ranges between 30° and 34° depending on the average grain size of the sediment. Sand dunes generally retain their characteristic profile as they migrate in relationship to prevailing winds. Where these winds are intense and unidirectional, sand dunes can move upward of 1m per day (Livingstone and Warren 1996).

Coastal Dune Research along the Lake Michigan Shoreline

Early Research

The coastal sand dunes along the eastern shore of Lake Michigan have been a focal point of geological and geographical interest for over 100 years. The earliest

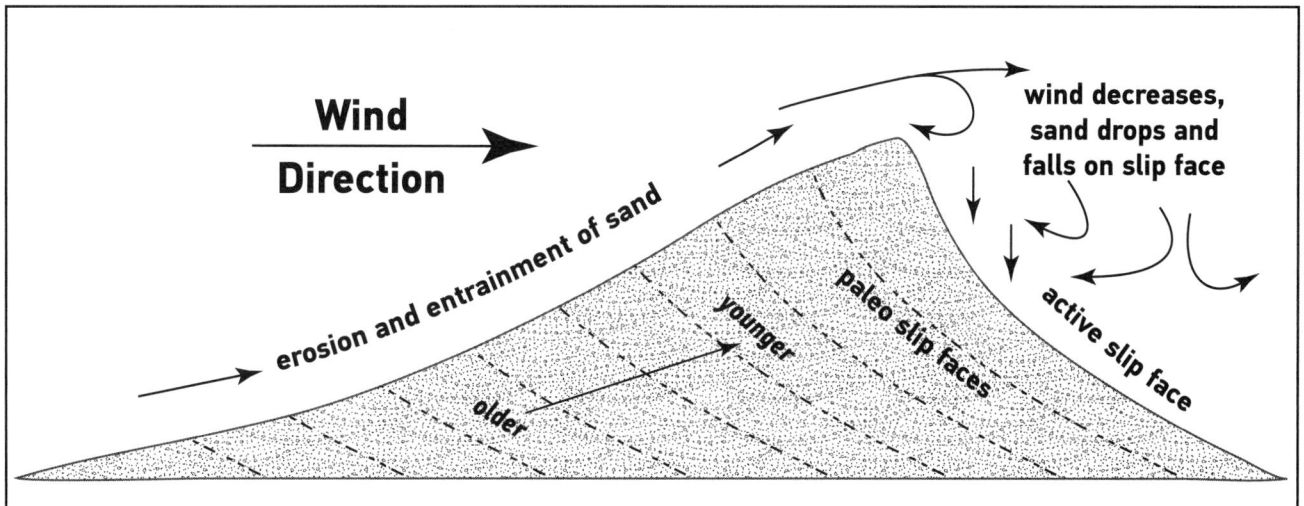

3-4. Cross-section of a typical sand dune. Sand moves up windward slope by saltation. Once the sand grain moves past the crest, it slides down the slip face by gravity. Dune migrates in direction of wind.

research on these landforms was conducted by Cowles (1899), who studied dune vegetation and how certain plant species relate to the geomorphology of the coast and the evolution of foredunes. Cowles (1899) divided coastal sand bodies into five geographic/geomorphic units and associated ecological stages: (1) the beach, (2) primary embryonic dunes, (3) secondary embryonic dunes, (4) wandering dunes, and (5) established dunes. This classification scheme became the foundation for further research on Lake Michigan coastal dunes.

According to Cowles (1899), the beach functions as the sand source for Lake Michigan foredunes when prevailing westerly winds flow across that surface. Under these conditions, eolian sand moves inland until it reaches an obstruction, such as a pocket of vegetation. Sand subsequently piles up quickly around the obstruction, forming a primary embryonic dune that has characteristic windward and leeward slopes. These dunes subsequently stabilize when sand supply diminishes and marram grass (*Ammophila brevigulata*) expands across the surface.

In contrast to primary embryonic dunes, secondary embryonic dunes are found farther inland and grow more slowly. Nevertheless, the crests of secondary embryonic dunes may reach such a height that the anchoring plants are unable to reach the water table, or they are exposed to the desiccating effects of wind exposure (Cowles 1899). These developments reduce vegetative cover and increase erosion, at which point the dunes may become "wandering dunes" that migrate with the prevailing winds in the predictable way. According to Cowles (1899), these wandering dunes are the building blocks for more complex dune systems. The final component of Cowles's (1899) classification scheme was "established dunes," which are fundamentally wandering dunes that have stabilized by the growth of climax plant species such as basswood (*Tilia americana*). Cowles acknowledged that established dunes are subject to erosion and may again become wandering.

Following the Cowles study, the next research on coastal dunes was conducted by Dow (1937), who investigated the formation of what he called *perched dunes* along the east coast of Lake Michigan. According to Dow (1937), perched dunes are large (> 30 m) parabolic dunes that mantle topographically high surfaces such as glacial headlands. Although such dunes occur in many places along the northeastern shore of Lake Michigan, he was particularly interested in determining the source area for the sand contained within the dunes that mantle the Manistee moraine. He hypothesized four possible sources, including (1) sand of beach origin that is driven directly upslope; (2) sand of beach origin that is driven upslope as lake level rises toward the base of the bluff; (3) sand derived from glacial sediments exposed in the bluff; and (4), a combination of the three sources. Based on field observations, Dow (1937) concluded that the dominant source is glacial sediment exposed in the bluff face.

In addition to determining the source of eolian sand in perched dune systems, Dow (1937) also introduced a model for the formation of these dunes. He suggested that perched dunes form during periods of high lake level when intense wave action erodes the bluff, which destabilizes the upper part of the slope such that sand can be liberated and blown to the adjacent plateau. He argued that sand supply to perched dunes is reduced when lake level falls and coastal erosion diminishes.

Following Dow's study, Scott (1942) gave a presidential address to the Michigan Academy of Sciences about his observations of Lake Michigan coastal dunes. He argued that foredunes are the primary dune form and that they develop during low lake phases. When lake level rises, these dunes are either eroded completely or modified to form secondary dunes that contain blowouts and/or have parabolic forms. Scott (1942) also noted that dune landscapes north and south of the isostatic hinge line are fundamentally different. South of the hinge line, dune complexes typically consist of closely spaced ridges that are frequently imbricated on one another. In contrast, dune landscapes north of the hinge line consist of well-defined ridges that are spaced apart over a broad zone. According to Scott (1942), this difference occurs because isostatic uplift continues north of the hinge line, resulting in progressively raised beaches that are the foci of deposition during low lake phases.

Shortly after Scott's (1942) observations, Tague (1946) used Cowles's (1899) findings to describe the dunes of the Grand Marais Embayment along Lake Michigan's southeastern shore. Like Cowle's investigation, Tague's study was a qualitative description of dune types and development. He argued that foredunes are the primary dune form along the coast and that sand is supplied to them from the adjacent beach. Foredunes may then erode or be modified in some fashion that leads to the development of large secondary coastal dunes. In contrast to Cowles, Tague was also interested in the evolution of the coastal dunes. In this context, he used the relative location of various dune forms to infer a chronology of dune evolution. He argued that the oldest dunes, which mantle older lake terraces, were probably constructed during the Nipissing high stand. The youngest dunes, in contrast, cover topographically lower surfaces that are closer to the lakeshore.

In the late 1950s a series of important studies were conducted by Olson (1958a, 1958b, 1958c) that contributed greatly to the growing understanding of Lake Michigan coastal dune evolution and the relationship to plant succession.

In the first of these studies, Olson (1958a) used plant structure and distribution to reconstruct changes in dune morphology. He specifically used the distance between stem nodes of dune-colonizing plants such as marram grass as an indirect measure of annual sand accumulation. Olson (1958a) argued that widely spaced nodes occur after rapid deposition of sand in the winter, whereas narrow node spacing reflects relative dune stability because the grass grows more slowly as sand accumulation decreases. Olson (1958a) suggested that active dunes accumulate approximately 1 foot of sand per year.

In his second study, Olson (1958b) proposed a new model for the evolution of Lake Michigan foredunes by suggesting that their formation is tied to lake-level fluctuations, with growth phases occurring during low-lake intervals. In this model, the foundation of a foredune is an offshore bar that forms in shallow water where waves break. When lake level subsequently falls, the ridge is exposed inland of a wide beach that functions as an eolian-sand source. This ridge then becomes the focal point of eolian sand deposition forming an incipient foredune. The foredune continues to grow as marram grass begins to colonize the dune, which promotes additional dune growth by trapping eolian sediment. This landform will continue to evolve as long as water levels remain low. In contrast, it will likely erode when a subsequently high lake phase occurs. Olson (1958c) hypothesized that lake-level oscillations sufficient to form these landforms probably occurred every 30 years (1958c).

Modern Dune Research

After Olson's work in the late 1950s, research on Lake Michigan coastal dunes was not pursued in any kind of systematic way until the 1990s. In the 1970s only two noteworthy references to coastal dunes were made. The first of these studies was by Dorr and Eschman (1970), who briefly discussed the basic geomorphology of the dunes in the *Geology of Michigan*, the second by Buckler (1979), who constructed a classification system for dunes along Lake Michigan. Dorr and Eschman (1970) referred to the large dunes as *high dunes* because of their immense size and generally assigned a Nipissing age to them. Subsequently, Buckler (1979) introduced the term *barrier dunes* because the large dunes form a significant topographic barrier between the lakeshore and the interior. He further classified groups of barrier dunes based on their morphologic characteristics, including categories such as blowouts, dune ridges, domal dunes, foredunes, overlapping dunes, and parabolic dunes among others (figure 3-5). In addition to providing this classification scheme, Buckler (1979) argued that the most extensive period of dune formation probably occurred during the waning phases of the Nipissing high stand, when sediment supply was increased due to wide beaches.

Following Buckler's (1979) report, the next research focusing on Lake Michigan coastal dunes was conducted by Snyder (1985), who investigated the perched dunes (e.g., Dow 1937) at Sleeping Bear Dunes National Lakeshore. These dunes mantle a tall (ca. 100 m) coastal bluff composed of glacial sediment contained within the Manistee Moraine. Snyder's (1985) study was possible because erosion of the bluff had revealed several buried soils that he used to reconstructed the chronology of eolian activity. Snyder was the first to acquire radiocarbon dates from coastal dunes in Michigan.

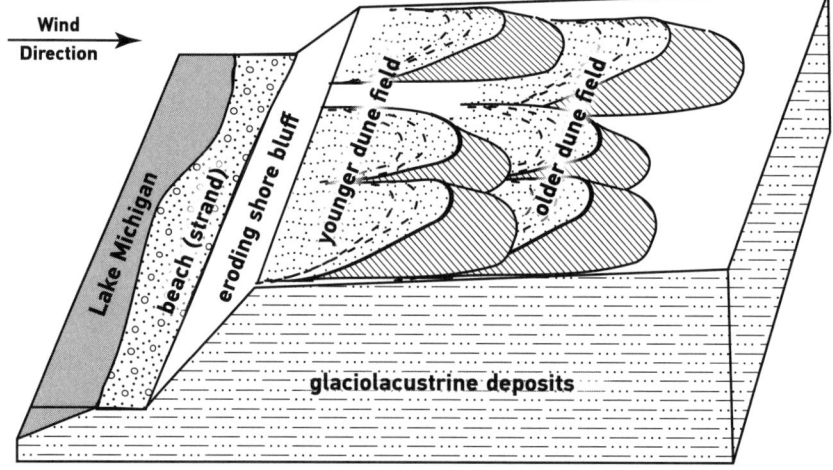

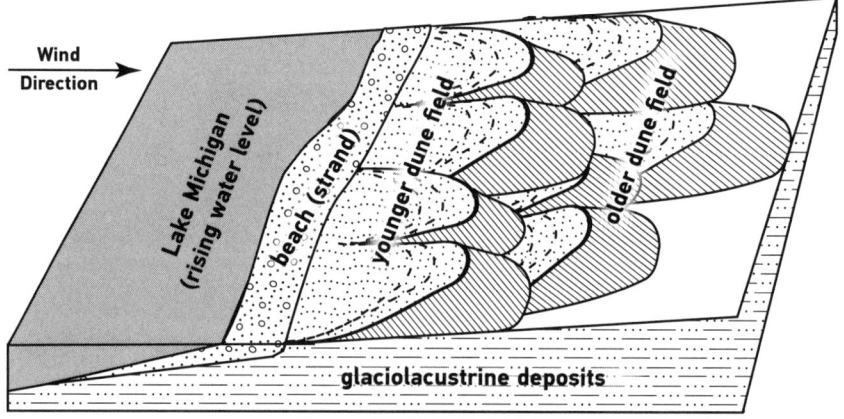

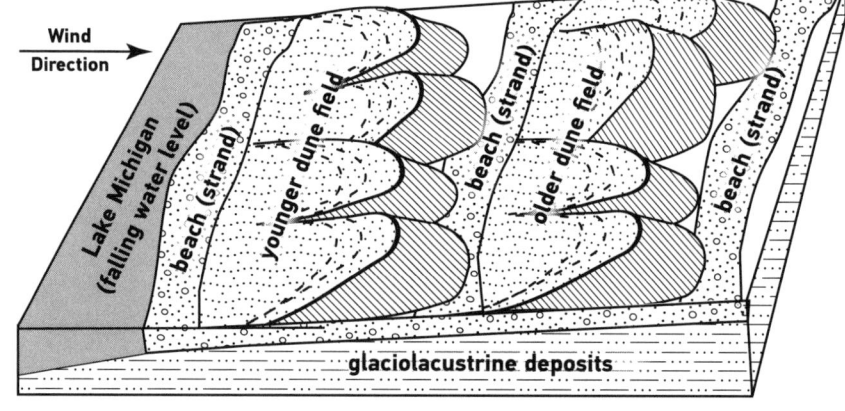

3-5. Classification of dune morphological types within typical Lake Michigan shoreline setting; after Wilson (2000).

The oldest date Snyder obtained was from a moderately developed paleosol formed in the uppermost part of the glacial sediment. This soil provided a date of 4559 ± 225 BP, suggesting it was buried by eolian sand during the Nipissing high stand. Snyder (1985) argued that the soil formed during the Chippewa low phase of Lake Michigan when the active lakeshore was far to the west. Initial mobilization of eolian sand subsequently began during the Nipissing high stand because wave erosion at the base of the bluff destabilized the upper bluff face so transport of eolian sand could occur. In addition to the basal paleosol, Snyder (1985) identified a pair of soils in the overlying eolian sands. The oldest of these soils yielded a radiocarbon date of 2781 ± 160 BP, whereas the younger soil provided a date of 688 ± 180 BP. Based on these dates Snyder believed that Dow's (1937) perched dune model best explains the growth of dunes at Sleeping Bear Dunes National Lakeshore.

Following the Snyder (1985) study, research on coastal dunes in Michigan surged in the 1990s. This wave of research focused on reconstructing the geomorphic evolution of dunefields through stratigraphic analyses and radiocarbon dating of buried soils. The first of these detailed geomorphic investigations was conducted by Anderton and Loope (1995), who investigated perched dunes at the Grand Sable dunefield on the southern shore of Lake Superior in Michigan's Upper Peninsula. The primary goal of this study was to test the validity of the perched dune model. Anderton and Loope (1995) conducted stratigraphic investigations at 17 sites, with each including at least one buried soil from which a radiocarbon date was obtained. The resulting dates were subsequently compared to the hypothetical lake hydrograph generated by Fraser et al. (1990).

Consistent with the observations at Sleeping Bear Dunes National Lakeshore (Snyder 1985), Anderton and Loope (1995) noted that the dunes at Grand Sable are directly underlain by a well-developed soil—in this case, a Spodosol—that formed in the uppermost part of the thick glacial sediments at the site. This soil was originally recognized by Farrell and Hughes (1985) and was called the *Sable Creek Soil* by Anderton and Loope (1995). Given the horizonation of the soil, Anderton and Loope (1995) suggested that it began developing immediately after glacial ice retreated from the Lake Superior basin about 9800 BP (Drexler et al. 1983). Radiocarbon dates from the soil indicated to Anderton and Loope (1995) that it was buried between 5170 BP and 4640 BP. Based on this time frame, Anderton and Loope argued that eolian sand deposition began during the Nipissing lake phase when sustained high water levels caused incision of coastal terraces and destabilization of bluffs.

In addition to the Sable Creek Soil, Anderton and Loope (1995) recognized a variety of other paleosols in dune exposures that are far less developed, indicating relatively brief periods of landscape stability and soil formation. Radiocarbon dates obtained from these soils indicated that episodes of dune growth occurred at circa 4,000, 3,550, 2,900, 1,500, 1,000, and 500 years ago. These ages fundamentally correlate with high lake phases in a hypothesized lake level curve (Fraser et al. 1990), which led Anderton and Loope (1995) to conclude that sand is supplied to the Grand Sable dunes in a manner consistent with the perched dune model (Dow 1937).

Following Anderton and Loope's (1995) study, Arbogast and Loope (1999) turned their attention to dunes along the eastern shore of Lake Michigan and investigated a variety of large (> 30 m high) dunes near Grand Haven and Muskegon. These dunes mantle proglacial lake terraces that, in comparison to the perched dunes, lie only a few meters above the lake. Given this landscape position, it had been generally assumed (e.g., Buckler 1979) that Olson's (1958c) foredune model best explains their formation, and that they developed shortly after the Nipissing high stand of ancestral Lake Michigan when beaches were wide and the eolian sand supply was high (Dorr and Eschman 1970).

Arbogast and Loope (1999) tested this hypothesis by radiocarbon dating soils formed in the uppermost lake sediments that underlie dunes at four locales. Whereas dates from three locales suggested that initial dune growth began shortly after the Nipissing stage (4,820–4,410 years ago), a date from a fourth locale indicated that the first pulse of eolian sand there occurred during the Algoma high stand (3270–2940 BP). According to Arbogast and Loope (1999), the results suggested that the onset of dune formation was not synchronous across the shore and that dune growth occurred during high lake stages. This latter finding was especially significant because it suggested that, despite the relatively low topographic position of the large dunes, they may have developed in a manner consistent with Dow's (1937) perched dune model.

At about the same time that Arbogast and Loope (1999) published their findings, a study by Arbogast (2000) investigated the timing of major perched dune growth along the southern shore of Lake Superior. This study focused on the Nodaway dunefield, which is perched about 90 m above Lake Superior about 20 km west of Sault Ste. Marie. This dunefield is isolated on a raised headland and contains a number of immense dunes that are > 50 m high. Arbogast (2000) tested the hypothesis that the dunes formed during the Nipissing high stand in a manner similar to the dunes at Sleeping Bear Dunes National Lakeshore (Snyder 1985) and the Grand Sable dunes (Anderton and Loope 1995).

To test this hypothesis, Arbogast (2000) used a combination of soils analysis and radiocarbon/OSL dating to determine the age of the dunes. A maximum-limiting age of circa 7,000 years ago was obtained from the edge of the bluff, which suggested to Arbogast (2000) that some dune building probably occurred during the Nipissing transgression. In addition to this age, an AMS radiocarbon age of 3480–3160 cal BP (see appendix A for a discussion of the conventions underlying this designation) was obtained in a soil formed in the top of the underlying glacial sediments beneath the lee slope in the southernmost part of the dunefield. This date, coupled with OSL ages of circa 3,700 and 3,100 years ago from dune crests elsewhere, indicated that the last major interval of dune growth occurred after the Nipissing high stand. Based on this chronology, Arbogast (2000) suggested that most dune growth occurred during the Nipissing stage and that the dunefield slowly stabilized subsequently.

As the surge in coastal dune research in Michigan accelerated, Loope and Arbogast (2000) conducted a study that more rigorously tested the hypothesis (e.g., Arbogast and Loope 1999) that large dunes on topographically low lake terraces in Lower Michigan enlarged during high lake stages. In this later study, Loope and Arbogast (2000) radiocarbon-dated 75 buried soils at 32 sites along the eastern coast of Lake Michigan. Given the large sample size, they statistically compared the radiocarbon distributions to the lake-level curve produced by Thompson and Baedke (1997, 2000). According to Loope and Arbogast (2000), the results of this analysis indicated that dune building episodes on the large dunes on lake terraces correlate with high lake intervals that occurred about every 150 years. These results support the hypothesis that the dunes formed via Dow's (1937) perched dune model.

In 2002, Arbogast et al. continued research on large dunes on the eastern shore of Lake Michigan near Holland, Michigan. This study assessed stratigraphic relationships at four locales and included several radiocarbon dates obtained from buried soils. These dates indicated that the dunes initially began forming during the Nipissing high stand (ca. 5,500 years ago) and that rapid and episodic dune growth subsequently occurred from circa 4,000 to 2,500 years ago. Arbogast et al. (2002) argued that this period of rapid dune formation may correspond to Olson's (1958c) foredune model in that sand was supplied as the lake regressed. Later periods of dune growth circa 3,200, 2,400, and 900 years ago appear to correlate with higher lake levels and thus support the perched dune model (Arbogast et al. 2002).

The stratigraphy and chronology of the dunes at Holland has been recognized at other locations along the southeastern coast of Lake Michigan. Van Oort et al. (2001) expanded on the research of Arbogast et al. (2002) at Van Buren State Park near South Haven. Van Oort et al. (2001) obtained a number of radiocarbon dates on a variety of buried soils that were exposed in a pair of excellent exposures. The basal soil at Van Buren is a peat layer that formed at the top of lake sediments. This peat provided maximum and minimum limiting ages of between circa 6,170 and 4,900 years ago, respectively, suggesting that a marsh existed for 700 to 1,000 years before the dune began to form.

Immediately above the peat is a sequence of stratigraphic units in eolian sand that contain weakly developed Entisols with A/C horizonation. These soils provided radiocarbon ages that ranged from circa 3,690 to 1,970 years ago, suggesting that episodic dune building lasted about 2,500 years. The best-developed buried soil at Van Buren occurs in the upper portion of the exposure and consists of a weakly developed Spodosol. A similar soil was also recognized in the Holland study (Arbogast et al. 2002). At Van Buren State Park, charcoal from the Ab horizon in this soil provided an age of circa 500 years ago (Van Oort et al. 2001). Given the date of circa 2,000 years ago from the paleosol immediately below, Van Oort et al. argued that the Spodosol formed over a circa 1,500-year period. Above this paleosol is a pair of Entisols that developed between about 300 cal BP and the present. Overall, Van Oort et al. (2001) concluded that the dunes at Van Buren are similar in both chronology and geomorphology to the dunes near Holland (Arbogast et al. 2002).

Research on the coastal dunes along the east coast of Lake Michigan has largely focused on the large dunes (or barrier dunes; Buckler 1979) that front the lake. Although these dunes are certainly the most prominent eolian landforms along the southeastern shore of the lake, a variety of smaller dune complexes occur in their lee. The evolution of these *backdunes* was the focus of a study near Holland (Hansen et al. 2003). In order to determine the chronology of these landforms, Hansen et al. (2002) obtained OSL samples from the uppermost sand deposits in the dunes. This study thus marked the first effort to systematically reconstruct dune history along Lake Michigan with this method.

Results from the study indicate that the backdunes near Holland are much younger than originally thought (e.g., Tague 1946). A basal sample provided an age range of 4,300 to 3,600 years ago, whereas those taken from the crests of the dunes range in age from 4,000 to 3,700 years ago. These dates indicate that the backdunes and the lake-fronting dunes were simultaneously active in the early phase of coastal dune formation. Hansen et al. (2002) suggested that the dunes were active as the lake fell after the Nipissing II high stand. In this scenario, falling lake levels exposed a wide beach to supply sand for dune building via Olson's model (1958c). In either case, OSL dates suggest that the dunes stabilized between 4,000 and 3,500 years ago. Hansen et al. (2002) hypothesized that sand delivery to the backdunes ceased because the lakeward dunes (Arbogast et al. 2002) grew sufficiently large to cut the sand supply off, allowing vegetation to stabilize the dunes (Hansen et al. 2003).

By 2004 several studies (e.g., Arbogast et al. 2002; Van Oort et al. 2001; Hansen et al. 2004) had reported the presence of numerous buried soils in coastal dunes on Lake Michigan's southeastern shore. The most prominent of these soils was the weakly developed Spodosol that is present in the upper third of many dunes from Indiana Dunes National Lakeshore in Indiana to Montague, Michigan. This relatively well developed soil was the focus of a study by Arbogast et al. (2004), who assessed its character and age in detail. The stratotype of this soil is a weakly developed Spodosol, exhibiting A-E-Bs-BC-C horizonation, near Holland. This degree of development clearly reflects an extended period of landscape stability when little if any eolian sand accumulated. At other locations, the paleosol has overthickened an A horizon or multiple thin A horizons. Regardless, the soil indicates a period in which sand supply to the dunes decreased or stopped altogether.

Arbogast et al. (2004) assigned an age to the soil on the basis of its development, radiocarbon dates, and data from Arbogast et al. (2002) and Hansen et al. (2004). Results from this data integration indicate that the soil formed in eolian sand that was deposited circa 3,000 to 2,500 years ago. It was subsequently buried circa 1,000–900 years ago in the Holland, Michigan area (Arbogast et al. 2002) and circa 400 years ago at Van Buren State Park. Due to its prominence along the lakeshore and obvious geomorphic significance, Arbogast et al. (2004) informally named this soil the *Holland Paleosol*.

At the same time the Holland Paleosol was studied by Arbogast et al. (2004), Van Dijk (2004) investigated the seasonal timing and patterns of sand deposition on foredunes at Hoffmaster State Park. These patterns were determined by using erosion pins, sand traps, and microclimate instruments to measure changes in dune morphology. Van Dijk found that the rate of sand accumulation on the dunes was greatest during the fall, winter, and early spring, when onshore winds are strongest. By late fall, the foredune had grown sufficiently in size to cut off the sand supply to the secondary foredune. Sand movement on the secondary foredune consisted mostly of reworking existing dune sands. Erosion of the secondary foredune and blowout were also observed, with those sediments being deposited in a leeward depositional area.

Van Dijk (2004) concluded that wind direction and speeds are important geomorphic variables that affect foredunes along the coast of Lake Michigan. Lake-level fluctuations and vegetation patterns also play important roles. Van Dijk (2004) verified Olson's (1958b) model that foredune growth is more likely during periods of lower lake levels that expose wide beaches. Decreased vegetation cover in the fall, winter, and early spring contributes to increased erosion and remobilization of dune sands.

Although most dune research has focused on the southeastern part of the Lake Michigan coast, a study by Cordoba-Lepczyk and Arbogast (2005) focused on the geomorphic evolution of an isolated dune complex at Petoskey State Park near the city of Harbor Springs, Michigan. In order to determine the chronology of the dunefield, they radiocarbon-dated buried soils and obtained OSL ages from selected sites. Five geomorphic map units were identified in the park, including (from most inland to lakeward, resp.): (1) lake terrace, (2) parabolic dunes, (3) onlap dunes (dunes that overlap the parabolic dunes), (4) linear dunes and shadow dunes, and (5) active dunes.

Results from this study indicate that the lake terrace probably corresponds with Nipissing lake levels and is overlain by large (> 50 m high) parabolic dunes on the eastern margin of the park. OSL dates from the parabolic dunes indicate that eolian sands first accumulated in these landforms circa 4,800 to 4,400 years ago (during the Nipissing II transgression) and accumulated for approximately 1,300 years.

As the name implies, the onlap dunes are inset on the western edge of the large parabolic dunes. These dunes are much smaller than the parabolic dunes, which implies that they took less time to form. One buried soil was identified in these dunes, one that provided an age of circa 2,800 years ago. Based on this age, Cordoba-Lepczyk and Arbogast (2005) proposed that eolian sand was probably first supplied to the onlap dunes shortly after the Nipissing high stand when beach sands were exposed to the wind. After the stable period in which the soil formed, another period of dune growth occurred after circa 2,800 years ago. Although ages were not obtained from the linear and shadow dunes, they probably formed shortly thereafter. Dates from the active dunes near the beach indicate three periods of stability at circa 515 to 315 years ago, 470 to 290 years ago, and 300 years ago to the present. These latter periods correlate reasonably well the patterns of dune growth along the southeastern shore of Lake Michigan (Arbogast et al. 2002; Van Oort et al. 2001). Cordoba-Lepczyk and Arbogast (2005) concluded that dune building at Petoskey State Park was cyclic and related to the high lake levels proposed by Thompson and Baedke (1997, 2000).

In the very recent past an effort has been made to reconstruct the history of eolian sand mobilization along Lake Michigan by assessing the character of lacustrine sediments that accumulated in small lakes in the lee of coastal dunes. The first such study was conducted by Fisher and Loope (2005), who analyzed lake sediment obtained from Silver Lake, Michigan. They plotted sand as percent weight within the sediments and demonstrated that high concentrations correlate with the elevation curve of Lake Michigan (Thompson and Baedke 2000) since the Nipissing phase. According to Fisher and Loope (2005), it appears that influxes of eolian sand into the lake are controlled by same climatic fluctuations that drive Lake Michigan lake-level changes and that dune instability occurs during high lake levels when coastal bluffs are destabilized.

In a subsequent study, Timmons et al. (2007) investigated eolian sand concentrations in two small lakes within a coastal-dune complex southwest of Holland. Four cores were obtained in the study. Each of these cores contained visible sand laminae and invisible sand peaks within mud and sapropel units. All but one of the episodes of dune construction reported by Arbogast et al. (2002) was identified in the cores. Additional sand peaks were preserved in the lake cores, however, suggesting to Timmons et al. (2007) that the sand record from small lakes gives a more complete record of eolian activity. In contrast to the lake study by Fisher and Loope (2005), no strong correlation was observed between rising prehistoric high phases in Lake Michigan (Thompson and Baedke 2000) and peaks in lacustrine sand concentration.

According to Arbogast (2009) most of the coastal dunes along Lake Michigan are transgressive dune systems that have migrated across previously vegetated or semivegetated surfaces (e.g., Hesp and Thom 1990). Although they may move and change form over time, transgressive dunes along the Lake Michigan coast contain records of eolian sand deposition extending back thousands of years (Arbogast and Loope 1999; Loope and Arbogast 2000; Van Oort et al. 2001; Arbogast et al. 2002; Hansen et al. 2004; Cordoba-Lepczyk and Arbogast 2005). These transgressive dunes can be further subdivided into high-perched dunes and low-perched systems (Arbogast 2009). High-perched dunes are typically tens of meters above current lake levels atop coastal bluffs composed of till or outwash. Low-perched dune complexes, in contrast, are typically only a few meters above current lake level resting on lacustrine plains or baymouth bars.

The most recent study of coastal sand dunes along the shore of Lake Michigan was conducted by Hansen et al. (2010). This research integrated new age data from P. J. Hoffmaster and Warren Dunes State Parks with results from previous investigations near Van Buren State Park (Van Oort et al. 2001), Holland (e.g., Arbogast et al. 2002; Timmons et al. 2007), and Silver Lake State Park (Fisher and Loope 2005; Fisher et al. 2007) to construct a comprehensive reconstruction of dune activity along the southeastern shore of Lake Michigan. Results indicated that dune growth began during the Nipissing high stand and continued for a time after peak water levels, forming broad fields of low dunes. After a hiatus during a low lake phase, large parabolic dunes began to form around 3,200 years ago. These dunes grew episodically, resulting in sequences of poorly developed paleosols. Dune activity slowed considerably from about 2,000 to 1,000 years ago, resulting in the prominent Holland Paleosol. Renewed dune activation began shortly after 1,000 years ago and has continued episodically until the present time.

Results of Current Research on Coastal Dunes in Michigan

Sampling Strategy for Dunes along the Lake Michigan Shoreline in Michigan

The sampling strategy for this project was governed by several important considerations and interrelated research goals. One of the primary factors that influenced our overall sample location selection, for example, was to reconstruct regional trends in the age of dune formation. We paid particular attention to sampling locales where very limited or no chronological data existed. Until this study, the focus of coastal dune research in Michigan was clearly the southeastern shore of Lake Michigan (e.g., Arbogast et al. 2002; Hansen et al. 2002, 2004; Arbogast et al. 2004; Timmons et al. 2007). As a result, a great deal was already known about the age and evolution of dunes in this part of the Lake Michigan basin and a definitive regional pattern has emerged for the southeastern end of

Lake Michigan. In contrast, few studies have been conducted in the northern or northeastern parts of the basin. Those that have been conducted have focused on isolated places like Sleeping Bear Dunes National Lakeshore (Snyder 1985) and Petoskey State Park (Cordoba-Lepczyk and Arbogast 2005). Given these local studies, a regional understanding of coastal dune formation and history in this part of the basin did not exist prior to this study. In this context, we focused our sampling strategy on the northern end of the basin so we could develop a regional model of dune formation that is as robust as the current understanding for the southeastern part of the shore.

After determining that the greatest need was in the northern end of the Lake Michigan basin, we further modified our sampling strategy to test areas that were representative of specific geomorphological contexts, particularly related to the middle and late Holocene evolution of Lake Michigan. We searched for areas in our preliminary reconnaissance that would yield important pieces to a developing puzzle about coastal dune evolution in the northern part of the basin. Some sites were chosen because they allowed us to test hypotheses relative to the timing of dune formation related to the Nipissing high-water phase. The Manistique Sand Pit was chosen as a sample site, for example, because eolian sand deposits directly overlie beach gravels. This stratigraphic relationship was considered important because it enabled us to directly place the evolution of the dunes within the context of known lake-level fluctuations. Further east along the north shore of Lake Michigan near St. Ignace, dunes in the Moran dunefield were sampled because they allowed us to assess the age of widely spaced dunes in what was believed to be a Nipissing stage coastal embayment. Other sites were chosen for sampling because they enabled us to reconstruct the relationship between dune growth and isostatic rebound. In this context, the sites sampled within the Torch Bay Nature Preserve and the Fisherman's Island State Park transects allowed us to assess the ages of the dunes formed on an emerging (i.e., isostatically rebounding) lacustrine surface in the late Holocene.

Another factor that influenced our sampling strategy was the overall stratigraphy preserved within the potential study sample sites. In this context, a particularly important variable was the presence of buried soils, which indicate episodic variation in the formation of dunes and reflect intermittent periods of landscape stability. Understanding where, when, and for how long such periods of dune activation and stability occurred is key to developing a temporal and spatial framework for archaeological site burial within the dune. Although buried soils are quite common in the southeastern part of the Lake Michigan coast (e.g., Arbogast et al. 2002, 2004), they are apparently relatively rare in the northern end of the basin. As a result, locales with paleosols buried within the dunes were considered a target of opportunity whenever they were encountered because they are such key stratigraphic markers from a paleoenvironmental perspective and provide the most likely contexts for potential buried archaeological sites. Four locations were chosen for detailed study and age determination because they contained buried soils. These included the Sturgeon Bay Point exposure, the Mt. McSauba archaeological site and dune, the Camp Miniwanca archaeological site and dune, and Ludington State Park. The Sturgeon Bay Point exposure was particularly important because it is the most northerly sample locale that contains buried soils that did not also include archaeological material or contexts. Mt. McSauba and Camp Miniwanca were significant because they contain paleosols that also include prehistoric pottery and archaeological features. The former location was known to exist prior to this study; the latter, however, was only found during our research. The exposure at Ludington State Park was significant because it contains five clearly visible and easily accessible buried soils. This sample locale gave us a unique opportunity to rigorously assess the timing of dune growth as it related to a distinct episodic history.

One of the unique aspects of our research program was to use a GeoProbe to assess internal stratigraphy of dunes and collect OSL samples to determine vertical rates of dune accretion. In most previous research programs throughout the world, dune sampling necessarily focuses on existing exposures, such as an eroded dune face, and as such spatial sample distribution is limited to only the places where chance occurrence has developed such exposures. Use of the GeoProbe, and its dual-tube sampling system, allowed us to collect samples from any place we could access the tops of dunes (see the discussion in appendix A concerning procedures used for sampling dunes with a GeoProbe). This procedure was particularly effective in our efforts to determine the basal and near-surface ages of tall dunes across the study area where good natural exposures were lacking, and to collect detailed changes in ages of dune formation for sequences of progressively younger dunes developed within Holocene embayments (i.e., Fisherman's Island State Park and Torch Bay Nature Preserve). In addition we could also sample more shallow archaeological sites with minimal impact to the subsurface (i.e., Winter and Ekdahl-Goudreau sites). At such locations even minor, shallow excavations would not be permitted.

In summary, several important considerations and interrelated research goals governed our sampling strategy. First, we focused on the northern end of the basin because little was known about northerly dunes. Enough chronological control already existed further south through which a solid model for dune formation in the southeastern part of the shore had been developed. We further focused on dunes that had key stratigraphic/geomorphic relationships, such as presence of beach gravels related to the middle Holocene Nipissing transgression and/or buried soils and landscape position from which to determine the spatial and temporal variations in dune formation. Finally, we employed equipment (i.e., GeoProbe) through which we could sample dunes to systematically determine their rates of accretion. In this fashion, given the funds available for dating, we believe we have tested locations that are key and representative of the range of coastal eolian setting and contexts in the northern part of the Lake Michigan basin.

Ages of Dunes along the Lake Michigan Shoreline in Michigan

As noted above, this study focused largely on the geomorphology of coastal dunes in the northern part of the Lake Michigan basin. Fifteen dune sample locales were investigated during this investigation; four of these included archaeological materials in stratified contexts (figure 3-6). Stratigraphic relationships were described as thoroughly as possible at each of these locales. A total of 32 samples were obtained for OSL age determination at these dune locales and other archaeological sites in the study area (see chapter 5 and appendix C for discussions of these sites and locales). The first part of the following discussion outlines the general results

Coastal Dunes, Processes, and Cycles

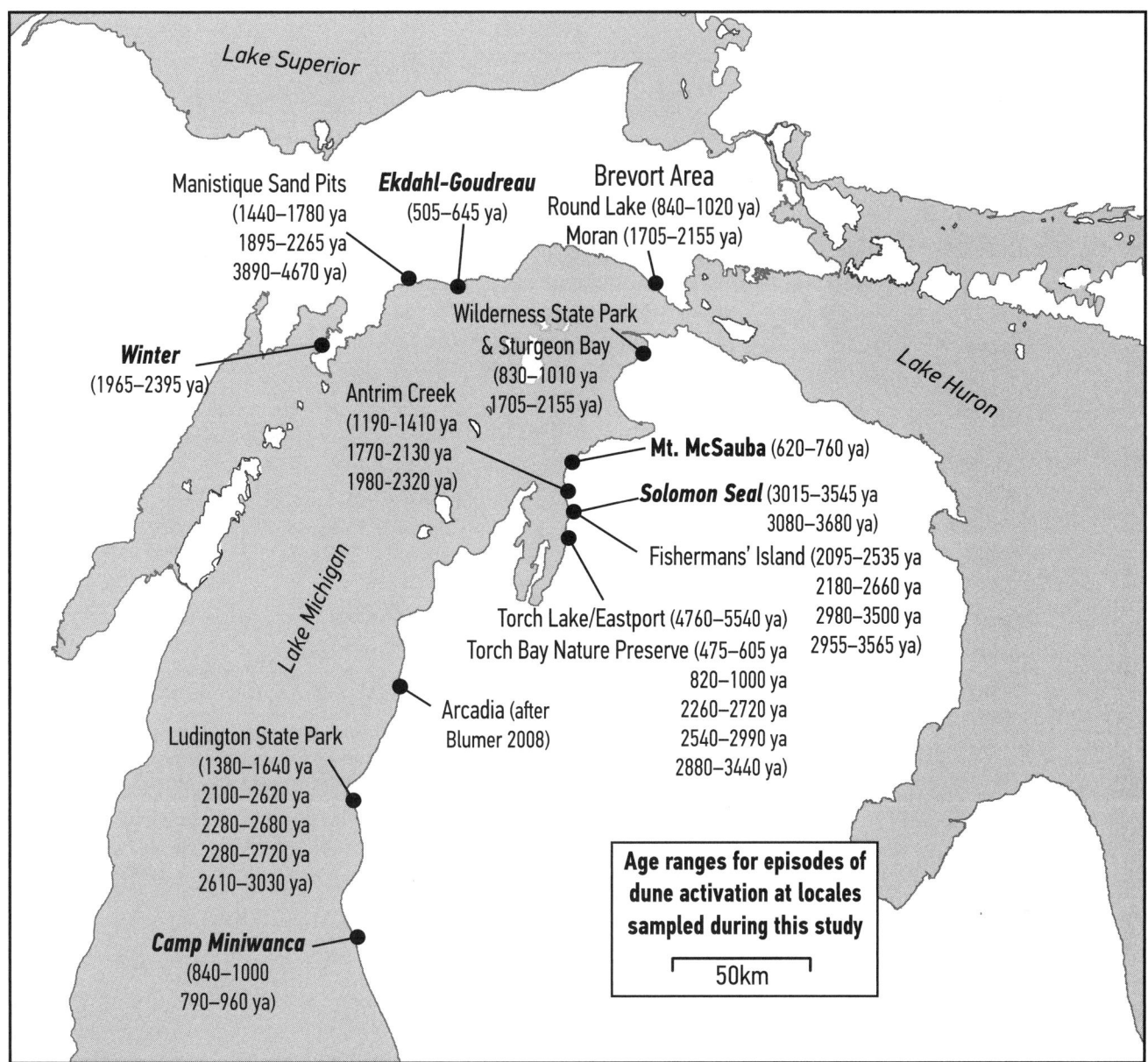

3-6. Locations of dunes and dunefields sampled and analyzed during this study. Age ranges of dune accretion labeled beside location; ages given in years ago (ya) based on OSL determination (see appendix B). Age range is 1σ error distribution. Names of sample locales with stratified archaeological materials are italicized and bolded.

associated with these geomorphic investigations. Detailed chronostratigraphic relationships of these sample sites are presented in chapter 5 and appendix C. The following discussion also reviews the general history of dunes in the southern part of the basin and compares it with the new results gleaned from this study.

In addition to the geography of dune locales investigated in this study, figure 3-6 also shows the geographical pattern of ages that were obtained from each of these locations. From a general perspective, the most obvious pattern is that virtually all

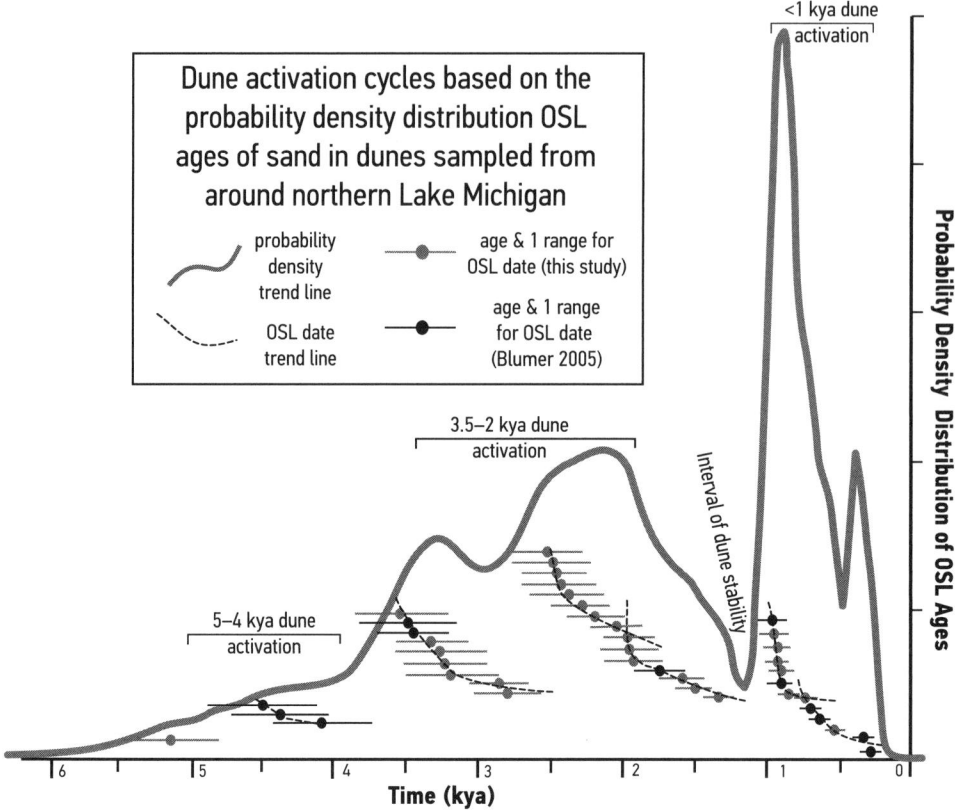

3-7. Temporal variation in dune construction in the northern half of Lake Michigan. Gray curve shows probability density distribution of OSL ages (kya = thousand years ago) from dunes in northern Lake Michigan; sample ages shown below curve, sample mean age shown by solid dot, 1σ range shown by lines. Dotted line shows the temporal trend in rates of dune formation.

dunes in the northern part of the Lake Michigan basin provided age estimates of less than 3,500 years ago. In fact, only one sand dune provided an age older than 3,500 years ago. This dune is located east of the Nipissing bluff in the village of Eastport (figure 3-6) and provided an OSL age of circa 5,000 years ago. Although an age older than 3,500 years ago (i.e., 4,280 years ago) was obtained from the Manistique Sand Pit, this sample was collected from apparent beach sands that are sandwiched between gravel beds. These sediments are likely beach deposits and thus not eolian. All other ages are less than about 3,500 years ago, with no clear geographical pattern as far as the northern part of the basin is concerned.

The best way to view the age chronology in the northern part of the Lake Michigan basin is shown in figure 3-7, which illustrates the cyclic nature of dune activity based on the probability density distribution (PDD) of OSL ages. This distribution describes the variability in the density (or likelihood of clustering) of OSL ages through time and normalizes this function to unity (i.e., a value of 1). It was developed for figure 3-7 by integrating the OSL ages and their 1σ error over the circa 5,000 years of recorded dune activity. Thus, the PDD axis (i.e., *y*-axis on figure

3-7) does not actually represent "probability" (as would be the case for a normal curve). It does show, however, that the OSL ages are not uniformly distributed through time but rather that some OSL ages are more likely than others. Taken together, these groupings have higher density distribution values and show up as "peaks" in the PDD curve. For example, if certain OSL ages do not ever occur (such as either 6,000 or 0 years ago), then the "density value" for these ages would equal zero. Conversely, the more OSL ages that cluster within groupings, the higher the "density" value of that age cluster, and the more pronounced the peaks of the density distribution become. These peaks and valleys in the PDD show periods of dune construction (peaks) and intervals of dune stabilization (valleys).

Intuitively, these "peaks" in dune activity can be observed by comparing the distribution of OSL ages collected during this study, which are shown below the PDD curve. These ages (shown by circles) and their 1σ range (shown by horizontal lines through the circles) were simply "stacked" on top of each over through time regardless of geographic position and show that five or six clustered groupings exist. These clusters are characterized by the occurrence of relatively greater numbers of dates initially that then taper off through time. The temporal pattern of these groupings is highlighted by a dashed line connecting the initial cluster of dates with those that taper off from the dashed line. The initial cluster, which generally occurs under a "peak" in the PDD curve, marks the ages of active dune building and occurs because a greater volume of sand is active. The tapering off of ages, on the other hand, represents periods of dune stability, when relatively little sand is moved to build the dunes. When we integrate these groups of date clusters, a temporal pattern of dune activation emerges.

The PDD curve clearly indicates that several cycles of dune activity occurred in this part of the Lake Michigan basin. The earliest period of /dune growth occurred about 5,000 years ago and is represented primarily by the dune within the village of Eastport (figure 3-6). This initial episode of dune growth was a relatively minor event in this part of the basin, however. This observation is significant and suggests that lake-level fluctuations associated with the Nipissing transgression and regression did not directly drive dune formation. This finding is surprising, as prior to the study we expected that a significant amount of dune building occurred during the Nipissing high stand.

The first major period of dune activity in the northern part of the Lake Michigan occurred between 3,500 and 2,000 years ago. During the early part of this episode the largest ($\geq$ 30 m high) dunes found in the northern part of Lake Michigan were constructed in positions that are now the most inland from the modern shore. As this interval of sand mobilization continued, progressively younger dunes were deposited in time-transgressive fashion. These dunes become progressively younger toward the current coastline. Dunes subsequently stabilized, or the supply of eolian sand diminished markedly, until circa 1,000 years ago, when another major period of dune growth transpired. Since the onset of this latter cycle, the supply of sand has apparently slowly diminished to the present time. As with the earlier cycle of dune growth, this latter interval has resulted in the development of progressively younger dunes toward the modern shore.

The reconstructed record of dune formation in the northern part of the Lake Michigan basin differs in some significant ways from the history of dune construction along the southeastern shore of the lake. An important difference between the two ends of Lake Michigan is that extensive dune growth during the Nipissing high stand (ca. 5,000 years ago) has been recognized along the southeastern shore (e.g., Hansen et al. 2002), whereas it is largely absent in the north. This period of dune growth built large (ca. 10 m high) backdunes in what is now the lee of the large dune complexes along much of the southeastern shore. As noted above, a dune of Nipissing age was recognized at only one sample locale in the northern end of the basin (Torch Lake Township), and this dune is relatively small, only about 3 m high.

Another significant difference between dunes at each end of Lake Michigan is that dunes along the southeastern shore tend to consist of large, overlapping parabolic dunes with active blowouts in many places. In contrast, northern coastal dunes consist largely of dune ridges that contain few parabolic forms. This lack of parabolic dunes in the north suggests that dunes in this part of the basin formed quickly and have not been subsequently reworked. Dunes along the central part of the coast, such as at Ludington, appear to represent a geomorphic transition, with distinct dune ridges that contain parabolic forms. Dunes in the southern end of the basin tend to contain several paleosols, indicating that they have grown upward in an episodic fashion through time. Aside from the exposures in the southern parts of Ludington and Wilderness state parks (figure 3-6), few such soils have been recognized within the coastal zone of Lake Michigan. This dichotomy reflects the fact that northern dunes have grown time-transgressively, with progressively younger dunes closer to the shore. Yet another geographical difference is that dunefields along the southeastern part of the shore tend to line the coast for many miles, forming the extensive barrier dunes described by Buckler (1979). In contrast, dunefields around the northern shoreline of the basin are generally confined to embayments that have apparently trapped littoral sand that has ultimately been reworked to form dunes.

In spite of these significant differences, important similarities in the timing of dune growth exist between the northern and southern ends of the Lake Michigan basin. The major interval of dune growth recognized in this study between circa 3,500 and 2,000 years ago correlates very well with the most intensive period of eolian sand deposition in the southern end of the basin (e.g., Arbogast et al. 2002). Similarly, the hiatus in dune formation observed between circa 2,000 and 1,000 years ago in the north has also been recognized along the southeastern shore. A primary difference between the two ends of the basin is that this hiatus in dune growth resulted in the formation of the well-developed Holland Paleosol in dunes along the southeastern shore (Arbogast et al. 2004), whereas it is absent north of Ludington. This soil is probably absent in the north again because dunes in this part of the basin have grown time-transgressively, with younger dunes formed on new, uplifted shoreline and older dunes displaced more inland away from active coastal processes. Along the southeastern shore, conversely, dunes typically grew upward through time, which resulted in episodes of growth becoming stacked vertically. Regardless of the presence or absence of the Holland Paleosol, the supply of eolian sand in both ends of Lake Michigan increased dramatically about 1,000 years ago. This increase in sand supply buried the Holland Paleosol (as late as ca. 500 years ago in some places) within dunes along the southeastern shore and resulted in the formation of a younger set of dunes lakeward of older dune ridges in the northern end of the basin. Since that time, the supply of eolian sand seems to have slowly dwindled in both the northern and southern ends of the basin. Dunes along the southeastern shore of Lake Michigan appear to have been extensively eroded during the past few hundred years, resulting in steep dune exposures immediately facing the lake. Such exposures are rare in the north. As discussed in chapter 4, the geographic differences in the characteristics of coastal landforms and dune growth probably related to the rate and amount of uplift in the northern and southern parts of Lake Michigan. In the north, particularly at the Straits of Mackinac, uplift was, and continues to be, relatively rapid. In the south, however, uplift of the coastline was minimal or may even have subsided since the middle Holocene.

Summary

A great deal of research has been conducted on Lake Michigan coastal dunes in the past 100 years. Given the lack of absolute dating techniques, research in the first half of the twentieth century was largely descriptive in nature and focused on basic geographic and geomorphic relationships. The concept of perched dunes originated at this time to describe dunes that mantle high headlands in the northern part of the basin (e.g., Sleeping Bear Dunes). Rigorous qualitative research began in the 1950s with Olson's (1958a, 1958b, 1958c) work, which led to enhanced understandings of coastal dune processes and plant succession. Following this seminal work, dunes were largely ignored until the 1990s. The major exception during this hiatus was work by Snyder (1985) at Sleeping Bear Dunes National Lakeshore. Since the middle 1990s a great deal of geomorphic research has been conducted on Lake Michigan coastal dunes. Until the present study, this previous research focused largely on the southeastern part of the lakeshore, where a number of excellent exposures occur, ranging generally from the Muskegon area to Indiana Dunes National Lakeshore (figure 3-7). Contained within these dunes are a number of buried soils that are clearly visible along the coast. The stratigraphic relationships of these dunes have been vigorously studied (e.g., Arbogast et al. 2002, 2004; Fisher and Loope 2005), with a particular effort to radiocarbon-date organic material contained within buried soils. These studies generally indicate that dunes in that part of the basin began to grow during the Nipissing high stand and that they grew episodically, with brief periods of stability, for the next circa 2,500 years. An extended period of stability apparently occurred in most dunes from circa 2,000 to 1,000 years ago, one that resulted in the prominent Holland Paleosol (Arbogast et al. 2004). This soil was subsequently buried between circa 1,000 and 500 years ago, depending upon the location, during a period of dune growth that has gradually diminished to this day.

Given the extensive work previously conducted along the southeastern shore of Lake Michigan (e.g., Arbogast et al. 2002; Hansen et al. 2002; Fisher and Loope 2005) this study focused on dunes in the northern part of the Lake Michigan basin, ranging from the Garden Peninsula to essentially Little Point Sable (figure 3-6). Stratigraphic relationships were investigated at the dunefields along the lake's northern rim using OSL ages to estimate when eolian sands were most recently deposited (figure 3-6). Results from this study indicate that the first major pulse of dune growth occurred circa 3,500 and 2,000 years ago. This interval generally resulted in the largest dune ridges in the northern end of Lake Michigan and also correlates reasonably well with episodes of dune growth within the southern end. Following this period of dune growth, the supply of eolian sand diminished and dune growth stopped. A similar period of stability occurred along the southeastern shore, which resulted in the formation of the Holland Paleosol in this area. This soil is not recognized north of Ludington. The supply of eolian sand increased again in the northern part of the basin circa 1,000 years ago, which also correlates reasonably well with dunes along the southeastern shore of the lake. Following this spike in eolian activity, the supply of sand to dunes has apparently diminished in a gradual way in the past 1,000 years.

CHAPTER FOUR

Middle and Late Holocene Lake-Level Variation, Isostatic Rebound, and Environmental Changes in the Upper Great Lakes

Geomorphology, Lake Level, Dunes Development, and Human Settlement

Introduction and Geomorphological Background

The overarching goal of this book is to formulate a temporal and spatial framework delineating the natural and cultural processes that control the locations and burial of archaeological sites within coastal dunes settings. The purpose of this chapter is to outline some of the major factors that influence the development and distribution of both dunes and archaeological sites through space and time (figures 4-1 and 4-2). Broadly, these factors include:

- The basic geomorphology of the Lake Michigan coastline, which relates mainly to the deglaciation history of the basin
- Regional climate change, particularly during the late Holocene
- Lake-level variations related to environmental change
- Variability in isostatic rebound rates and directions

These factors and influences are not independent of each other and include both secular and cyclical variations and long-term, directional developmental changes. Although most of the direct influences relate mainly to Holocene development, the geomorphology around Lake Michigan and its basic configuration is derived primarily from the late Wisconsin retreat of the Laurentide Ice Sheet between 15,000 and 10,000 years ago.

Although Lakes Michigan and Huron are called separate lakes, they actually form a single, hydraulic lake basin because they share the same water plane (about 176–177 masl) and are connected at the Straits of Mackinac. This lake system ultimately discharges through the St. Clair River at Port Huron (figures 4-1 and 4-2). Depending on the interaction of factors noted above, at various times during

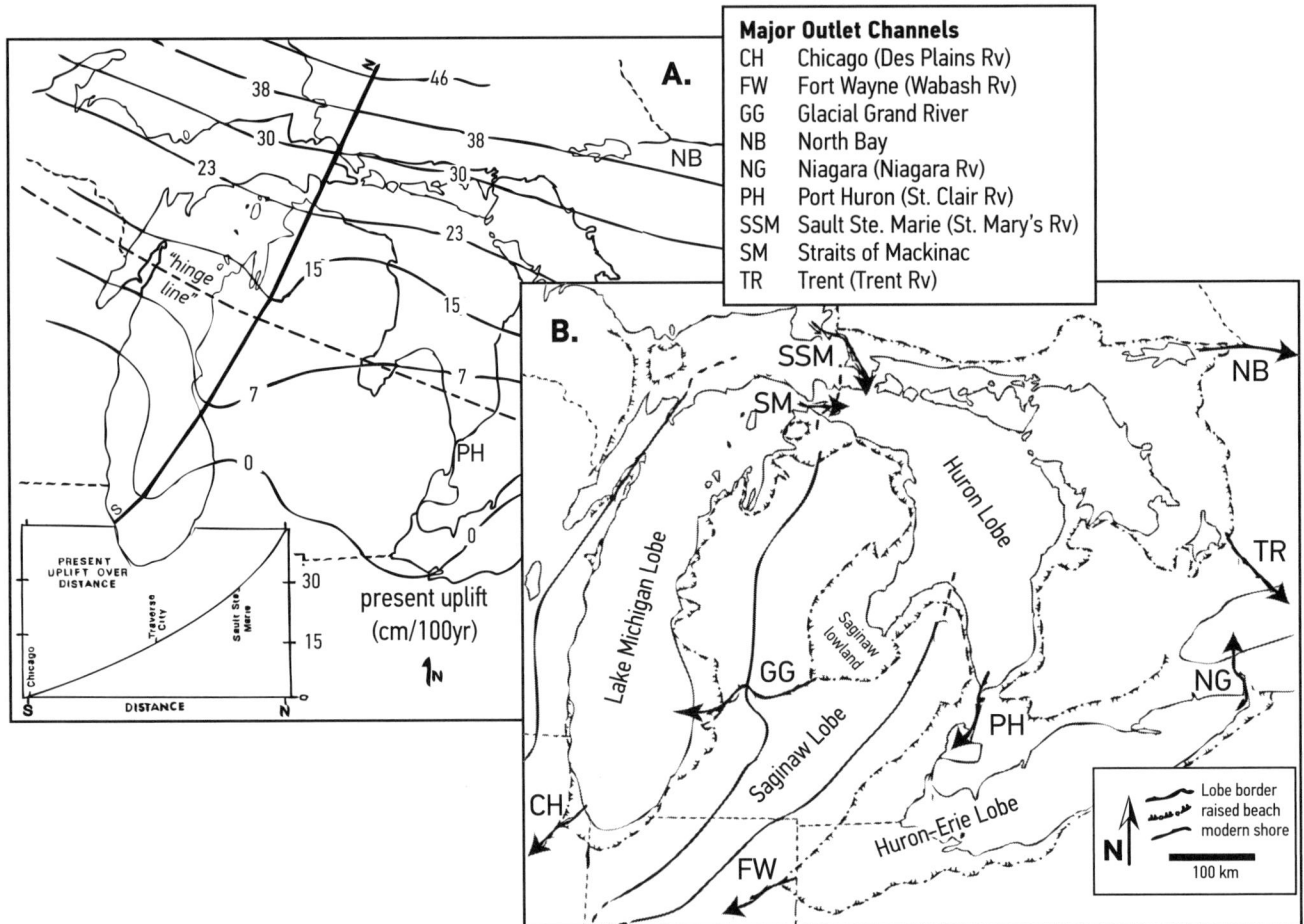

4-1. Maps of the upper Great Lakes. A. Map showing the modern rate of isostatic rebound within the Great Lakes region. Insert profile shows general north-south variation in uplift (after Larsen 1985a and Monaghan and Lovis 2005). B. Geomorphological map of the region. Location of major outlets used labeled; abbreviations shown above map; Monaghan and Lovis (2005).

the Holocene, the entire upper Great Lakes system (Lakes Superior, Michigan, and Huron) formed several different, separate lakes as well as a single hydraulic lake system. For example, Lake Superior now exists as a separate lake basin at an elevation of about 183 masl and discharges southward over a bedrock threshold near Sault Ste. Marie, thence through the St. Mary's River into Lake Huron. As recently as circa 2,000 years ago, however, Lake Superior was also directly connected to Lakes Michigan and Huron to form a single lake. Clearly, the modern configuration of the upper Great lakes is a relatively recent manifestation and has only existed since circa 2,000 years ago (Monaghan and Lovis 2005).

The modern water levels for the upper Great Lakes (Lakes Michigan, Huron and Superior) are controlled by a balance between the amount of water input to the basin and the amount that discharges through the outlet. Currently, the outlets at Port Huron and Sault Ste. Marie (Michigan-Huron and Superior, respectively)

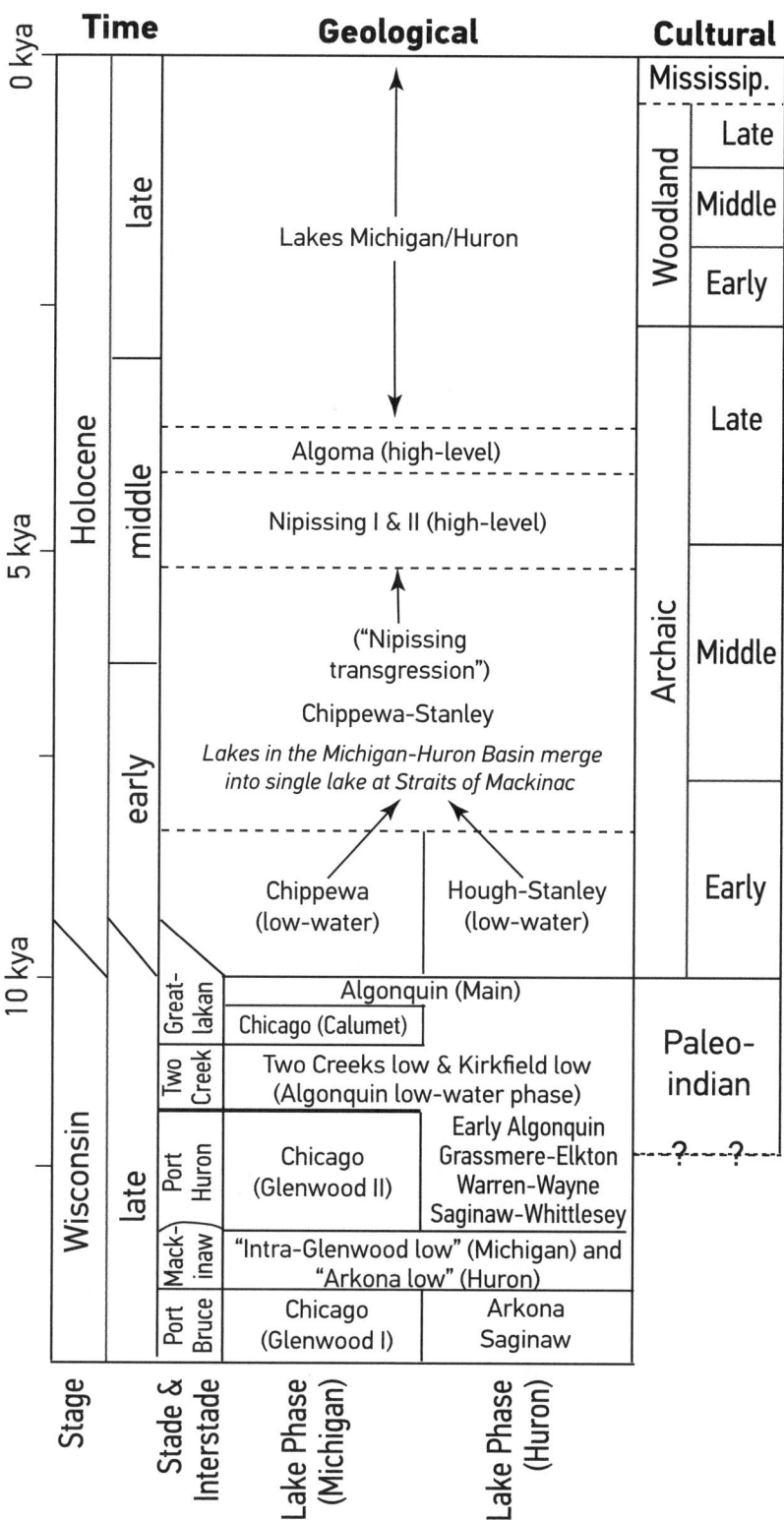

4-2. Time stratigraphy for geological and cultural events during the late Wisconsin and Holocene in the upper Great Lakes; after Monaghan and Lovis (2005).

both appear to be stable (i.e., they are not undergoing long-term, measurable erosion), which apparently has been true since circa 4,000 years ago (see Larsen 1985a; Monaghan and Lovis 2005). In fact, over the past 15,000 years the upper Great Lakes basins have included both high-level (water plane higher than modern level) and low-level (water plane lower than modern) lakes that had vastly different water levels, existed for a matter of years or for several centuries, and also used several geographically diverse outlets (figures 4-1 and 4-2).

The present topography and geomorphological configuration of the Lake Michigan basin were dictated mainly by erosional and depositional events associated with the Lake Michigan lobe of the Laurentide Ice Sheet. The Lake Michigan lobe, which was one of three main ice streams that covered Michigan during the late Wisconsin stage (ca. 18,000 to 10,000 years ago; figure 4-2), flowed south through a series of topographical lows now occupied by Lake Michigan (figure 4-1). Consequently, it probably regularly advanced into proglacial lake systems that occupied the basin lows. As it advanced, the Lake Michigan lobe filled the basin and then spread south and eastward into western Michigan and northern Indiana and at its maximum extent extended well south and west into eastern Wisconsin and northern Illinois (figure 4-1). In Michigan, its extent is marked by sequences of mainly north-south trending morainal highlands, which include the prominent Sturgis, Kalamazoo, Valparaiso, Lake Border, and Manistee morainal systems (Leverett and Taylor 1915).

The easternmost maximum extent of the Lake Michigan lobe in Michigan generally follows a line from South Bend, Indiana, through Kalamazoo, Grand Rapids, Cadillac, Grayling, and Cheboygan and was at this extent circa 15,000 years ago (Leverett and Taylor 1915; Monaghan et al. 1986; Monaghan and Larson 1986; see figure 4-1). During the next circa 5,000 years, the ice margin underwent an overall withdrawal from the Lake Michigan basin that was characterized by a series of ice-marginal retreats and readvances, with each readvance less extensive than the last. The oscillations of the Lake Michigan lobe during this period created a sequence of prominent morainal systems and also resulted in the complex sequence of glacial lakes that are the hallmark of the Lake Michigan lobe (figure 4-1). The fronts of the morainal systems formed during this general retreat commonly include extensive outwash systems, fans, and sandar. Sometimes the ice margin also terminated within extensive glacial lakes creating sub- and proglacial fans and deltas. Consequently, sand is an important constituent of the Lake Michigan lobe drift. Ultimately, it was this sand that was eroded from upland landforms and shoreline bluffs, accumulated on beaches and bars in Lake Michigan, and became the source for the numerous dunes that characterized the Lake Michigan shoreline. Understanding and contextualizing the geomorphological processes and geological history that move this sand to places where it could be eroded, transported, and reworked to build dunes during the Holocene are essential in formulating the temporal and spatial framework for dune construction or reactivation and burial or stratification of any associated archaeological sites within the dunes.

Dune Formation, Lake-Level Variations, and Human Settlement

Although many influences affect eolian landform construction, at the most basic level dunes form because of three major factors that must occur simultaneously.

1. Enough wind must exist to form sand into a dune.
2. An abundant sand supply must exist to maintain dune construction.
3. This supply of sand must also be poorly vegetated and thus erodible.

Unless each of these necessary factors occurs at the same time, then regardless of sand volume or wind energy, dunes will not form or at best will only occur as localized sand sheets. Although this research is mainly focused on the age and construction of the more prominent coastal dunes, sand-sheet deposition may also have import for understanding stratification of cultural deposits as well as some of the more fine-scale formation processes within archaeological sites found in dune contexts (see discussions of this phenomenon in chapter 5).

To varying degrees, the three factors mentioned above each commonly occur along all shorelines, but vary in intensity across time and space. For example, some sand is always present on beaches as part of the near-shore and littoral processes. Likewise, wind is a common component along most shorelines because of both large-scale atmospheric circulatory processes and diurnal changes in land/water temperatures, shoreline aspect, and other such conditions. Additionally, the long fetch of the prevailing west and northwest wind over Lake Michigan and the generally north-south shoreline may be particularly important for the shore of eastern and southern Lake Michigan. Regardless, the formation of large or pervasive dunes only occurs when an abundant and sustainable supply of sand coincides with a regular and persistent source of energy to erode and transport the sand. Many of the factors that allow dune growth are similar to, but occasionally opposite from, those that influence variation in the Great Lake water levels. Thus it is important to understand how these two systems interact.

Cyclical and long-term directional fluctuations in lake level greatly influence the development or reactivation of shorelines and associated coastal dunes. Shoreline processes are the most obvious agents for increasing (or reducing) the supply of sand to beaches, and an abundant supply of sand is vital for the formation of dunes. The vigor of such processes directly affects sand supply and is largely impacted by lake-level variations, although sometimes in unexpected and complex ways. For example, even though beach and shore bluff erosion are generally greatest during high-water stages, which should increase sand supply to the beach, transgressions also reduce beach widths, and consequently the new supply of sand may not be readily available to blow into a dune until the water level falls during a subsequent regression. Such regression creates a wider beach and increases the supply of available sand. Depending on specific circumstances, however, a low water level may also reduce the supply of sand to the beach and, thus, not resupply sand for sustained dune construction. For example, assuming that transgressions in the upper Great Lakes are related to cooler, wetter periods and that regressions occur during warmer, dryer intervals (Larsen 1985a; Monaghan and Lovis 2005), storminess may be more common during high-water intervals and less so when the water level is low. If so, although abundant sand may have been available on wider beaches to form dunes during low water level, less storminess may have also resulted in relatively few dunes being formed because of a lack of sufficient and sustained winds or because sand supply was not rejuvenated on the beach. In short, the interaction of variables needed to form dunes is complex.

Human economic subsistence potential and cultural ecology has also been impacted by the Holocene development of the upper Great Lakes, particularly during the late Holocene. How humans utilize coastal landforms influences the numbers, ages, and types of archaeological sites that may be present within dunes. Moreover, some of the same environmental factors that control dune growth or reactivation also influence human settlement and subsistence. Even minor variations in water level can dramatically influence the types and abundances of plant and animal resources that may be available to humans. For example, lake transgressions will raise water tables, flood wetlands, rework lake bottom conditions to enhance (or destroy) fish spawning grounds, and promote beach and bluff erosion. During regression events many of these same processes will act on the shoreline. Regardless of trajectory, subsistence possibilities can be either enhanced or made poorer as a result of water-level variations. Interestingly, from the standpoint of long-term human subsistence patterns, even if resource productivity is not highest during stable intervals, the plant and animal resources that do occur may be more reliable from year to year. This is to say that during these intervals the seasonal occurrence of specific resources is probably more predictable and dependable at a specific place and, therefore, can be more reliably scheduled as part of the seasonal rounds for a hunter-collector economy. Understanding how all of the human and natural factors intersect within a dynamic coastal environment is necessary to formulate the temporal/spatial framework that controls the locations and burial of archaeological sites within dunes settings.

Late Wisconsin and Holocene Lakes in the Upper Great Lakes

Overview of Lake Stages in the Upper Great Lakes

The configuration of the Lake Michigan lobe, as well as more generally the Laurentide Ice Sheet, had a significant impact on the developmental history of Lake Michigan and its ancestors. The direction and rate of its retreat, the ice volume and weight of the lobe, and the length of time it sat over particular areas all had significant impact on the later geomorphology and development of the ancestral and modern lakes in the basin. The process and timing of deglaciation also dictated the levels of lakes, their duration, and ultimately the position of outlets for these lakes. For example, early in the deglaciation of Michigan, when the three lobes (Lake Michigan, Saginaw, and Huron/Erie; figure 4-1A) were near their maximum extent in Michigan, drainage was generally south and west, away from the ice margin and eventually into the Mississippi River. As the ice margins retreated back into the lake basins, however, meltwater was trapped between morainal uplands that rimmed the basins and isostatically depressed areas near the ice margin (i.e., Lake Michigan basin, Saginaw Bay lowlands, and Lake Erie basin; figure 4-1A). This resulted in the formation of the extensive proglacial lake system that is the hallmark of the Wisconsin geological history in the upper Great Lakes (figure 4-2). The initial lakes were all "high-level" (i.e., significantly higher than modern Lakes Michigan and Huron) and overflowed through sags in the rimming morainal uplands (Monaghan and Lovis 2005, chapters 2 and 3). These high-level lakes drained generally westward and southwestward from one lake to the next and ultimately discharged through the Chicago outlet and into the Mississippi valley (figures 4-1 and 4-2). When the ice-margin retreat in the upper

Great Lakes was great enough to expose isostatically depressed, lower outlets through northern Michigan or the Algonquin highland of Ontario, the lake level fell and the drainage pattern reversed to eastward or northeastward discharge, only to reestablish the high-level lakes and be forced back to westward discharge during subsequent readvances of the ice. This pattern probably repeated itself throughout the Wisconsin, but only the lake sequence developed during the last general retreat (post-15,000 years) was broadly preserved.

Although oversimplified, the pattern of lake sequences noted above (higher, westward-flowing lakes during stadial intervals and lower, eastward-flowing lakes during interstadials) was characteristic in the upper Great Lakes during the late Wisconsin and Holocene interval. More detailed summaries and discussions of the lake sequences and their significance are presented in Monaghan and Lovis (2005), which is based on more than 100 years of research by a wide variety of scientists (e.g., Spencer 1881, 1888, 1891; Taylor 1894, 1897; Leverett 1897; Goldthwait 1891, 1908; Wright 1918; Leverett and Taylor 1915; Stanley 1936, 1938; Bretz 1951a, 1951b, 1953, 1955, 1963,1964; Hough 1955, 1958, 1963, 1966; Dreimanis 1958, 1969; Lewis 1969, 1970; Dreimanis and Goldthwait 1973; Farrand and Drexler 1985; Fullerton 1980; Karrow et al. 1975; Karrow 1980; Monaghan et al. 1986; Monaghan and Larson 1986; Hansel et al. 1985; Larsen 1985a, 1985b, 1987; Eschman and Karrow 1985; Calkin and Feenstra 1985; Hansel and Mickelson 1988; Lewis and Anderson 1989; Clark et al. 1990; Monaghan and Hansel 1990; Kaszicki 1985; Finamore 1985, Thompson 1992; Thompson and Baedke 1995). This chapter, however, will focus mainly on providing background concerning the spatial and temporal contexts of the Holocene development of ancestral Lake Michigan phases and the geomorphological interaction between this lake sequence and the formation of coastal dunes.

The Holocene, which began in the upper Great Lakes circa 10,500 years ago, includes several related lakes formed after the ice margin had completely retreated from the Lower Peninsula of Michigan. Broadly, most of the Holocene lake sequence can be described through a simple "outlet-controlled" framework in which dramatic changes in water level and drainage are controlled largely by changes in outlet configuration. This framework generally fits the broad developmental outline of low-level lakes related to isostatically depressed outlets in the northeastern part of the upper Great Lakes and higher lakes associated with less depressed southern or western outlets described above. The early and middle Holocene in the Lake Michigan basin is epitomized by one of the largest magnitude and longest duration transgressions ever recorded in the region. It is informally referred to as the "Nipissing transgression." The first major Holocene event in the upper Great Lakes was initiated when the front of the Laurentide Ice Sheet retreated north of North Bay, Ontario, and exposed an extremely low-level outlet into the Ottawa valley. Subsequent water level in the Lake Michigan basin was controlled mainly by the outlets used and by the rate of its isostatic uplift.

When it was uncovered, water levels in the Michigan, Huron, and eastern Superior lake basins fell as drainage was transferred to the North Bay outlet. This initiated the Chippewa-Stanley phase. Initially, water level in the basin fell to an extremely low altitude and probably created a series of very low-level, connected lakes in both the Michigan and Huron lake basins. Two lakes were believed to have formed in the Huron lake basin, Lake Stanley in the main part of the basin and

Lake Hough in the Georgian Bay region (Lewis 1969; Sly and Lewis 1972; Eschman and Karrow 1985). The lowest level achieved by these lakes was < 80 masl and may have been as low as 45 masl (Sly and Lewis 1972). Lake Chippewa formed in the Michigan lake basin and was connected to Lake Stanley through a channel at the Straits of Mackinac (figure 4-1). The lowest level of Lake Chippewa was at least 117 m and may have been as low as 107 m (Hough 1958; Hansel et al. 1985; Buckley 1974). Early in its history, Lake Chippewa was actually two separate lakes, one that formed in a "southern" basin and another that formed in a "northern" basin. The southern basin existed south of a line from Ludington, Michigan, to Milwaukee, while the northern basin existed between this line and approximately Beaver Island, Michigan. The level of the southern basin was controlled by a relatively deep canyon through subcroppings of Paleozoic limestone located offshore from Milwaukee. Water level in the northern basin, on the other hand, was constrained by a deep bedrock canyon that forms the Straits of Mackinac.

The youngest ages from Lake Algonquin deposits and bogs on the Algonquin plain, as well as from stumps and peat deposits below the current level of Lakes Michigan and Huron, all indicate that the Chippewa-Stanley low phase began circa 10,000 years ago (Monaghan and Lovis 2005). Based on ^{14}C ages near the outlet, Lewis (1969) indicates that the outlet at North Bay was deglaciated prior to circa 9,800 years ago. During the next 5,000 years or so, isostatic uplift of the North Bay outlet progressively raised the water level in the basin. At some point, North Bay was raised higher than the Straits of Mackinac, and lakes in the Michigan and Huron lake basins were again confluent. Somewhat later, circa 7,500 to 7,000 years ago (Saarnisto 1975; Farrand and Drexler 1985), North Bay was also raised above the St. Mary's River, and all three basins (Superior, Michigan, and Huron) were confluent. Although a whole series of lake levels and phases must have existed during this period, any record of them is now submerged below Lakes Michigan and Huron or was effectively destroyed around the periphery of the lakes by the later Nipissing transgression. From a practical standpoint, Chippewa-Stanley probably had no actual stable level and in reality represented a circa 5,000-year-long transgression (informally the Nipissing transgression).

Continued uplift of the North Bay outlet during the Nipissing transgression eventually raised North Bay above the altitude of the previously abandoned southern outlets at Port Huron and (possibly) Chicago, and by circa 5,000 years ago drainage was transferred to the southern outlets. Lake Nipissing formed at this time and through erosion of the St. Clair River, Port Huron in southern Lake Huron became the sole outlet (figures 4-1 and 4-3). Like Chippewa-Stanley/Lake Nipissing probably did not have a long-term stable level but was rather characterized by a several-hundred-year-long regression to the level of modern Lakes Michigan and Huron. This regression resulted from the erosion of the Port Huron sill and may have included a brief stabilization, Lake Algoma (figure 4-3). By circa 4,000 years ago, erosion at Port Huron apparently ceased and the modern level of Lakes Michigan and Huron was achieved, at least in the southern end of the basin. Importantly, at this time the entire upper Great Lakes (Michigan, Huron, and Superior) existed as one lake that drained through Port Huron. The outlet for this lake was maintained at its essentially modern level until circa 2,000 years ago, when isostatic rebound raised the rapids in the St. Mary's River at Sault Ste. Marie above Port Huron and separated Lake Huron from Lake Superior.

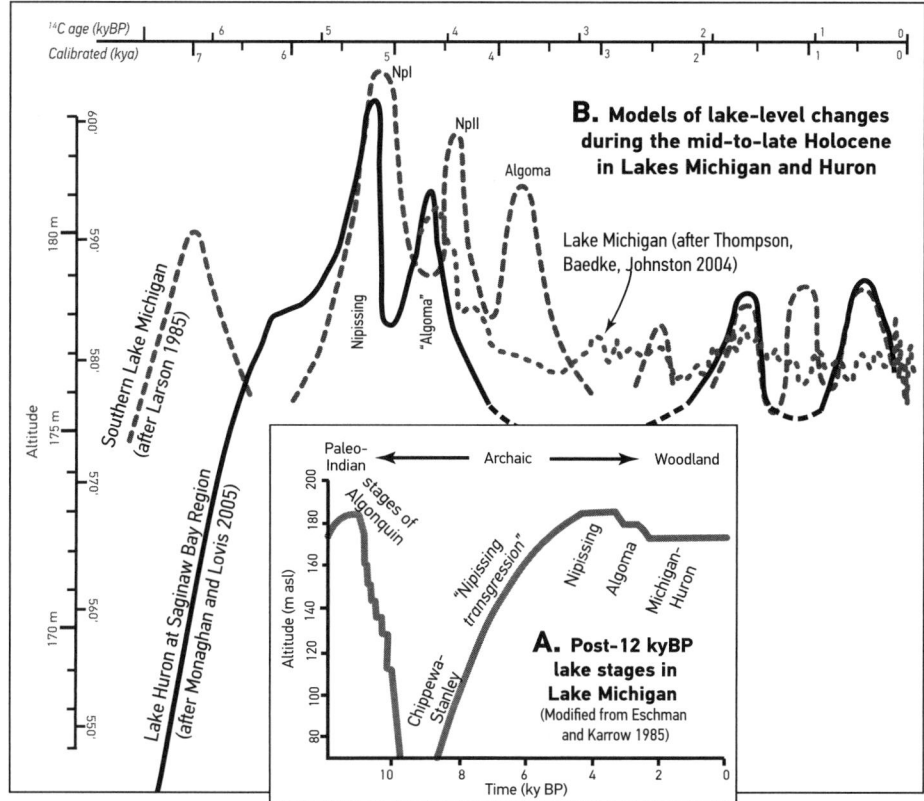

4-3. Lake-level curves for Lakes Michigan and Huron. A. Post-12,000 BP lake stages in Lakes Michigan and Huron; modified from Eschman and Karrow (1985). B. Three different models for middle and late Holocene lake levels in the upper Great Lakes region; modified from Larsen (1985a), Monaghan and Lovis (2005), and Thompson and Baedke (1999).

Once these Port Huron and Sault Ste. Marie outlets became broadly stable and the modern configuration of the upper Great Lakes was achieved, the water level within these basins began to be controlled mainly by variation in how much water entered versus left the lake (i.e., input/output). The "outlet controlled" framework does not adequately describe the hydraulic system for the late Holocene upper Great Lakes. Rather, water level in each basin was controlled mainly by secular variations in climate (i.e., precipitation and temperature; Larsen 1985a; Monaghan and Lovis 2005). These lakes, modern Michigan, Huron, and Superior, can be more usefully thought of as "climate-controlled" lakes (Larsen 1985a), which is a conceptual model that more realistically depicts both the broad and the smaller-scale lake-level variations during the past 2,000–3,000 years (figure 4-3).

Controls for Lake Level and Configurations in the Upper Great Lakes

Factors Controlling Lake Level and Configurations

The above description of late Wisconsin and Holocene lakes in the upper Great Lakes region depicts a relatively simple sequence that has fluctuated by over 100 m over the past circa 10,000 years (figure 4-3). The sequence generally consists of a circa 5,000-year-long transgression followed by a circa 1,000-year-long regression and then stabilization. Although simple, even this brief outline shows that several complex circumstances interact to create this system. Specifically, spatial and temporal factors interact with these broad-scale, lake level changes to produce additional geographic variability. Variation in the water level for these lakes is related to adjustments in the outlet utilized by each lake. Specific lake levels are generally controlled by either the specific altitude of the outlet channel for that lake or by the amount of water entering and leaving the basin (the water budget for the lake). Consequently, the sequences of lakes that evolved within the upper Great Lakes reflect four major processes (Hansel et al. 1985; Monaghan and Lovis 2005):

1. Fluctuations of the ice lobe margin that either uncovered lower outlets during retreats or blocked them during advances
2. Differential isostatic changes (postglacial rebound) in the altitudes of different parts of the lake basins or outlet channels themselves
3. Erosion of outlet channels and sills
4. Significant changes in the volume of water entering or leaving the basin, through climatic changes and changes in the hydraulic configuration of the outlet channel

Although all of these processes probably played some role in the development sequence of proglacial lakes that occupied the basin during the late Wisconsin, only the last three (outlet erosion, isostatic rebound, and climatic changes) influenced the Holocene lake phases on which this research focuses. Factors 1 and 2 (i.e., uncovering or blocking outlets due to ice-margin fluctuations and isostatic rebound) can result in dramatic changes in lake level of several tens of meters (figure 4-3). In the case of ice-margin fluctuations, such changes can be catastrophic, whereas isostatic uplift is usually gradual. Smaller-scale variation in water level of a few to several meters, however, generally results from either downcutting of outlet channels or from changes in the basin-wide water budget.

The influence of the four factors mentioned above is significant for understanding the middle-to-late Holocene sequence of lakes. Channel floor erosion is probably relatively gradual and is controlled both by the lithology of the substrate comprising the channel floor and by the velocity of water discharging through the channel. For example, the channel floor of the Chicago outlet remained stable after the Calumet phase, since by then it had been eroded to bedrock (Hansel et al. 1985). Although Lake Nipissing may have used both the Chicago and Port Huron outlets simultaneously, because at Chicago the outlet is floored by bedrock while more easily eroded, unconsolidated drift underlies Port Huron, the 7 m drop in lake level from Lake Nipissing to modern levels apparently resulted solely from erosion of the St. Clair River. The average rate of erosion for this drop was circa 45 cm per century, which is actually similar to the uplift of North Bay that controlled the Nipissing transgression (Monaghan and Lovis 2002). Conceptually, the causes and implications of lake-level change related to outlet erosion and climate change are relatively straightforward. The issues and implications for isostatic depression and uplift, however, are more complex because of the north-south geographic variation in amount of movement and the temporal variation in rate. Isostatic uplift continues to influence shorelines around the northern half of Lake Michigan. Furthermore, modern, sophisticated measurements of uplift collected recently in the upper Midwest and Great Lakes regions indicate that the southern half of Lake

Michigan is not isostatically stable, as has long been postulated, but is actually subsiding, particularly relative the Port Huron outlet. As noted throughout this book, the process of isostatic depression and recovery has important direct and indirect influences on shoreline development and coastal dune formation processes.

Isostatic Depression and Recovery in the Upper Great Lakes

Isostatic depression of the earth's crust resulted because viscous mantle material below the earth's crust was displaced by the mass of the Laurentide Ice Sheet (1–2 km thick in the northern parts of the upper Great Lakes). Assuming that the greatest amount of depression occurred beneath the greatest ice thickness (i.e., the point of maximum load) and decreased closer to the ice margin, relatively greater crustal depression occurred generally north-northeastward from the most southern ice margin. As the ice thinned during northward retreat between 15,000 and 10,000 years ago, the load depressing the crust was removed and displaced mantle material began to flow back beneath the crust. As a consequence of this process, the earth's crust began to gradually rise in areas under the former ice sheet, which results in "isostatic rebound" of the surface.

Within the Great Lakes region the maximum amount of crustal depression, and consequently the maximum potential amount of isostatic rebound, occurred in the most north-northwestern part of the basin. Studies have shown that the rebound rate varies through time by a factor of e^{-kt}, where e is the base of natural logarithms, k is the decay constant, and t is the years since deglaciation (Andrews 1970; Fillon 1972). This relationship indicates that the greatest rate of uplift occurred directly following deglaciation and that it also logarithmically decreases through time (Farrand 1962; Broecker 1966; Andrews 1970; Walcott 1970). Based on data from lake-level gages collected from around the Great Lakes basin, the region has not yet achieved isostatic equilibrium, and differential north-south rebound is still occurring throughout the basin (figures 4-1 and 4-3; Clark and Persoage 1970; Larsen 1985a, 1985b; Clark et al. 1990).

The process of isostatic rebound is particularly significant for understanding the development, correlation, and age of Holocene beaches. The regional and temporal differences in uplift also have important implications for beach/bluff erosional characteristics, fluctuations in sand supply, and other spatial geomorphological variations around Lake Michigan. First, consider at the most basic level how isostatic differential rebound influences beaches, their regional correlation and formation processes. Given that water planes of lakes are horizontal, when any associated beach was formed and developed it must also have been horizontal and, therefore, must have occurred at the same altitude throughout the basin. Isostatic rebound dictates that such a beach, formed at a specific time in the past, does not occur today at the same altitude at which it was originally formed. Rather, the altitude of older beaches varies throughout the basin in a regular and predictable fashion such that the beaches in areas of high rebound are raised higher than those in areas of lower rebound. Specifically, because the Lake Michigan isostatic rebound rate is greatest in the northeast (i.e., Straits of Mackinac) and lowest in the southwest (i.e., Chicago), ongoing uplift means that the altitude of fossil beaches will be highest in the north and gradually fall to the south where rebound rates are lowest (see figure 4-1). Consequently, given that the amount

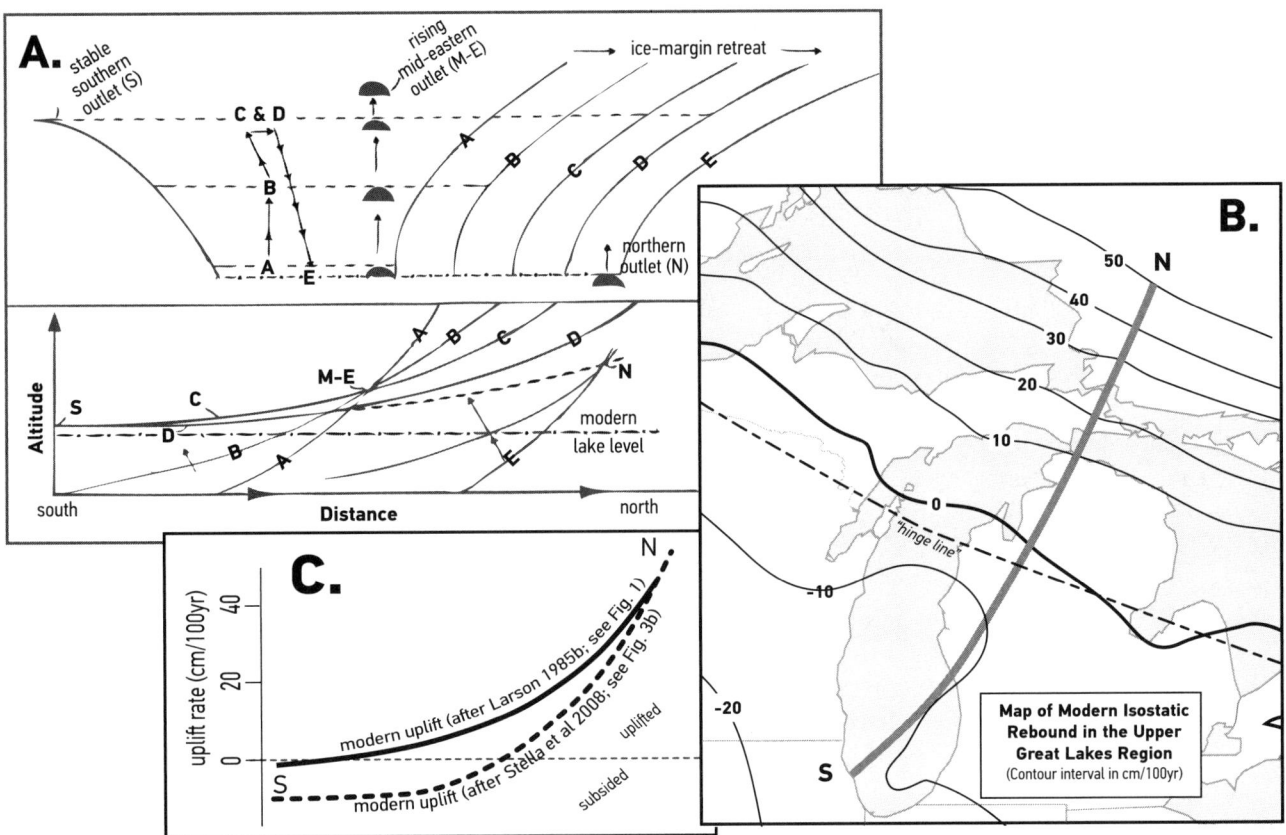

4-4. Isostatic uplift and subsidence in the upper Great Lakes. A. Diagrammatic models of the theoretical affects of uplift based on variation in outlet use and differential uplift. B. Map showing present rates of uplift in the upper Great Lakes region based on GPS and tidal buoy measurements (after Sella et al. 2007). C. Profiles of uplift. One is along line S–N in 4B and the other is from profile line shown in figure 4-1. Uplift diagram/rates modified from Larsen (1985a), Clark et al. (1990), and Sella et al. (2007).

of uplift generally increases northward, segments of the beach formed in more northern parts of the basin will be raised relatively more than in the southern part. From a modeling standpoint, differential uplift varies exponentially with distance and is approximately curvilinear (Larsen 1985a, 1987; see figures 4-1 and 4-3). Moreover, because the rate of uplift decreases exponentially through time, beaches formed 10,000 years ago will be raised significantly more than those formed 5,000 years ago even if the actual altitude of the outlet and consequently lake level remained the same.

The process of fossil beach formation and their regional correlation is complicated by the geographic and temporal variations in outlets and their differential uplift. For example, although the basin-wide water level achieved in a specific ancestral lake is controlled largely by the altitude and hydrological properties of the outlet channel, differential isostatic rebound of the outlet compared to other parts of the lake basin affects both the preservation of beaches observed today and the rate of rise in lake level through time. This process is diagrammatically illustrated in figure 4-4, which shows how differential rebound affects beach preservation as the outlet varies in geographic position in relationship to rebound rate. It specifically shows the generally accepted model of lake-level change in the Huron lake basin for the Algonquin-Stanley-Nipissing sequence of lakes. Three hypothetical outlets, southern (Port Huron), middle (Kirkfield/Fenelon Falls), and northern (North Bay), each have variable rates of rebound with the greatest rate associated with the most northern (see figure 4-1 for outlet locations). As the ice margin retreats to position A, it uncovers the middle outlet and lake level falls to level A as drainage is transferred to this outlet. During retreat of the ice to positions B and C the outlet rebounds and lake level progressively rises from level A to levels B and C. When level C is achieved the middle outlet has risen to the same altitude as the southern outlet and drainage is transferred back to it.

Although a simple transgressional sequence might be predicted as lake level rises from A to C, in fact isostatic rebound causes both "transgressional" and "regressional" beaches to form in different parts of the basin. This sequence of beaches, as they would appear today, is shown on the lower half of figure 4-4A. Because the rate of isostatic rebound increases northward in the basin and lake level is assumed to be controlled by the rate of uplift at the outlet, the rise in lake level would be greater than the rebound of the land surface south of the outlet. Beaches in this area would represent "transgressional" sequences as older beaches are reworked and submerged by younger, relatively higher-level lakes, and thus the youngest beach would occur at the highest altitude. North of the outlet, however, where the rate of isostatic rebound is greater than that at the outlet, the land surface would rise relatively faster than the level of the lake. The sequence of beaches formed in this region would appear "regressional" with the oldest beaches occurring at a greater altitude than younger. The result of such an uplift-controlled lake level is that once discharge ceases through the outlet, a single beach would be preserved in the south, while the northern beaches would "split" and preserve a series of progressively younger and lower beaches. Many modern researchers (Larsen 1985a, 1985b, 1987; Clark et al. 1990) believe this apparent splitting of beaches is actually the origin for the "hinge line" proposed by Goldthwait (1908) and which has been broadly employed by Leverett and Taylor (1915) and Hough (1958, 1966) to explain the evolution of the Great Lakes.

The concept of a hinge line, which is simply a point of "0" isobase and in the upper Great Lakes region marks the northern limit of crustal stability (Clark et al. 1990), has been integral to the debate on the evolution of the Great Lakes for more than a century. Most contemporary reviews of hinge lines, however, suggest that they are more apparent than real and may even be contrary to principles of solid-earth geophysics (Clark et al. 1990). Such modern critiques suggest that "hinge lines" derived from incorrect correlation of differing-age beaches. In such a view, the hinge line does not actually mark a change from crustal stability to isostatic uplift but rather marks an inflection from crustal uplift in the north to crustal subsidence in the south. This idea was first proposed by Spencer (1881) and can be illustrated from figure 4-4. For example, the point "M–E," which appears to be a place where beaches "bifurcate" on figure 4–4A, represents an apparent hinge line. South of that point, levels "C," "A" and "B" on figure 4–4A all appear to merge to define an area of crustal stability. By time of level "D," the "M–E" outlet has risen above the southern outlet and drainage is transferred south. However, because the "stable" (i.e., least warped) southern outlet controls lake level after level C on figure 4-4A, continued retreat of the ice margin to position D results in no actual change in water level within the lake. Depending on rate of differential uplift or ice-margin retreat, very little difference may exist between C and D south of the M–E point, creating what appears to be a hinge line with lakes draining through the southern outlet.

Ultimately, when the ice margin retreated to position E, the northern outlet was uncovered and consequently allows the lake level to drop again (figure 4-4A). Through time, lake level will rise at a rate controlled by the rebound of the outlet. Because the northern outlet rebounds fastest, however, the change in lake level will always be greater than the land surface south of the outlet. This results in a completely transgressional sequence throughout the basin and will continue until the altitude of the northern outlet reaches that of the southern (dashed line in lower part of figure 4-4A). At this point drainage is again transferred south. Importantly, in the south end of the lake basin, this lake level will come close to the altitude of beach D and will create another apparent "hinge line" with a relatively simple history. In fact, the sequence of lakes was vastly more complex than is apparent. The evidence for the complexity, however, was destroyed by the several transgressive sequences. Lacking direct evidence, or only having more ambiguous data, researchers unraveling the lake history have more recently focused on collecting direct measurements of uplift, more detailed chronology of beaches, and geomorphological or bathymetric data to develop realistic models of uplift.

Shoreline Subsidence in Southern Lake Michigan

When the mantle is displaced from beneath the Laurentide Ice Sheet and forms a crustal depression, it flows away from the maximum load to areas where the ice is thinner or absent near the glacial margin. This process causes the crust to rise and form a "bulge" in front of the ice sheet, an effect that is global in scale (Peltier 1994, 2002, 2004; Sella et al. 2007). Through time, as the mantle material flows back to areas where it was displaced, the bulge readjusts and begins to sink. This readjustment is ultimately a global process, but more sophisticated measuring equipment has made it possible to recognize the process in the Midwest region well south of the Laurentide Ice Sheet margins (Clark et al. 1990; Sella et al. 2007). Unlike most of the Great Lakes and upper Midwest regions where isostatic rebound occurs, areas where the bulge occurred are now experiencing crustal subsidence. More significantly for this book, recently collected and more refined data on crustal movement collected from very sensitive global positioning satellites (GPS) and lake-gage arrays suggest that the area of subsidence actually extends into southern Lake Michigan (figure 4-4). These data have important implications for the sequence of beaches related to the middle and late Holocene levels of Lakes Nipissing through modern Lake Michigan, as well as to our understanding of regional differences in the long-term effect of coastal processes. The GPS and tidal gage records of uplift indicate that the location at which isostatic rebound changes from positive to negative (i.e., from uplift to subsidence) occurs generally on a northwest-southeast trending line that crosses Lake Michigan from about Frankfort, Michigan, to Sturgeon Bay, Wisconsin. Probably not coincidently, this line also occurs at the approximate position of the so-called hinge line discussed above (figure 4-4).

The concordance of the area south of the hinge line in Lake Michigan, which is believed not to have been warped during the Holocene, with the area of subsidence has important implications for the correlations of lake stages, raised fossil beaches, and long-term geomorphological processes that control beach development, shoreline erosion, and sand supply. These geomorphological processes, particularly in relationship to the rate of uplift, also affect coastal dune formation and development. For example, if correct, the constant shoreline subsidence in the southern end of Lake Michigan effectively creates a continual transgressive sequence during the Holocene because, as the shoreline subsides, lake level will appear to rise relative to the beach. Through time this process will amplify beach erosion and accentuate or accelerate bluff erosion and recession. Depending on specific conditions at a given locale, such a process may create a relatively continuous supply of sand from which to build dunes. However, the same process that erodes bluffs will also erode the newly formed dunes. This is shown in a large regional study (Buckler and Winters 1983) that measured bluff recession at sites south of the Door and Leelanau peninsulas (Wisconsin and Michigan, respectively) but regrettably included no sites from the northern third of Lake Michigan. These data indicate that bluff recession is generally highest during high water stands, is lowest along dune-dominated shorelines, and is generally higher in the southern than northern parts of the basin. Specific estimates of bluff recession in western Michigan (i.e., east shore of Lake Michigan) average 40 cm yr^{-1} but are 40–50 percent lower in dune-dominated shorelines (Buckler and Winters 1983). In Wisconsin, bluff recession averages 74 cm yr^{-1} in the north and 22 cm yr^{-1} in the south, but is more irregular in Michigan. These rates suggest that, depending on the specific geological conditions at a given locale, during the past 4,000 years 1,600–880 m of shoreline has been eroded from the southeastern shore of Lake Michigan (Michigan) and as much as 3,000 m may have been eroded from bluffs along the southwestern shore (Wisconsin and Illinois).

The high rates of bluff recession noted above suggest that considerable shoreline has been lost since Lake Nipissing was abandoned circa 4,000 years ago, but that this is also an expected outcome within a subsiding coastal environment. The opposite outcome, considerably lower bluff-shoreline recession-erosion or even coastal accretion, should occur for the northern end of Lake Michigan where uplift dominates the coastal zone. Because no study of long-term beach erosion similar to that of Buckler and Winters (1983) has been reported from the northern or northeastern shore of Lake Michigan, we suggest the following logical model to explain this phenomenon. In an area of uplift of Lake Michigan, such as the Straits of Mackinac where uplift rate is highest, continuous uplift effectively creates shoreline regression because during uplift, water level appears to fall relative to a beach. This situation is illustrated in figure 4-4, which shows a sequence of recessional beaches that developed after the Lake Nipissing fell. Unlike the southern end of Lake Michigan, where subsidence prevails and beaches undergo relatively continuous erosion, the beaches and shoreline environments developed in the northern part of the basin are generally uplifted and preserved. This includes landforms associated with the uplifted shoreline. From the standpoint of shoreline dunes, the net effect is that shoreline or coastal dunes are more likely to be preserved in the north than in the south. As a consequence, any associated coastal dune archaeological sites are also more likely to be preserved in such areas.

The results of the process of uplift and preservation of shoreline environments can be observed along the northeastern shoreline of Lake Michigan within one of the sample areas between Brevort and St. Ignace in the Upper Peninsula (figure 4-5). The original interpretation of this area suggested that the sets of prominent dunes that occur 2–3 km inland from Lake Michigan (labeled "2 kya dune complexes"; figure 4-5A) were middle Holocene in age and probably formed related to the maximum level of Lake Nipissing. The sets of prominent beach ridges (shown by dashed lines; figure 4-5A) formed as Nipissing fell to Algoma and modern Lake Michigan. The high dunes closer to Lake Michigan (labeled "1 kya dune complex"; figure 4-5A) were believed to have formed related to receding Lake Algoma or to modern Lake Michigan. As shown on the map (figure 4-5A), however, most of these inferences were wrong. Based on ^{14}C and OSL ages of dunes, shoreline features, and wetland deposits, most of the parts of the landscape are considerably younger than previously believed. These data indicate that the "Nipissing" dune complex formed just after circa 2,000 years ago and that the younger, "Algoma" dunes formed after circa 1,000 years ago. At least some of the beach ridges are probably also of different ages than originally supposed. For example, the beach ridges with an embayment likely formed less than 2,000 years ago and are probably the shorelines to which the 2,000- and 1,000-year-old dune complexes relate.

The formation, infilling, and preservation so far inland of these relatively young beach sequences probably resulted because uplifting shorelines and lowering water level combined to effectively remove older beaches from erosion and support the development of increasingly younger sequences. This process is illustrated in figure 4-5B, which is a lake hydrograph that shows the fluctuations in lake level through time for Lake Michigan. The lake level curve shown is based on Monaghan and Lovis (2005: figure 4–3) and has been adjusted for isostatic recovery in the Brevort–St. Ignace area based on models presented by Larsen (1985b), Clark et al. (1990), and Sella et al. (2007; figures 4-1 and 4-3). This adjustment was undertaken to directly relate lake level, which has been controlled by the Port Huron outlet since circa 5,000 years ago, to uplift in specific areas. The lake-level curves were adjusted by placing a "0" uplift isobase through Port Huron and normalizing other isobases to Port Huron. Several other lake hydrographs were similarly prepared for sites in northern Lake Michigan (figure 4-6). Locales that were on similar isobases at opposite sides of the lake (i.e., Manistique-Petoskey and Winter-Charlevoix) were grouped. To test whether the modeled uplift rates and lake hydrographs accurately reflected local conditions, age controls for lake level (i.e., indicators such as wood in beach, peat under/over beach, OSL age of beach, age of occupation horizons) from these areas were the altitudes from which they occurred at each locale. These data suggest that the calculated uplift and the projected water-planes at least broadly reflect reality and that we can use these data to understand how uplift and lake level locally interact to create different transgression-regression rates through time and space around Lake Michigan.

The calculated lake hydrograph from the Brevort–St. Ignace area (figure 4-5B) shows that lake level fell after circa 4,000 years ago and rose again circa 2,000 years ago. This rising water level probably filled the embayment (labeled

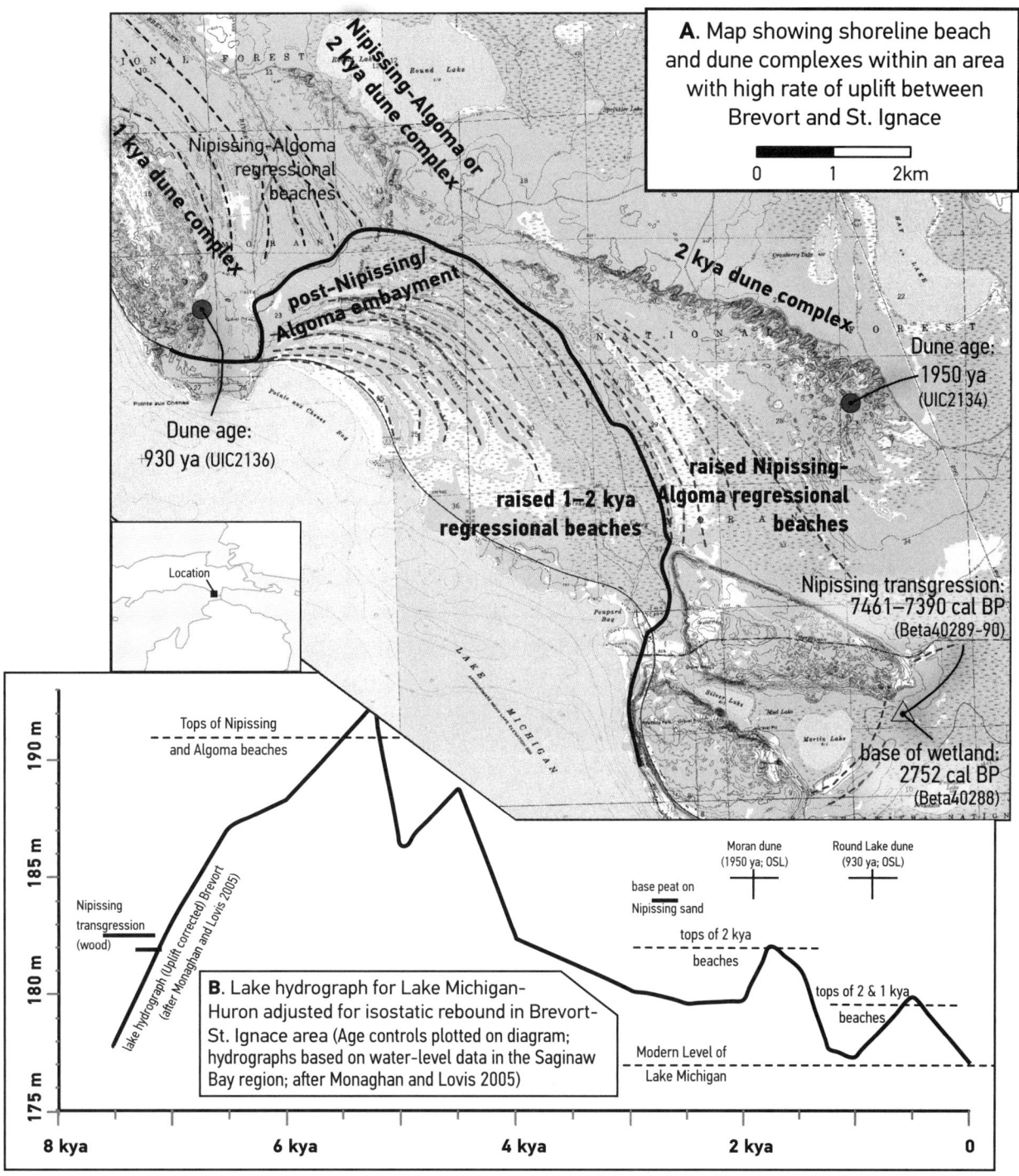

4-5. Geomorphology of area between Brevort and St. Ignace. A. Beach ridge and dune complexes. Age estimates for dunes and beach deposits shown by circles and triangle (respectively); age results and method of dating label for points. B. Lake hydrograph for Lake Michigan-Huron adjusted for isostatic rebound in Brevort–St. Ignace area. The general levels of beach ridge tops derived from figure 4-5A. Age controls plotted on diagram; hydrographs based on water-level data in the Saginaw Bay region; after Monaghan and Lovis (2005).

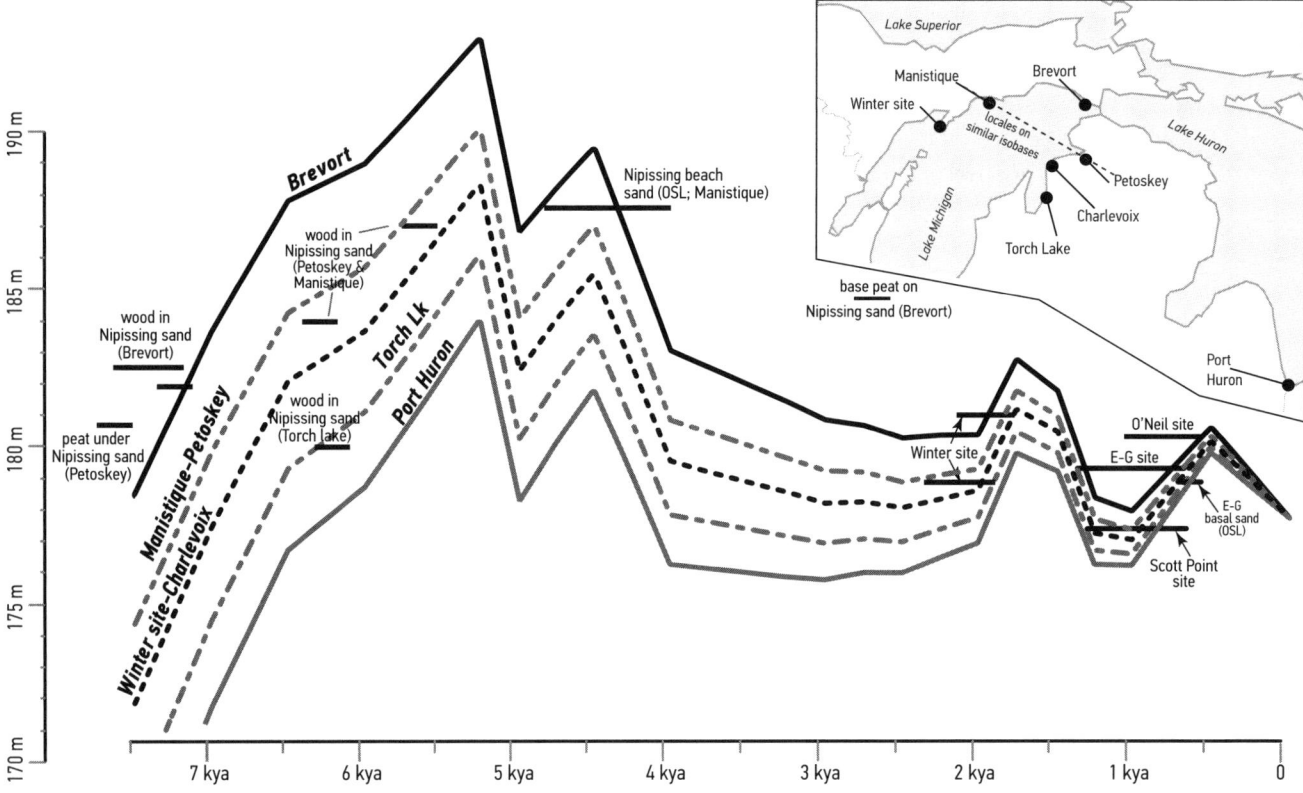

4-6. Lake hydrographs for Lake Michigan adjusted for isostatic rebound in areas of northern Lake Michigan. Age controls plotted on diagram; colors of age controls match colors of hydrographs; hydrographs' curves based on water level in the Saginaw Bay and southern Lake Huron; after Monaghan and Lovis (2002, 2005).

"raised 1–2 kya regressional beaches"; figure 4-5B) and instigated a sequence of beach ridge formation and dune development ("2kya dune complex"; figure 4-5B). Following another regression, lake level again rose after circa 1,000 years ago and initiated a younger episode of beach ridge and dune construction. The preservation of these episodes is a direct result of the uplift process that removes shorelines from erosion but is also accented by the cyclical nature of late Holocene transgressions and regressions. Moreover, the north-south variations in differential uplift are also apparent through the comparison of the predicted changes in water level between circa 4,000 years ago and 2,000 years ago from the least warped beaches (Port Huron) to the most warped (Brevort). These data show that while water level probably remained relatively stable near Port Huron during this interval, the relatively high rate of uplift near Brevort created an apparent regressive sequence that would have preserved older beaches by raising them to higher altitudes.

The interactions between uplift or subsidence, cyclical, and climate-related changes in lake level related to erosion are complicated and can mask some important relationships. Much of the discussion above concerning the general processes and expectation of beach and dune formation and their preservation or erosion as a result of differences in isostatic rebound is presented within the context of a static lake level. As shown in figure 4-3, however, Holocene lakes within the Great Lakes were not static. Lake level actually varied greatly even during the late Holocene, as a result of outlet uplift and erosion as well as regional climate fluctuations. figures 4-5 and 4-6 clearly show some of the temporal and spatial complications in how uplift and lake-level excursions interact to limit or enhance the magnitude, rate, and effects of transgression and regressions around Lake Michigan. How the transgressions and regressions created by the actual changes in water level interacted with those that are only apparent and related to isostatic uplift or subsidence is the focus of the remainder of this chapter.

Climate Controls for Holocene Lakes

Lake-level changes in the upper Great Lakes were affected by regional climate variations and cycles. This was most significant during the middle and late Holocene (figure 4-4). These cycles, the duration and magnitude of which are by no means universally agreed upon, are believed to be intimately related to regional climate variations, particularly temperature and precipitation events such as the Little Ice Age and Medieval Warm intervals (e.g., Larsen 1985a; Monaghan and Lovis 2005). Regardless of event correlation, most researchers agree that late Holocene water levels in the upper Great Lakes fluctuated related to basin-wide variations in precipitation and/or temperature (Fraser et al. 1975; 1990; Larsen 1985a, 1985b; Monaghan and Lovis 2005). For example, reduced precipitation during relatively dry periods resulted in lower lake level because less water was input to the basin. Similarly, particularly wet intervals resulted in increased lake levels because more water was input to the basin.

Changes in temperature also affected lake level by increasing evaporation during warm periods or reducing evaporation during cool periods. Relatively long-term increases in evaporation will, of course, lower the lake level. Changes in water level due solely to variations in temperature and/or precipitation could result in only a few meters difference in the level of the Great Lakes. In fact, the data from the Great Lakes region indicates that climate change, on the magnitude observed during the middle and late Holocene, probably caused only minor changes in lake level during the late Wisconsin, but is probably the main factor controlling the middle and late Holocene and historic changes in lake level (figures 4-2, 4-3, and 4-4). For example, Larsen (1985a) has applied this model of climate-controlled lake-level fluctuation to the middle and late Holocene lake sequence. He suggests that the level of Lakes Michigan and Huron varied 2–3 m from the mean lake level and that this variation occurred over a 200- to 300-year interval (figure 4-3). Similarly, Monaghan and Lovis (2005) used data from archaeological and alluvial records preserved in streams near the margins of Saginaw Bay to construct a curve of water-level variation (figure 4-3). They indicate that these relatively minor transgressions and regressions resulted because of regional, millennial-scale variation in warm/dry and cool/wet climate, which structured human settlement and archaeological site burial, stratification, and preservation. Similarly, such lake level changes, in conjunction with climate cycling, probably also played a significant role in structuring episodes of dune formation or reactivation.

CHAPTER FOUR

Outlet Controls for Holocene Lakes

Although climate played a major role in controlling the relatively minor lake-level fluctuations observed during the late Holocene, the most dramatic changes in water level occurred because of either isostatic rebound or erosion of outlet channels. For example, the 5,000-year-long Nipissing transgression resulted because the isostatically depressed northernmost outlet in the region at North Bay, Ontario, gradually rebounded after it was uncovered circa 10,000 BP (North Bay outlet: NB; figure 4-1). Interestingly, because of the geographic configuration of the upper Great Lakes, the North Bay outlet rebound rate is significantly higher than that anywhere in Lake Michigan (figure 4-7A). Consequently, the rate of the Nipissing transgression in Lake Michigan is completely controlled by the rate of uplift of the North Bay outlet (figures 4-3 and 4-7B; Monaghan and Lovis 2002). The transgression culminated in the formation of Lake Nipissing, which is generally considered to begin when the Nipissing transgression rose above the altitude of the modern Lake Huron water level. Lake Nipissing, which was 6–7 m above the modern level of Lake Michigan in the least uplifted southern end, developed when the northern outlet at North Bay rose above the sill at Port Huron, Michigan (PH, figure 4-1) and transferred drainage to this southern outlet.

Lake Nipissing covered all of what today are Lakes Michigan, Huron, and Superior, and was abandoned circa 4,000 years ago because of accelerated erosion of the outlet sill at Port Huron (Monaghan and Lovis 2005). By circa 4,000 years ago water levels in southern ends of the Lakes Huron and Michigan stabilized at approximately modern levels, which is a proposition that has only been advanced during the past decade and is discussed in more detail below (Monaghan and Hayes 1998, 2001; Monaghan 2002; Monaghan and Lovis 2002, 2005). The rate of this regression was controlled mainly by the rate of erosion of the Port Huron outlet (Monaghan and Lovis 2002, 2005). Although the essentially modern lake level may have been achieved very early in the southern part of the upper Great Lakes basins (i.e., southern Lakes Michigan and Huron), because the northern part of the upper Great Lakes was still significantly isostatically depressed, the actual separation of Lake Superior from the lower lakes (Michigan and Huron) and formation of the modern configuration of the Great Lakes (Michigan-Huron and Superior) did not occur until circa 2,200 to 2,000 years ago (Monaghan and Lovis 2002, 2005; figures 4-5 and 4-8). At this time, the rapids at Sault Ste. Marie finally rose to the level of the Port Huron outlet and began to control the level and hydrologic properties of Lake Superior, as it does today.

Summary

From the standpoint of this project, several important conclusions can be reached concerning the interaction between lake-level variations, the effects of differential uplift around Lake Michigan, and dune formation. The factors and influences for dune formation and preservations are numerous, complicated, and often not independent of each other. At the most basic level, in order to form and sustain dune development and growth, sufficient sand must be present, must be erodible, and, in order to build large dunes, must be replenished. Sufficient wind must also be present. While many of additional factors, ranging from basin aspect

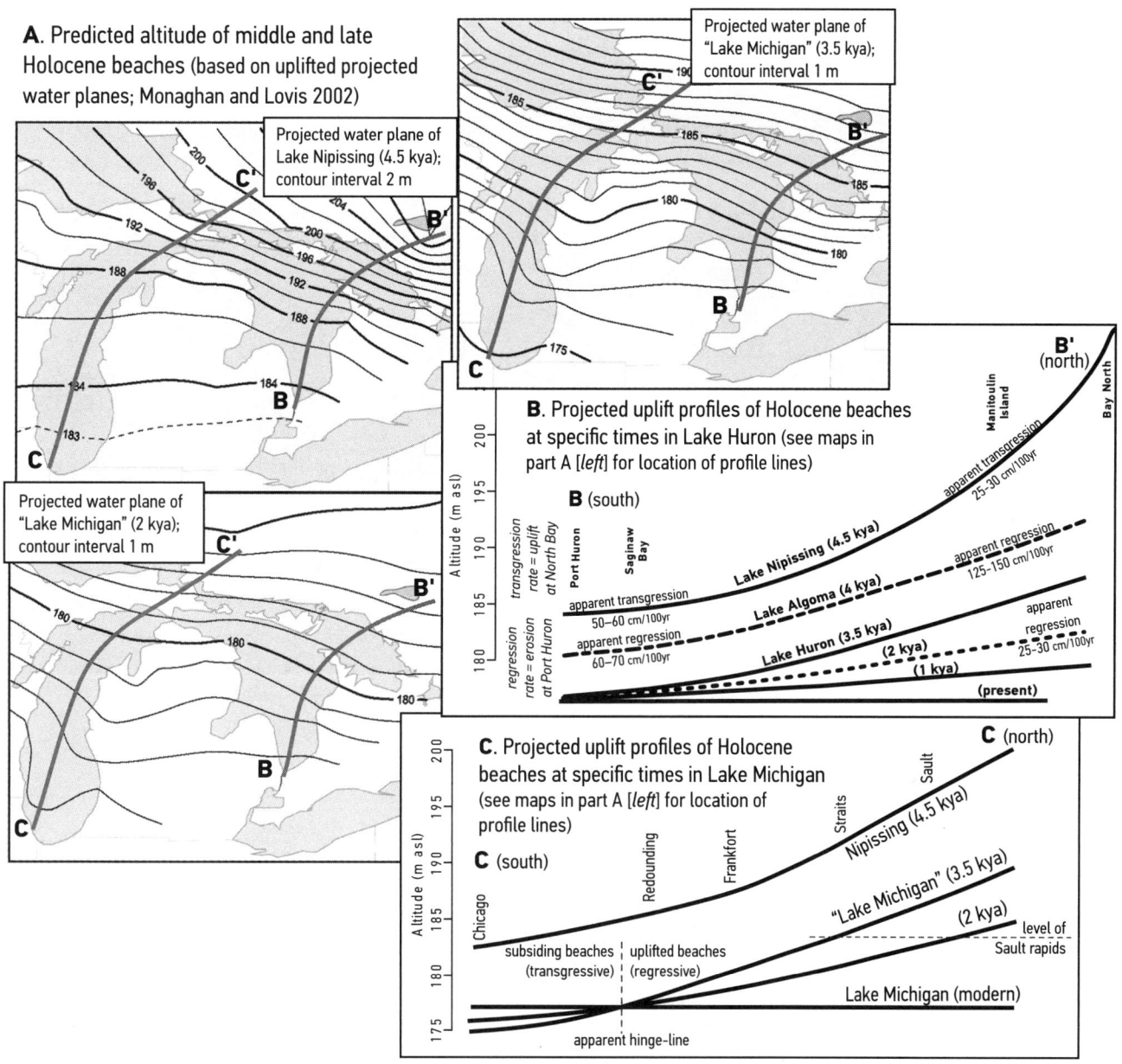

4-7. The effects of differential uplift and lake-level changes around Lakes Michigan and Huron. A. Predicted altitude of middle and late Holocene beaches based on uplifted projected water plains (after Monaghan and Lovis 2002). B. Projected uplift profiles of Holocene beaches at specific times in Lake Huron. See maps in part A (left) for location of profile lines. C. Projected uplift profiles of Holocene beaches at specific times in Lake Michigan. See maps in part A for location of profile lines. Lake-level curves after Monaghan and Lovis (2002).

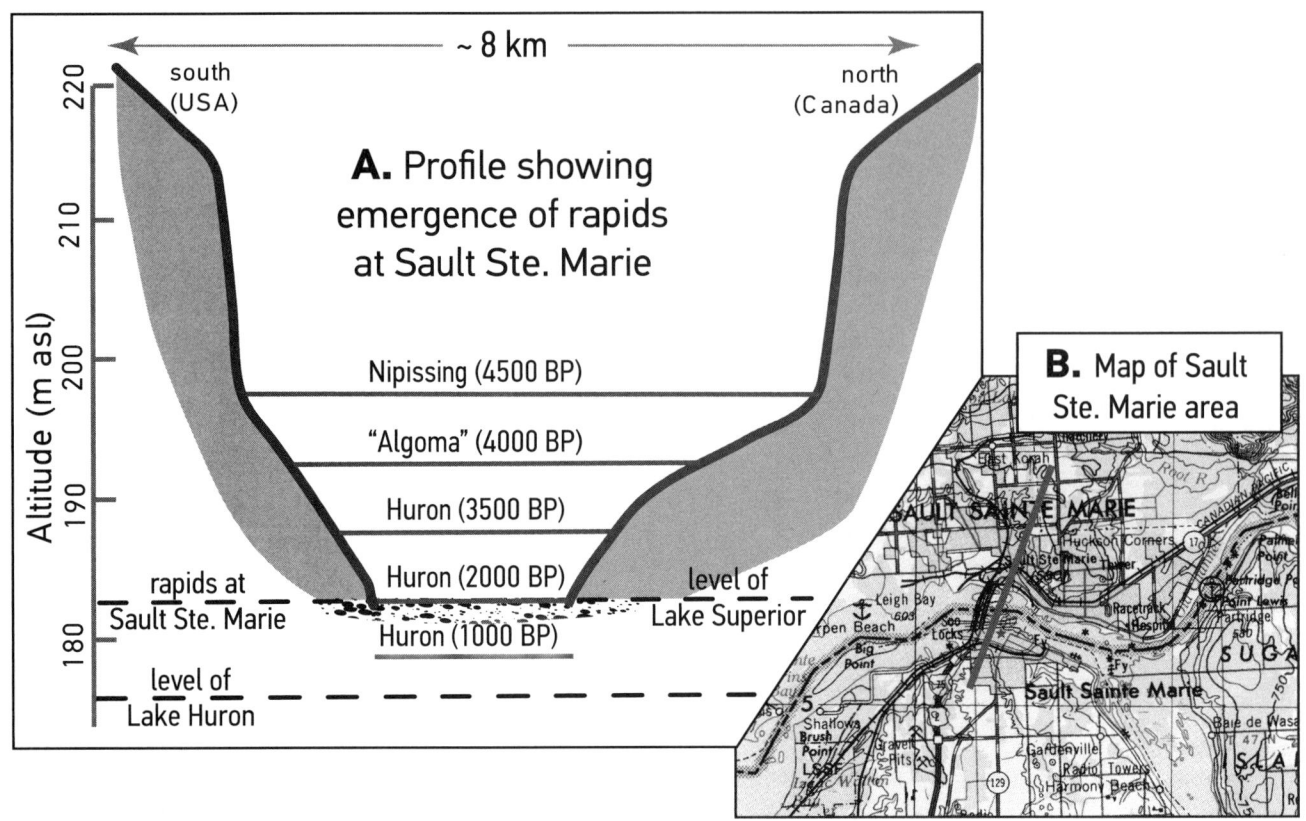

4-8. The effects of differential uplift and lake-level changes around Lakes Michigan and Huron. A. Predicted timing for emergence of rapids at Sault Ste. Marie and separation of Lake Superior from Huron. B. Location of emergent profile. Lake levels shown based upon uplifted projected water planes in Lake Huron; after Monaghan and Lovis (2002).

to late Wisconsin deglaciation sequences, influence dune formation, we believe that the most important factors relate to differential uplift and the Holocene transgressions and regressions of Lake Michigan and its ancestors. The most important characteristics of these factors are summarized below. Variation of their patterns through time and space provides the comparative basis to delineate the controlling events that initiate the formation, reactivation, and preservation of coastal dunes. Developing a clear understanding of these processes, however, is complicated because they include both secular and cyclical systemic variability as well as long-term, directional developmental changes.

Lake level has important influences on dune formation and has varied greatly during the Holocene. Several frameworks for changes in lake levels are summarized in figure 4-3, which shows three different lake-level curves developed by Larsen (1985b), Thompson and Baedke (1999), and Monaghan and Lovis (2005) for the middle and late Holocene. These curves are based on different types of data and derive from different areas in the Lake Michigan and Lake Huron basins. They also have different resolution, and emphasize different aspects of lake level. Consequently, because of the vagaries of preservation of fossil beaches or organic material, the imprecision of ^{14}C age estimates, and difficulties in sedimentological interpretations, the Holocene lake-level curves shown are general and do not always directly correspond with each other (figure 4-3). The lake level curves, however, are conceptually useful and taken together show some important broad trends. The most significant from the perspective of dune formation or reactivation are listed below.

1. High water levels (i.e., a transgressive sequence) occur throughout the upper Great Lakes during the middle Holocene (ca. 6,000 to 4,500 years ago). These are related to the well-established Nipissing phase, which is the largest and most significant transgression during the Holocene in the upper Great Lakes.
2. Low-water-level lakes (i.e., a regressive sequence) generally follow Lake Nipissing and represent a significant regression to modern lake levels near Port Huron. The fall from Nipissing to modern levels occurred between 4,000 and 3,500 years ago and is mainly a consequence of erosion of the outlet channel at Port Huron, but some lake-level variation may also relate to climate cycles (e.g., Nipissing II and Algoma).
3. Initially, the modern upper Great Lakes formed a single lake in the Lake Michigan, Lake Huron, and Lake Superior basins. It drained through Port Huron from circa 3,500–4,000 years ago until circa 2,000 years ago. At this time, isostatic rebound raised the rapids at Sault Ste. Marie above Port Huron and separated Lake Superior from Michigan-Huron, and the modern configuration of the upper Great Lakes was achieved.
4. A series of high-low water levels (i.e., transgressions-regressions) occurs within Lakes Michigan and Huron after circa 2,000 years ago. These cycles were individually several hundred years in duration and are generally believed to relate to climatic cycles. High-water (transgressive) phases occurred during cool/wet intervals and low-water (regressive) phases related to warm/dry conditions (e.g., Little Ice Age and Medieval Warm intervals, respectively; Monaghan and Hayes 1998; Monaghan and Lovis 2005)

The pattern of lake-level variation described above provides some clear temporal and spatial parameters through which the relationship between transgressive-regressive events can be compared to episodes of dune formation and reactivation.

Patterns to differential uplift that likely also influenced the timing and character of dune formation and preservation also exist around the Lake Michigan shoreline. The dominant trend is north to south; the lowest rates of uplift are in the southwest part of Lake Michigan (i.e., north of Chicago), while the highest rates are in the northeast (i.e., near the Straits of Mackinac; figures 4-1 and 4-4). Although this pattern has been broadly understood for over 100 years, direct measurements of isostatic recovery recently collected in the upper Midwest and Great Lakes regions suggest that some of the earlier conclusions concerning basin-wide rebound are probably incorrect. These new data indicate that the southern end of Lake Michigan is actually subsiding (figures 4-4, 4-6, and 4-7). The original hinge-line concept proposed by Goldthwait (1908) and Leverett and Taylor (1915) does not indicate an area of isostatic stability but rather probably demarks that place in the basin that separates regional uplift (north) from regional subsidence (south) (see figures 4-4 and 4–7). Most of the recent research about this issue has concluded or implies that isostatic recovery exists across Lake Michigan and that the region of isostatic stability south of the hinge line does not exist. The major point of disagreement surrounds the magnitude and spatial distribution of subsidence, and not whether the basin is actually subsiding.

Regardless of the different interpretations or spatial configuration concerning the magnitude and distribution of uplift and subsidence around Lake Michigan, these data have some important, general implications for modeling the timing and character of dune formation or reactivation within the basin.

1. The southern end of Lake Michigan is currently undergoing relative subsidence, while the northern end of the basin is uplifting.
2. The rate of differential isostatic recovery has lessened through time, which makes the relative differences in rebound across the basin greater during the middle Holocene than now. Consequently, the effects of uplift or subsidence on dune formation (discussed below) was accented between 2,000 and 4,000 years ago and became progressively less significant after circa 2,000 years ago. This fact is illustrated in figures 4-5, 4-6, and 4-7.
3. In areas of subsidence, beaches will generally appear to undergo long-term transgression, which will in turn promote shoreline and bluff erosion. This outcome has been observed by Buckler and Winters (1983), whose data suggest that in western Michigan up to 800–1,500 m of shore and bluffs have been eroded since the middle Holocene. The lower estimate derives from dune-dominated shorelines. Regardless of whether the recession rate is regular, has slowed down, or has sped up through time, the net result is that older coastal dunes will tend to be eroded and destroyed, or only partly preserved, along the south shore of Lake Michigan.
4. Within areas of relatively higher differential uplift (i.e., the northern and northeastern shore of Lake Michigan) on the other hand, beaches will appear as a general regressive sequence. Over time, this pattern will result in a series of raised beaches, each older beach positioned further inland from the younger beach. Unlike what occurs in areas of subsidence, such a pattern will also tend to preserve coastal dune sets that, like the beaches, are offset progressively more inland as that shoreline is raised (see figure 4-5).

Placing these and other factors into their proper context is critical to the formulation of a temporal/spatial framework, and our ability to delineate the natural and cultural processes that control the locations and burial of archaeological sites within coastal dunes settings.

CHAPTER FIVE

Archaeological Sites in Dune Contexts around Lake Michigan

This sample locale description and summary compendium contains many of the fundamental primary observations on the six major archaeological locales newly located, newly sampled, reassessed through field investigation, and/or dated and redated with new samples from 2006 to 2008 as part of the ISTEA buried site taphonomy/dune activation and cycling project. As we have done elsewhere, the sequential presentation is organized geographically around the Michigan locations in the Lake Michigan basin. The first site summarized, the Winter site, is the most western site the project visited along the northern coastline of Lake Michigan. Site descriptions then proceed eastward to the Straits of Mackinac, and then southward to our most southern site in Oceana County. As chapter 2 revealed, no buried and stratified coastal dune archaeological sites have yet been discovered south of this point on the eastern shoreline of Lake Michigan.

As will be readily evident, the specific content of each site or locale summary is variable, although each contains a synopsis of past and current research, a detailed statement of the depositional sequence and formation processes, and any associated absolute ages (AMS, ^{14}C, OSL) attached to either the geological deposits (paleosols, sands) or cultural residues (features, artifacts). Dependent on the availability of reports or published materials, other aspects of the summary may vary. For example, the Ekdahl-Goudreau and Mt. McSauba sites, among others, have never been published, and in some cases the only information available is from institutional field notes. These sites are less well described. At the opposite extreme the Winter site has been summarized extensively in major papers and a master's thesis. In our presentation we chose not to be overly redundant with the available literature.

Finally, it is essential to note that the sites described here are not the only sites incorporated into our analysis. Contributory sites are presented in appendix C, along with the many nonarchaeological dune locales field-sampled between 2006 and 2008.

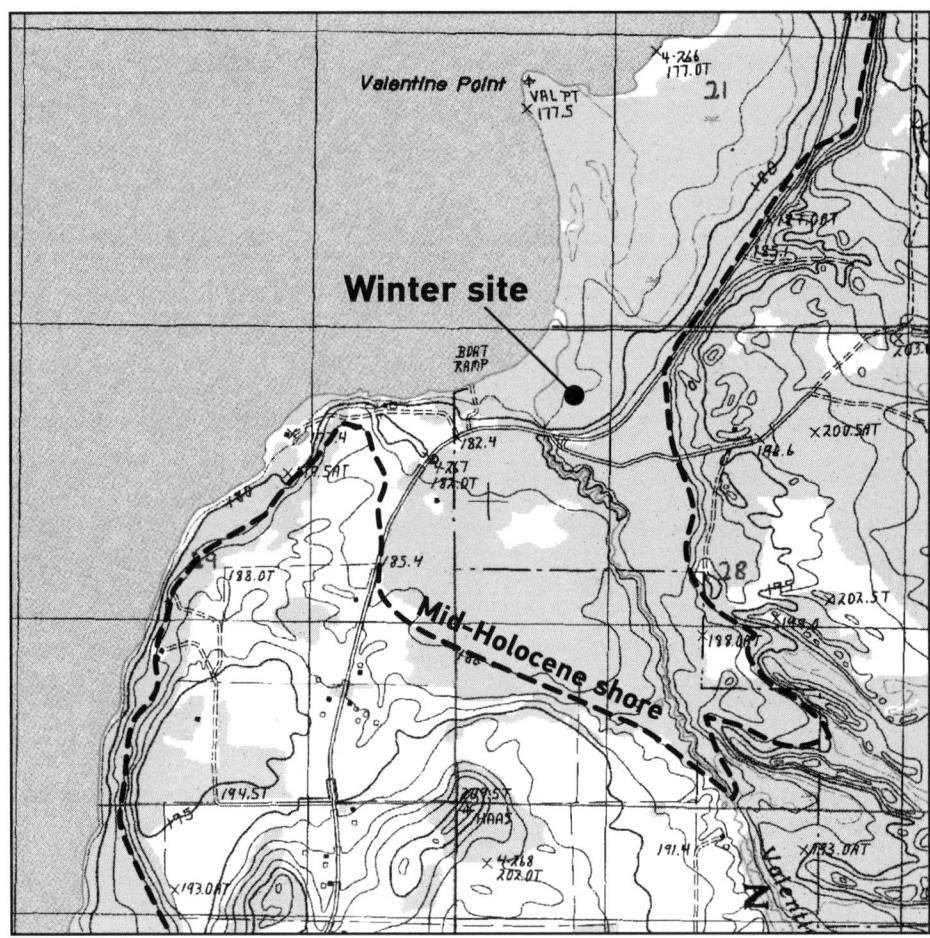

5-1. The Winter site location.

Winter Site (20DE17), Delta County

Location and Description

The Winter site (figure 5-1) is situated on the western shore of the Garden Peninsula, between the northern end of Green Bay and Big Bay de Noc. The site is located on the east bank of Valentine Creek, in the NW ¼ of the NW ¼ of Section 28, Fairbanks Township, T39N R17W, Delta County. The site was originally recorded by Thomas Bianchi in 1968, and was excavated by Jeffrey Richner, then a graduate student at Western Michigan University, in the summer of 1972.

Two major pieces of analysis and reporting resulted from this work. In 1973 Richner produced a MA thesis titled "Depositional History and Tool Industries of the Winter Site: A Lake Forest Middle Woodland Cultural Manifestation," and Bianchi subsequently in 1974 produced a major and unpublished description of the ceramic assemblage titled "Description and Analysis of the Prehistoric Ceramic Materials Recovered on the Winter Site." In 1980 Terrance Martin identified and reported on the limited faunal assemblage from the two major components at the site. The conclusions and interpretations resulting from these analyses were summarized in a recent comprehensive overview of northern Michigan Middle Woodland societies (Brose and Hambacher 1999). The primary reports by Richner (1973), Bianchi (1974), and Martin (1980) also form the basis for much of our ensuing summary.

Our revisit to the site 17 June 2007, by a team consisting of the three authors, confirmed that the depositional history of the site as reconstructed by Richner was fundamentally accurate. However, neither Richner nor Bianchi had access to absolute dates for his analysis and interpretation. Indeed, the site remained sans absolute chronology other than the relative sequence revealed by stratigraphic superpositioning of the occupations and comparative ceramic cross-dating for 25 years, until the current project rectified the situation.

Stratigraphy and Dating

The Winter site is situated on an isostatically uplifted beach or storm terrace on the eastern shore of Big Bay de Noc, and is consequently situated inland of the current lakeshore at an altitude of 179–80 masl. Richner correctly argues that the site would have been on the lakeshore during its occupation. Bianchi (1974), following Richner, describes the stratified deposits at the Winter site in terms of a "lower," a "middle," and an "upper" midden; these are occupations in stable soils here translated as A2/B or Ab horizons (figure 5-2; table 5-1). Richner observes that the "upper" and "middle" middens diverge from one another with intervening thin lenses of eolian sand, and that the "middle" midden periodically ends abruptly—they appear to be part of the same and very short-term depositional and stabilization sequence. The "lower" midden is stratigraphically more distinct, and separate from the "middle" and "upper" middens. Again according to Richner, the "lower" midden is underlain in places by gravel and clay deposits, and he favors an interpretation, which at the time related to Mason's observations at the Mero and Port de Morts sites on the Door Peninsula of Wisconsin, that these are the result of a Middle Woodland high-water stage dated to the second century A.D. More recent research has vindicated this Michigan-Huron basin-wide phenomenon (Monaghan and Lovis 2005).

TABLE 5-1. Correlation of 1972 Site Stratigraphy with the Current Study

RICHNER 1973	CURRENT STUDY
Sod	A horizon
Upper midden	A2/B horizon; 2σ 2152—1949 cal BP
Eolian sand	Unit 3; eolian sand
Middle midden	2Ab horizon
Eolian sand	Unit 2; eolian sand
Lower midden	3Ab horizon; 1883–1708 cal BP and 1949–1769 cal BP
Gravel	Unit 1; coarse lacustrine sand/gravel; 2605–1964 ya
N.a.	Red clay/Greatlakean till
N.a.	Limestone bedrock

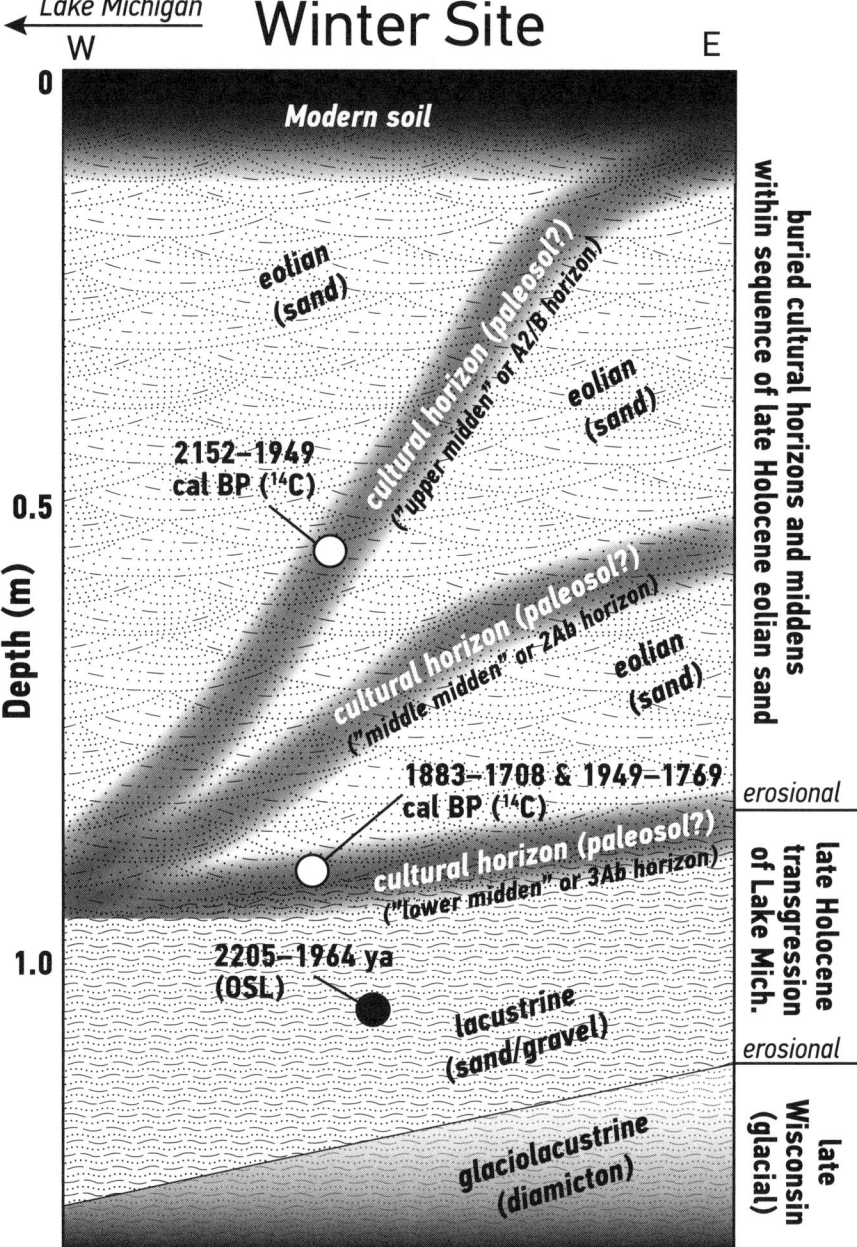

5-2. Generalized, composite stratigraphic column of the Winter site from Richner's archaeological units and as noted from the GeoProbe core. Ages of various units shown. See appendix A for discussion of how ages are reported. Unit depths and boundaries are approximate.

Our work at the Winter site was limited and very focused. We first relocated and identified the datum for the 1972 excavations and then reconstructed the site grid. We then placed three adjacent GeoProbe-generated cores in the vicinity of S11W5 on the excavation grid in an attempt to find the same stratigraphy recorded and published in Richner's thesis. Core 1 was fully exploratory to assess whether we had in fact relocated the stratigraphy recorded by Richner in 1972. Core 2 specifically targeted what we termed the "upper" end of the sequence, terminating in gravel deposits. The final core, Core 3, was designed specifically to sample the organic sequence in the lacustrine sands overlying the basal red clay. The clay formed a sharp contact in the sequence at a depth of 1.75 m below the surface. Correlation of Richner's stratigraphic observations and labels with those derived from our own work at the Winter site is presented in table 5-1. Winter site composite stratigraphy is depicted in figure 5-2.

The Winter site had not been absolute-dated prior to our research. We employed AMS and OSL dating in tandem. Based on the work of Bianchi (1974) we identified several rimsherds from the lower, middle, and upper occupations at the Winter site that had adhering carbonized food residues. The lower occupation was AMS-dated on North Bay Punctate and North Bay Plain sherds, and the middle/upper occupation AMS-dated with residues from a North Bay Vertical Corded vessel. The North Bay Vertical Corded vessel dated to 2090 ± 40 BP (Beta-237019; 2σ 1949–2152 intercept 2050 cal BP, Stuiver and Reimer 1993). The North Bay Punctate vessel from the lower zone dated to 1920 ± 40 BP (Beta-237017; 2σ 1736–1967 cal BP, intercept 1870 cal BP, Stuiver and Reimer 1993), and the North Bay Plain vessel from the lower zone dated to 1860 ± 40 BP (Beta-237018; 2σ 1708–1883, intercept 1820 cal BP, Stuiver and Reimer 1993). The dates from the lower zone are statistically identical at 95 percent ($t = 1.125$, $df = 2$), with a mean pooled age of 1890 ± 28 cal BP. The dates from the lower and upper zones are statistically different, although they are close in age. The basal sands at the Winter site were dated by OSL to 2180 ± 215 (UIC-2133, 1σ). This date is, as expected from its position underlying all of the cultural strata, earlier than any of the AMS dates. Here we are not concerned with the 190-year inversion of the AMS dates. Rather, what is of significance to our research is the rapidity of deposition between the basal sands and the occupations. There is only a 360-year difference between the intercepts of the youngest and the oldest ages from all of the dated strata.

Occupations at the Winter site

The Winter site is a sequentially stratified series of Initial Woodland occupations. Richner's speculation of general contemporaneity with second century A.D. occupations at the Port des Morts and Mero sites on the Door Peninsula has been vindicated by two radiocarbon dates from the so-called lower midden, or 3Ab horizon, although the date from the upper midden suggests that some sporadic and less intensive use of the site may have occurred earlier. Regardless, the occupations span the brief period of circa 2,150 to 1,710 years ago, and the basal lacustrine sands were deposited circa 2,180 years ago. The lower midden, or 3Ab horizon, demonstrates much more limited occupation densities than the middle and the upper middens. Due to the fact that the upper and middle middens at the Winter site were either separated by a thin zone of eolian sand, or alternatively merged and converged within the deposits, both Bianchi and Richner grouped the occupation materials recovered from the middle and upper middens in their analyses, and retained the lower midden as an independent analytic unit. Bianchi referred to the lower midden as the Winter I assemblage, and the combined middle and upper middens as the Winter II assemblage. Disparities in frequencies recorded by both analysts between the earlier and later occupations are largely an outcome of this decision to group the two upper middens.

Given the age and location of the Winter site, the lithic assemblage is unsurprisingly dominated by small end scrapers of varying size, a substantial bipolar core series, and expanding stemmed points common to the Garden and Door Peninsulas, the latter bearing close resemblances to North Bay series projectiles from the Door Peninsula and materials from nearby Spider Shelter. In the upper and middle middens, bifaces in varying stages of reduction, hammerstones, anvils, pigments, copper tools, and an abundance of ceramics including miniature vessels attributed to children lead to an interpretation of the Winter site as having functioned as a residential base camp in a mobile seasonal round.

Analysis of the faunal assemblage from the upper and lower occupations by T. Martin (1980) reveals an abundance of mammals and a paucity of fish. The earlier occupation is dominated by white-tailed deer, wapiti, and beaver, and contained a single identified lake sturgeon. The more recent occupation is likewise dominated by white-tailed deer, beaver, large cervids most likely moose and wapiti, and black bear, although in this instance fish including walleye, white bass, drum, round whitefish, or shallow water cisco have also been identified. Both occupations probably took place during the warm season, summer through autumn.

The temporary nature of the occupation, however, is well reflected in the lack of evident architecture or structures and few features present, as well as the lack of storage facilities. The lower midden has even fewer traces of such residential activity. Several fire-cracked rock features were recorded, and one clay-lined pit.

Of particular interest is the evidence in the lithic raw material for wide(r) ranging social contacts or exchange relationships. Augmenting a local raw material inventory of pebble and tabular cherts and quartzites are Knife River Chalcedony artifacts and debitage, and a limited inventory of very small obsidian flakes. A widely banded gray chert descriptively resembles Norwood chert from the northern Lower Peninsula of Michigan. In concert with certain ceramics similar to those from the Door Peninsula such as Dane Incised, and certain Havana-like Middle Woodland attributes on some vessels, the obsidian and chalcedony are attributed to more southern interactions with Middle Woodland populations.

Discussion

The Winter site is the westernmost sampled site in our research. The site landform is situated relatively far inland from the coastline as a result of uplift, and the consequent shifts in altitude of the site surface. Of particular interest here is the apparently brief period of time during which both active sand sheets, and brief periods of occupation associated with periods of stabilization, took place. The Winter site stratigraphy demonstrates increased amounts of eolian deposition as one moves inland from the shoreline. The multiple paleosols evident on the site diverge from each other by increasingly greater depths of sand deposits as one moves inland. The paleosols were presumably occupied, given the association of cultural materials primarily with the organic horizons, during periods of stabilization. The occupations are Laurel- or North Bay–related Middle or Initial Woodland, dating between circa 2,000 to 1,500 years ago and were short term at best, although there is evidence for residential occupation, which would suggest mixed gender and age composition. The faunal assemblage has been interpreted as a warm season occupation extending through early or mid autumn. Notably, this seasonality of coastal occupation is consistent with many models of upper Great Lakes seasonal mobility based on postcontact ethnographic data.

Ekdahl-Goudreau Site (20ST1), Schoolcraft County

Location and Description

Often known as the Seul Choix site, after the prominent geographic feature and lighthouse southeast of the actual location of the site, the Ekdahl-Goudreau site is located in Section 21, T41N, R13W, Mueller Township, Schoolcraft County (figure 5-3). A brief history of research at the site provided by Binford and Quimby (1963, 282) reveals the Ekdahl-Goudreau site was discovered in 1962 by George Quimby and James R. Getz. This field party of two made a surface collection of the site area, recovering an assemblage consisting largely of lithic debris, which was the subject of the 1963 paper. This may be the most comprehensive published work on the Ekdahl-Goudreau site. Subsequent to Quimby and Getz's visit bulldozer construction was initiated for a house foundation on the property, which led to extensive systematic surface collections and limited excavations at the site (UMMA field notes). The excavations occurred on both a level sandy bench several hundred feet removed from the lakeshore, and in an adjacent eolian dune deposit; the latter revealed stratified Middle and Late Woodland occupations (UMMA field notes). The results of this fieldwork have not yet been published, although a pair of radiocarbon dates has been obtained on the occupations (Crane and Griffin 1962), and the assemblage has been generally referred to in summary volumes (e.g.,

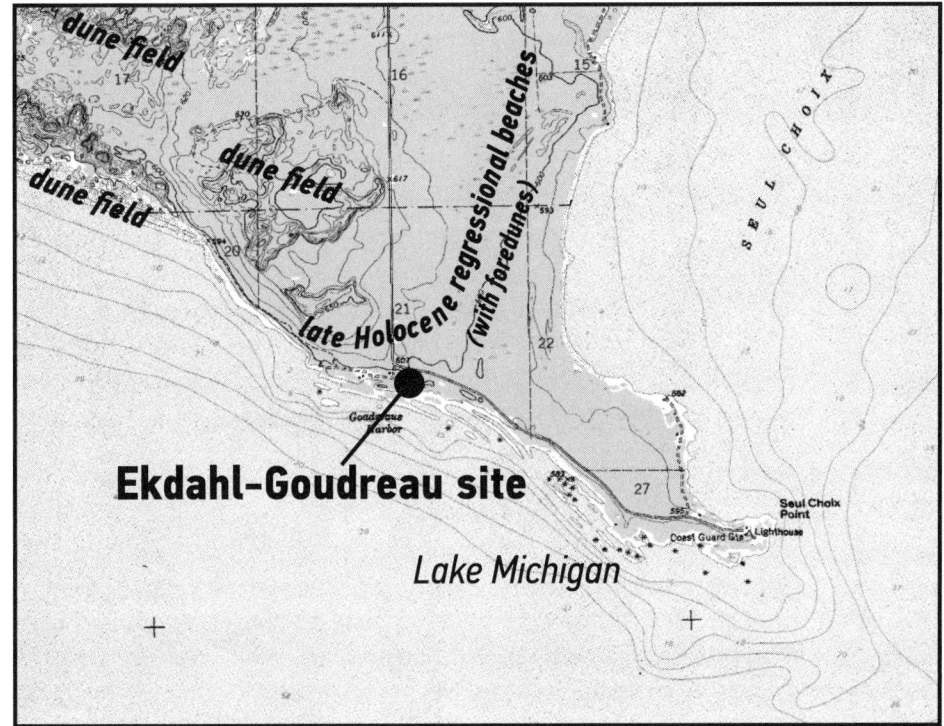

5-3. The Ekdahl-Goudreau site location.

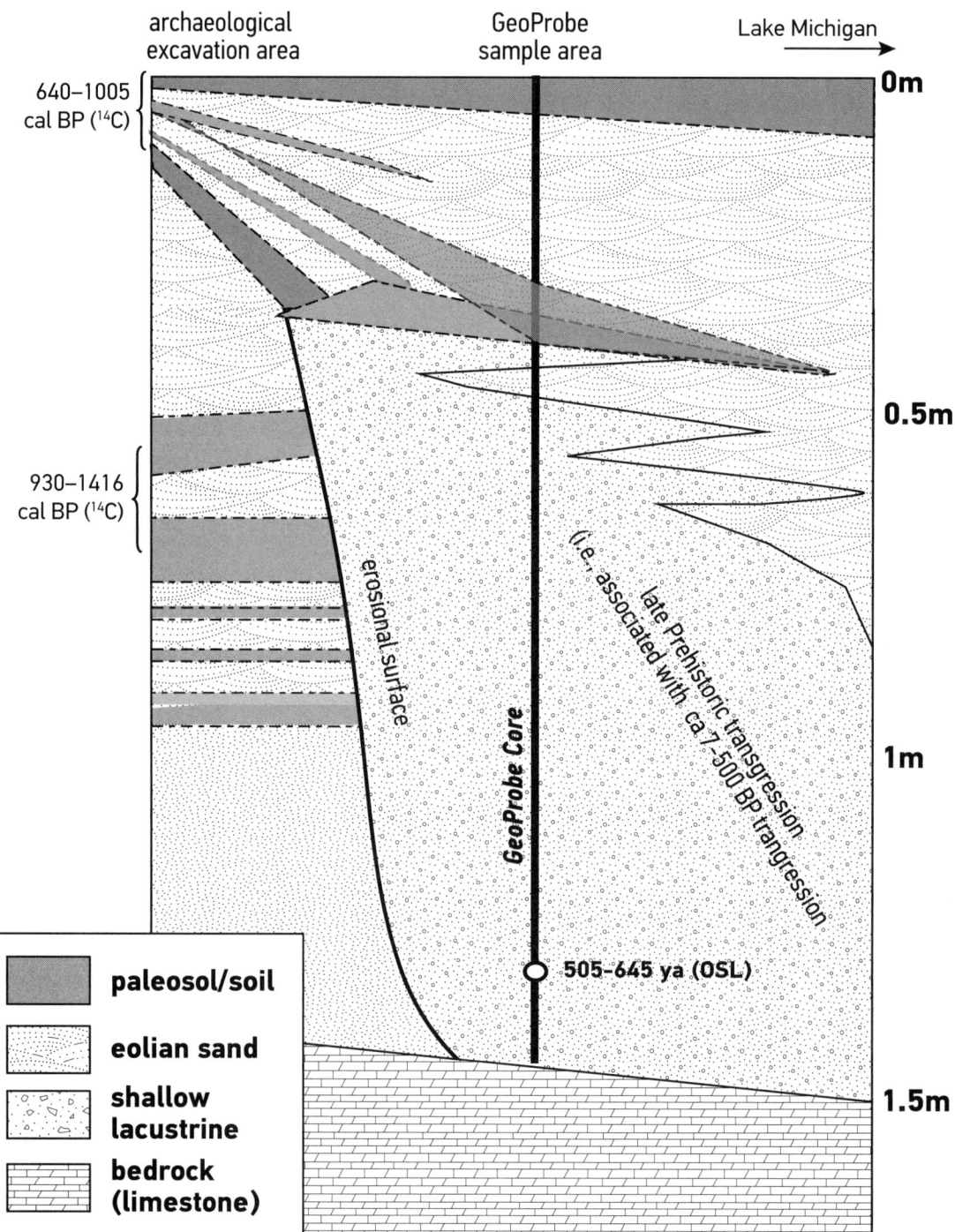

5-4. Generalized, composite stratigraphic column of the Ekdahl-Goudreau site from UMMA field notes and as noted from the GeoProbe core. Discrepancy probably related to imprecise location of the archaeological excavations in the early 1960s. Ages of various units shown. See appendix A for discussion of how ages are reported. *Note:* unit depths and boundaries are approximate.

Fitting 1970, 1975). Lack of published information, however, has resulted in only limited passing reference to the site in the most recent compendium statement on the Middle Woodland (Brose and Hambacher 1999).

The Ekdahl-Goudreau site was an elusive target for our field research. The State of Michigan Archaeological Site Files gave three locations and two names for the site, removed from each other in some cases by great distances. An initial field visit by Lovis and Elizabeth Bogdan-Lovis in 2006 ultimately confirmed the location of the site, and allowed us to contact the new property owners, who graciously gave permission for us to conduct fieldwork. Lovis, Monaghan, and Arbogast visited the Ekdahl-Goudreau site in 2007 and conducted a GeoProbe coring operation designed to relocate stratigraphy similar to that described in the field records, but to little avail. It is possible and even likely that much of the deeply stratified area of the site has been excavated and destroyed by construction.

Stratigraphy and Dating

Inspection of field notes from the UMMA excavations, housed at the Great Lakes Division, suggests that the deepest parts of the site have been excavated or destroyed. The unit labeled "Dune Strat Cut" is the most revealing of the stratigraphic sections from Ekdahl-Goudreau. This excavation is in excess of a meter in depth; approximately 110–120 cm based on scaling of photographs (figure 5-4). The top 30 to 40 cm consists of an internally stratified series of at least three A horizons separated by lighter-colored sand and positioned directly above an eolian sand deposit. Approximately 30 cm below the contact with the eolian sand is a faint organic horizon labeled "veg horizon," and about 30 cm below this is another or second well-defined organic horizon. The intervening deposits are eolian sand. Some 20 cm below the second horizon is a third well-developed organic paleosol with at least 20 cm of additional organic rich material underlying it. This reconstruction suggests that there are at least three and perhaps more buried paleosols in the eolian sands underlying the internally stratified A horizon.

A single OSL sample from the basal sand deposits produced an age of 575 ± 70 ya (UIC-2135). This age is younger than either of the ^{14}C dates from the site (appendix B). The Late Woodland, or "top level," associated with Juntunen-like ceramics is dated to 870 ± 120 BP (M-2311, Crane and Griffin 1972; 2σ 640–1005 cal BP; Stuiver and Reimer 1993). The basal Middle Woodland deposits are dated to 1290 ± 130 BP (M-2312, Crane and Griffin 1972; 2σ 930–1416 cal BP, Stuiver and Reimer 1993). The basal deposits could date as early as circa 1,416 years ago, which is substantially earlier than our OSL age. However, Earl Prahl comments in Crane and Griffin 1972 that "the date is late" by comparison with those from Summer Island. One interpretation of the depositional framework for this complex site is shown in figure 5-4.

Ekdahl-Goudreau approaches the complexity and depth of the deposits at the Scott Point site to the east. The stratigraphy is far more complex than simply an upper Late Woodland and basal Middle Woodland occupation, and appears to have varied considerably in depth across the site. In our opinion, it is likely that our fieldwork did not encounter the same stratigraphy or soil profile that was present in the vicinity of the house foundation excavations, but rather most likely sampled younger sand deposits inland of the UMMA excavations.

CHAPTER FIVE

Scott Point Site (20MK22), Mackinac County

Location and Description

The Scott Point site is located in Section 8, T41N, R11W, Newton Township, Mackinac County (figure 5-5). As previously noted, the Scott Point site also goes by several other names depending on which literature one accesses, including Point Scott, and the geographically incorrect Point Patterson. The Scott Point site was the subject of a series of visits by the UMMA in the early 1960s during which several surface artifact collections were made (Binford and Quimby 1963; Peske and Kent 1963). Binford and Quimby describe the Scott Point site admirably (1963, 284–285):

> The site is situated some 300 or 400 feet from Lake Michigan in a small bay just west of Scott Point. . . . Much of the Scott Point site has been exposed in large sand blows covering an area 200 feet by 300 feet or more. Numerous clusters of fire-cracked rocks indicate the locations of former dwellings and/or hearths. Pottery sherds, flint materials, and hammerstones lie on the surface in great abundance. . . . Faunal remains include deer, moose, beaver, and considerable quantities of fish, among which are sturgeon. There are large piles of fish remains two and three feet deep in various parts of the site.

While perhaps there may no longer be piles of fish bone as deep as Binford and Quimby claimed were present 45 years ago, there is still an abundance of fauna and other cultural debris on the deflated surfaces of the Scott Point site. The Scott Point site has been visited by several other institutions since the initial field forays by UMMA in the 1960s. In 1970 a MSUM field party under Charles Cleland performed limited surface collection at Scott Point. In 1979, a NMU archaeological field school directed by Marla Buckmaster performed extensive systematic surface collection as well as major excavations. Much of what we now know about the site derives from this work, although it is still in the process of being brought to print. The information presented here derives from access to artifacts, field photographs, field notes, and drawings provided by Marla Buckmaster prior to her retirement from NMU, and inventories of ceramic identifications made available by Claire McHale Milner of Pennsylvania State University. The collections from the Scott Point site are currently housed at the Office of the State Archaeologist in Lansing. The Scott Point site is listed on the National Register of Historic Places.

Absolute Dating

The NMU excavations at Scott Point site revealed what previous research at the site hadn't; the site was both well stratified and deep. Excavations exposed up to 1.5 meters of intact stratified deposits. Investigation was halted due to the instability of the dry sand at the site, which rapidly attempted to reach its angle of repose and posed safety issues. Because the excavations were not taken deeper, it remained, and still remains, unknown whether there are additional cultural deposits or evidence of dune stabilization at greater depths.

Our intention was to perform systematic coring at the Scott Point site to resolve the issue of the depth and age of the basal cultural deposits and paleosols. To

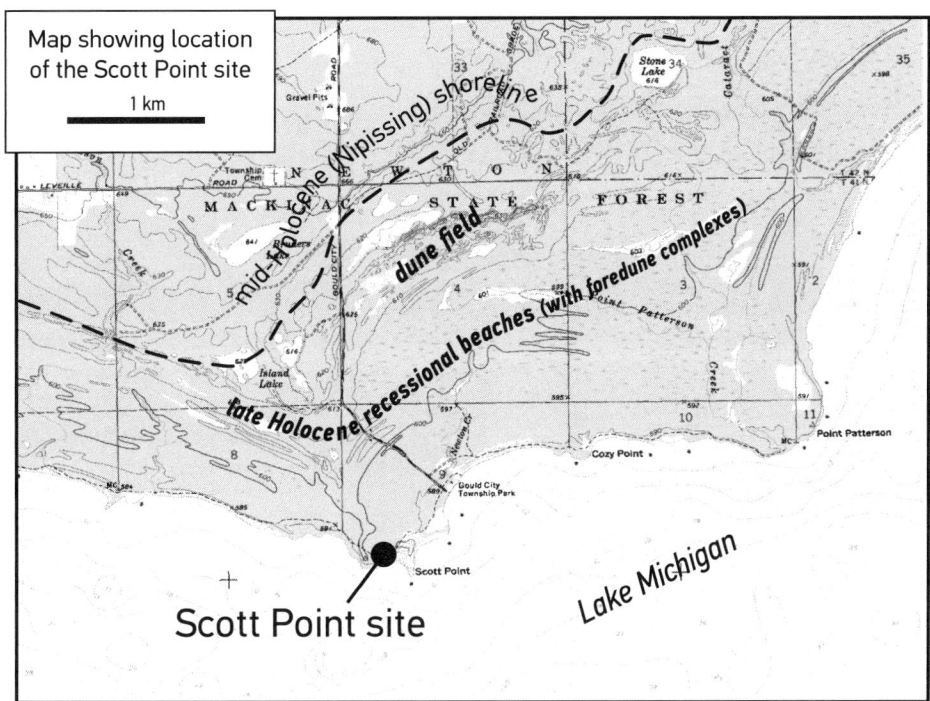

5-5. The Scott Point site location.

this end Lovis and E. Bogdan-Lovis visited the site in July 2006 to assess the field conditions and relocate the previous excavations. In 2007 permits were requested from the Department of Natural Resources, which due to the National Register status of the property required review by the Office of the State Archaeologist. The permitting procedure led to substantial delays, and we ultimately had to reapply for a permit, which we received in midwinter 2007 for the 2008 calendar year. As a consequence of its status as a critical dune, and the presence of endangered species, in addition to its National Register status, the permit had substantial restrictions on our activities to the degree that we made arrangements to undertake hand coring to a depth of several meters. However, our attempts to access the site both in the spring and in the fall of 2008 were unsuccessful. The Scott Point site has not, therefore, been cored by this project, and the depth and age of the earliest deposits remain unknown.

Although we were unable to conduct formal field investigations and coring, we are able to reconstruct the stratigraphy of the site from field drawings, color positive field photographs, and synthetic profile sections provided to the project through the courtesy of Marla Buckmaster. These proxy sources indicate that the stratigraphy of the Scott Point site is complex, with multiple units of eolian sand and associated paleosols evident across the excavated areas. These paleosols merge and diverge across the excavated area; in some locations as many as seven separate organic zones are evident, whereas in others only two or three might be present.

In an effort to estimate the age of these deposits, we elected to date the exposed "package" of depositional units by the AMS dating of carbonized residues from the interiors of typologically distinctive ceramic rimsherds. Two rimsherds were selected from the uppermost 30 cm of the NMU excavations; one sherd of Juntunen

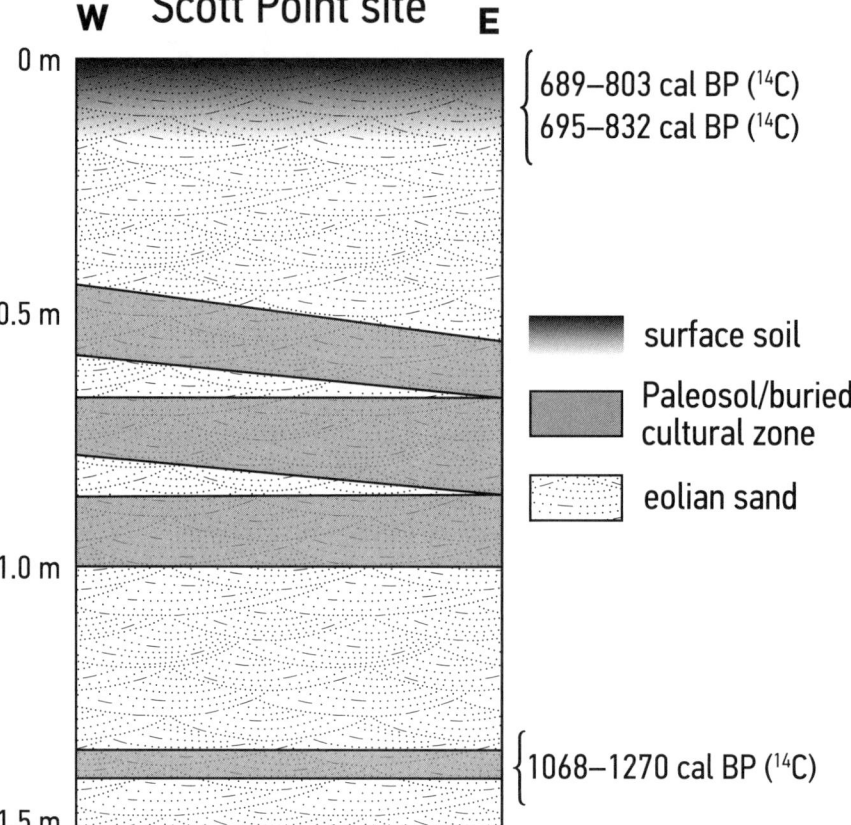

5-6. Generalized, composite stratigraphic column of the Scott Point site. Ages of various units shown. See appendix A for discussion of how ages are reported. Unit depths and boundaries are approximate.

Drag and Jab, and one rimsherd of plain shell tempered Oneota pottery. These yielded ages of 860 ± 40 BP (Beta-237014; 2σ 689–803, 808–831 and 851–906 cal BP, intercept 750) and 870 ± 40 BP (Beta-237015; 2σ 695–832 and 844–907 cal BP, intercept cal BP 780). These dates are statistically identical at 95 percent (df = 1), with a mean pooled age of cal BP 865 ± 28. A Mackinac Ware rimsherd was selected from near the base of the deposits, and produced an age of 1240 ± 40 BP (Beta-237016; 2σ 1270–1068 cal BP, intercept 1180). Thus, the 1.5 meters from top to bottom of the excavated deposits range from circa 1,180 to 865 years ago.

Site Stratigraphy

The representative section we employ here derives primarily from the north profile or section of unit N0 W0, which was excavated to a depth of 1.4 m (figure 5-6), supplemented by a complete north-to-south and east-to-west profile section, and field photographs made available by Marla Buckmaster of NMU. This stratigraphic reconstruction and nomenclature assumes that the basal eolian sand deposits are the first eolian depositional unit in this sequence.

- *Unit I—eolian sand.* This deposit is of unknown maximum depth.
- *4Ab—formed in Unit I.* This is a several-centimeter-thick organic horizon with

cultural material displaying little evidence of soil development. The interface between the organic and inorganic sand within Unit I is sharp. This deposit is horizontal from west to east. The Mackinac Ware rimsherd dated to circa 1,180 years ago derives from this unit.
- *Unit II—eolian sand.* This is a rather thick bed of eolian sand averaging 33–35 cm in thickness.
- *3Ab—formed in Unit II.* This is also a several-centimeter-thick organic horizon, thicker than the 4Ab horizon, and in some locations up to 7 cm in thickness. This horizon is also horizontal from west to east.
- *Unit III—eolian sand.* This deposit is a wedge of sand thicker at the west and narrowing in thickness to the east, at which point it only lies a couple of centimeters above the 3Ab. Consequently, the overlying 2Ab horizon formed within this unit slopes downward to the east.
- *2Ab—formed in Unit III.* The 2Ab as noted slopes downward to the east. This deposit is similar in thickness to the 3Ab, up to 7 cm or so. Slight leaching is evident at the base of the 2Ab where it interfaces with the inorganic deposits of Unit III. The fact that the 2Ab slopes downward to the east reveals that a shallow basin was beginning to develop in this vicinity of the site.
- *Unit IV—eolian sand.* Much like Unit III, this is a wedge of sand thinning to the east, albeit thinner. This deposition enhances the basin present in this vicinity of the site.
- *1Ab—formed in Unit IV.* This is a variably thick paleosol similar in thickness to underlying 2Ab and 3Ab horizons. Due to the slope of the deposits to the east, the separation of the 3Ab, 2Ab, and 1Ab horizons is minimal at the easternmost point of this section. The Juntunen Drag and Jab and shell tempered Oneota Plain rimsherds, with a mean pooled age of 865 cal BP, derive from this uppermost buried horizon.
- *Unit V—eolian sand.* This is the final large eolian deposit in this vicinity of the site. During this depositional episode the basin on the east end of the section is infilled by sand, and the surface of the eolian sand unit is once again horizontal. Consequently, this unit too is a wedge shape, with the base dipping to the east; it is thinner to the west, and up to 30+ cm in thickness to the east.
- *A horizon*—Current surface deposit formed at top of Unit V, in some places 8–10 cm in thickness.

Mt. McSauba Site (20CX23), Charlevoix County

Location and Description

Mt. McSauba is a large sand dune about 1 km north of Charlevoix in northwestern Lower Michigan that straddles Sections 14 and 23 in T34N, R8W, City of Charlevoix, Charlevoix County (figure 5-7). The northern half of the dune consists of two large mounds that are both about 40 m high. A shorter (ca. 20 m high) dune ridge extends south/southwestward for about 500 m from the most southern of the two mounds. Overall, the dune is about 1.5 km long from its northern to southern margins. The most notable feature in the dune is a large blowout in the southernmost large mound. This blowout is significant because it exposes the sediments that underlie the dune. In addition, the northern wall of the blowout

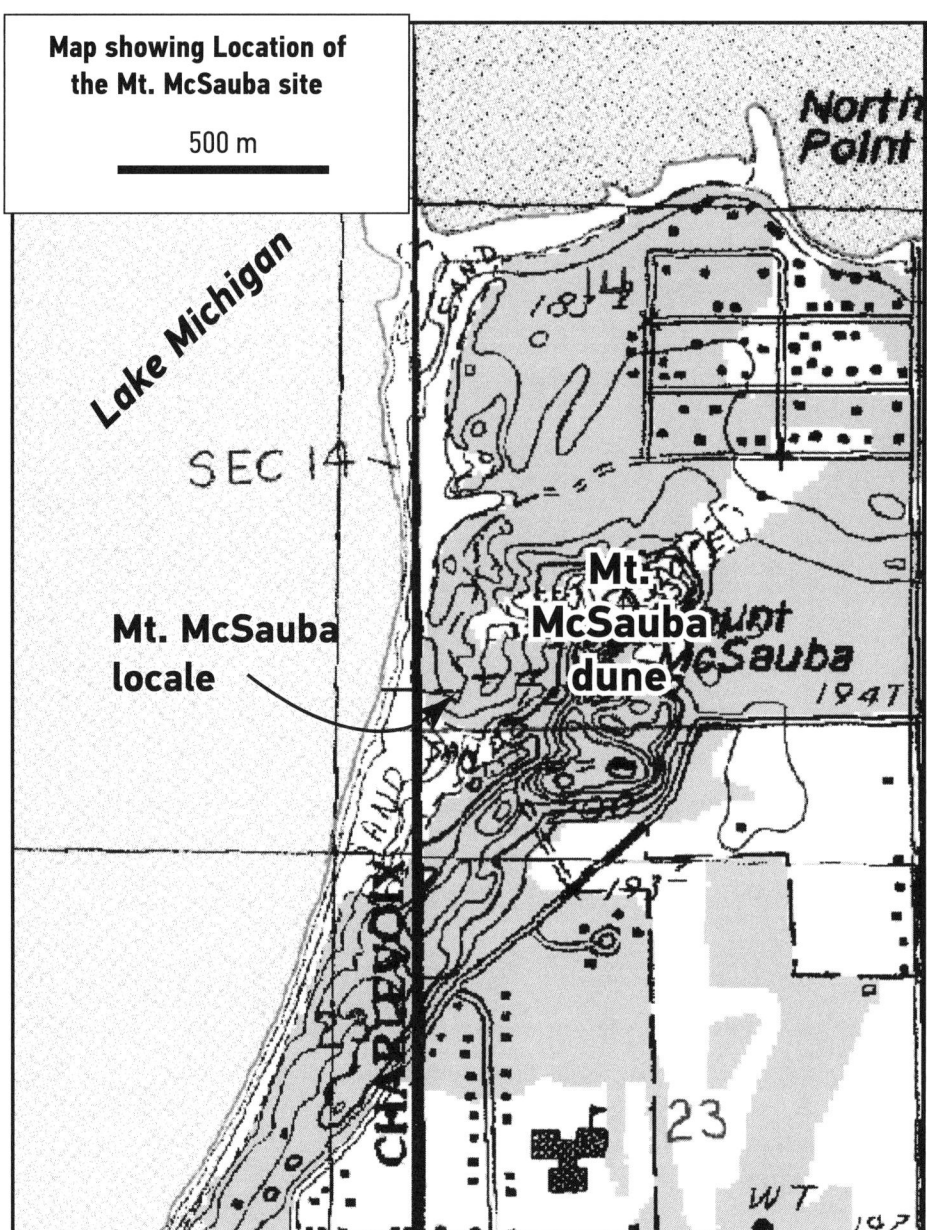

5-7. The Mt. McSauba site location.

exposes the stratigraphy of a portion of the dune. Embedded within this dune stratigraphy is an archaeological site known as the Mt. McSauba site.

A field party consisting of Arbogast, Lovis, and Jennifer Holmstadt visited the Mt. McSauba site on 19 June 2006. During this visit we observed multiple prehistoric ceramic vessel bases in situ at the base of a preserved paleosol, and eroding from the vertical face of the dune resulting in a large downslope sherd scatter. Rimsherds were present and field-collected, but upon inspection did not contain adhering datable materials (figure 5-8). We obtained an OSL sample from directly beneath the occupation horizon using standard tube collection procedures. This sample was sent to M. Bateman at Sheffield University UK, and processed by

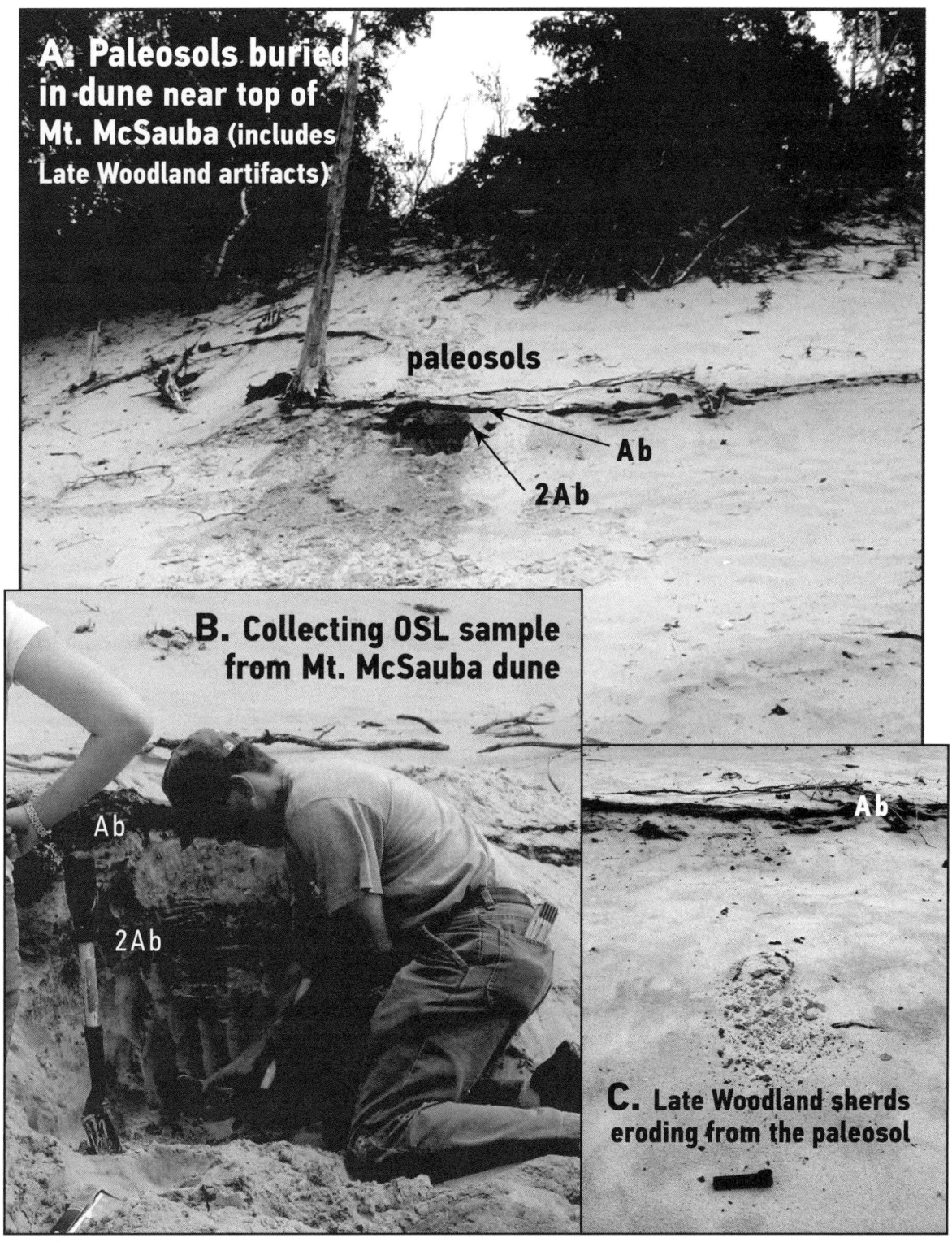

5-8. The Mt. McSauba site sample location. A. Overview of paleosol near the top of Mt. McSauba showing the Ab and 2Ab horizons. B. Samples being collected for OSL dating from minimally weathered dune sand below the 2Ab horizon. C. Fragments of Late Woodland sherds eroding from the paleosols in photo A.

the luminescence lab at the Sheffield Centre for International Drylands Research. It yielded an OSL age of 740 ± 70 ya (Shfd06139).

The Mt. McSauba site is actually a rather ephemeral site, since its visibility is conditioned by wind direction, seasonality, relative eolian dune activity, and intensity of human traffic. The actual preserved occupation surface or floor can, at times, be completely exposed as a pavement of natural and cultural materials on a windblown organic surface, with spatially discrete hearths, fragmented ceramic vessel segments, and activity areas readily visible. This is because the site is exposed within a parabolic dune hollow. The preserved occupation/paleosol surface is only shallowly buried by a thin veneer of sand that can be completely exposed or obscured in a short period of time. Those lakeward portions of the site that are more deeply buried can only be observed when both the buried paleosol and cultural material are eroding from the face of the dune. Consequently, surface artifact collection is difficult, and many visits result in the site not being evident from surface indications.

The multiple visits to the Mt. McSauba site by MSUM field crews over several decades resulted in the accumulation of a small assemblage of fauna from various site contexts. Since the occupations at Mt. McSauba appear to be more recent than circa 800 BP based on ceramic cross-dating and confirmed by the OSL date, the likelihood is that the faunal assemblage is broadly assignable to this time span. The faunal assemblage was analyzed by Beverley Smith (1983), and employed by S. Martin in her doctoral dissertation (1985). The faunal assemblage is reminiscent of other dune site assemblages, with beaver, woodchuck, goose, turtle, and white-tailed deer. Again, a warm season assemblage, ranging from spring through fall, would be an appropriate interpretation.

Direct dating of the human occupation at the Mt. McSauba site is difficult because charcoal exposed on the surface of the occupation is quickly damaged and dispersed as fine particles. The ceramics from Mt. McSauba did not contain sufficient adhering carbonized interior food residues for AMS dating. Thus, OSL has provided the only means of dating the occupation surface, which appears to be more recent than circa 750 years ago.

Dune Stratigraphy

Assessment of the dune stratigraphy indicates that that the dune landform directly mantles beach gravels that are visible in the base of the blowout (figures 5-8 and 5-9). Overlying this deposit in the northern wall of the blowout is about 9 m of eolian sand. Contained within this sand body are two eolian units that are separated by a buried soil. The lower unit (Unit I) is about 4 m thick. Formed in the uppermost part of this deposit is a soil that is about 40 cm thick. This soil has a distinct 2Ab horizon, which is about 7 cm thick and contains large woody fragments. The remainder of the solum is about 33 cm thick and consists of sand that is clearly less enriched in organic matter than the overlying 2Ab. The amount of organic enrichment appears to vary slightly with depth. Given the nature of this soil, it is characterized as having A/AC1/AC2/AC3 horizonation. In an effort to assess the age of this portion of Mt. McSauba, a sample was collected for OSL dating from a point directly beneath the paleosol at a depth of 6 m. This sample provided an age of 740 ± 70 ya (Shfd06139).

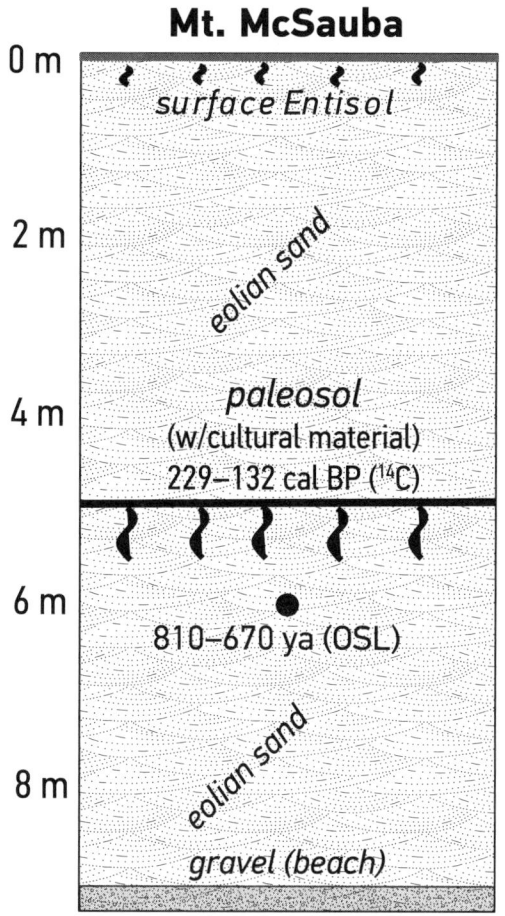

5-9. Generalized, composite stratigraphic column of the Mt. McSauba site. Ages of various units shown. See appendix A for discussion of how ages are reported. Unit depths and boundaries are approximate.

The stratigraphy exposed in the blowout at Mt. McSauba and the OSL age derived from the upper part of Unit I indicate the following history. The exposed basal gravel deposit indicates that the active beach was located at this point sometime in the past. Given the elevation (ca. 597 ft) of this deposit, it was most likely deposited circa 3,200 years ago during the recessional post-Nipissing phase of ancestral Lake Michigan. After the water receded from this point, dune construction probably began, resulting in the large mounds of eolian sand that lie to the east. The latter part of this depositional interval must have been episodic, as evidenced by the buried soil exposed in the north wall. The OSL age obtained from the underlying sands indicate that deposition of Unit I in the exposure occurred between 800 and 600 years ago. The overthickened nature of the buried soil formed in that deposit suggests that sand deposition did not cease entirely at first, but was sufficiently slow for slight (but variable) organic enrichment of the sands to occur. Deposition of eolian sand at this site completely stopped for a long enough span of time to allow the 2Ab to form, and for trees to grow on its surface. The uppermost sand unit probably accumulated in the past 500 years, as witnessed by the modern calibrated age of wood from the paleosol (i.e., 229–132 cal BP; figure 5-9, appendix B).

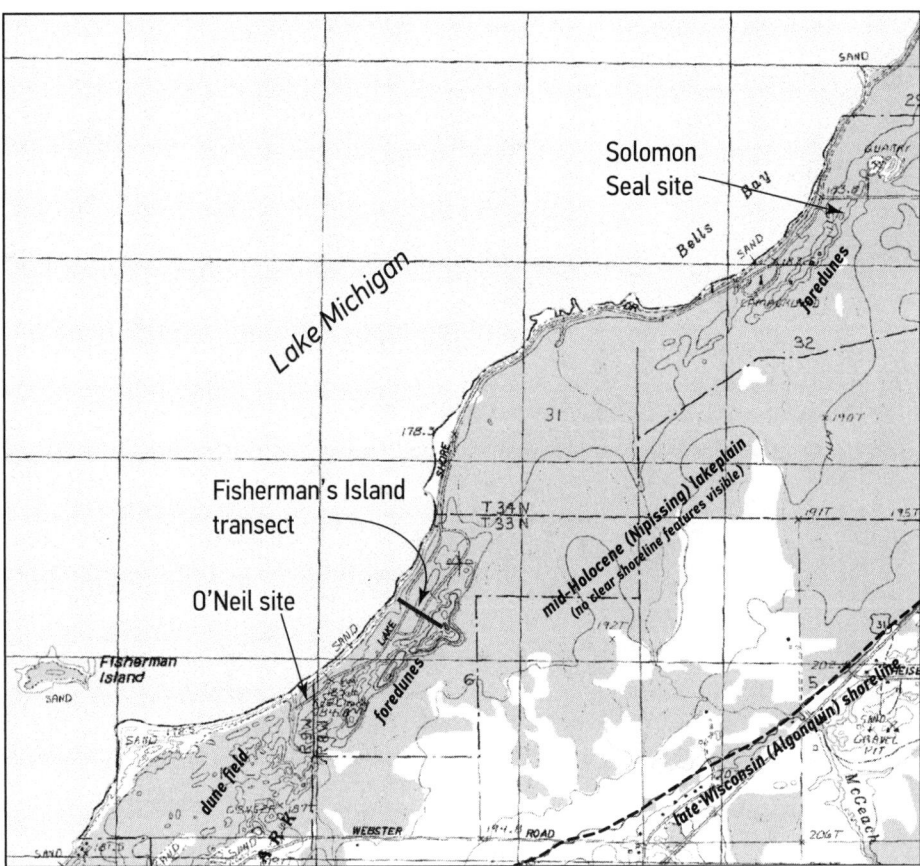

5-10. Fisherman's Island State Park archaeological sites and dune sample locations.

Solomon Seal Site (20CX42), Charlevoix County

Location and Description

The Solomon Seal site is a possible Late Archaic archaeological site that is contained within a coastal dune ridge that lies in the NW ¼, NE ¼, Section 32, T34N, R8W (figure 5-10), Charlevoix Township, Charlevoix County, currently within the boundaries of Fisherman's Island State Park Campground. The ridge is about 11 m high and extends for about 1.6 km in a generally southwest direction that follows the shore in a parallel fashion. It lies about 200 m inland and is centered in Bells Bay between two high headlands projecting into Lake Michigan.

The Solomon Seal site was initially recorded during a survey of Nipissing coastline features conducted by Janet Brashler (then of Michigan State University), who produced a brief descriptive report on the site (Brashler 1972). The site was subsequently revisited, and test excavated, as part of a management survey of the then Bells Bay State Park Campground, which was to be expanded and included within the proposed Fisherman's Island State Park. The larger management report included a description of the test excavations, the stratigraphy, and the resulting artifact assemblage (Lovis et al. 1976). The current project revisited the Solomon Seal site in July 2007 to perform a deep core through the dune and eolian sediments on which the site is situated, extract OSL samples, and attempt to put a *terminus post quem* on the occupation. The ensuing discussion is summarized from the two reports and the visit of the current project.

Solomon Seal is a buried site, situated near the crest of a stable sand dune some 350+ meters from the Lake Michigan shoreline, at an elevation of 625 ft (190.5 m) asl. The crest of the dune is some 3 meters greater in elevation; 635 ft (193–194 m). This sand feature was assumed to be Nipissing in age (i.e., ca. 5,000 years ago) because its elevation exceeded that of the 611 ft reported to have been attained in the local area after correction for uplift (Leverett and Taylor 1915, 458). The age and interpretation of this sand feature is the subject of additional discussion elsewhere in this book. The 1971 visit produced a limited assemblage of one hammer stone and three secondary flakes of local Norwood chert recovered from what appeared to be a buried feature or midden eroding from near the crest of the dune. The 1976 test excavation at the Solomon Seal site corroborated that this was indeed the case; the site preserved buried organic horizons and produced lithics including worked bifacial tools. All of the raw material was the local Norwood chert variety found nearby to the south.

The assemblage recovered from the 1976 test excavation is limited and not particularly diagnostic. The single five-foot by five-foot excavation unit was excavated to a depth averaging four feet (1.22 m) below surface, although given the steep slope of the dune crest one corner was at 3.5 feet (1.07 m) and the other greater than 4.7 ft (1.43 m) below surface. Cultural material was encountered to maximum depth when excavation was halted due to safety concerns. The ratio of 237 flat flakes and 333 blocky flakes recovered reveals substantial core testing, although of the flat flakes almost half are below 1.0 cm in length, suggesting they are the result of final reduction stages. The entire assemblage is a fine-grained and lustrous variety of local Norwood chert, and a small number of flat flakes reveal "pot lid" spalls perhaps indicative of heat treatment. Four formal tools were recovered: one complete biface, two biface fragments, and a single perforator, all on Norwood chert and temporally undiagnostic. Like many other sites in the vicinity of the Norwood chert quarry, Solomon Seal appears to be dominated by raw material reduction activities. A single unidentifiable bone fragment and 3,000 g of fire-cracked rock does not further elucidate site function. In 2007 the site was GeoProbe-cored to a depth of almost 28 ft (10 m). No cultural material was recovered from the core, but two OSL samples were obtained and sent to University of Illinois–Chicago for processing.

Dune Stratigraphy

In an effort to assess the chronostratigraphy of the dune (figure 5-11), a core approximately 9 m long was taken from the crest of the ridge in the vicinity of the site with a GeoProbe. The stratigraphy apparently consists of a single depositional unit of eolian sand, as no buried soils or obvious unconformities were seen. Unfortunately, the archaeological deposits were not encountered, suggesting that the remnants of the site are small and highly localized.

Two samples were derived from the core for OSL dating. The deepest sample was taken from a depth of about 8.5 m and provided an OSL age of 3380 ± 300 ya (UIC2139). Another sample was obtained near the crest from a depth of about 3 m and yielded an OSL age of 3280 ± 265 ya (UIC2138). This latter OSL date is

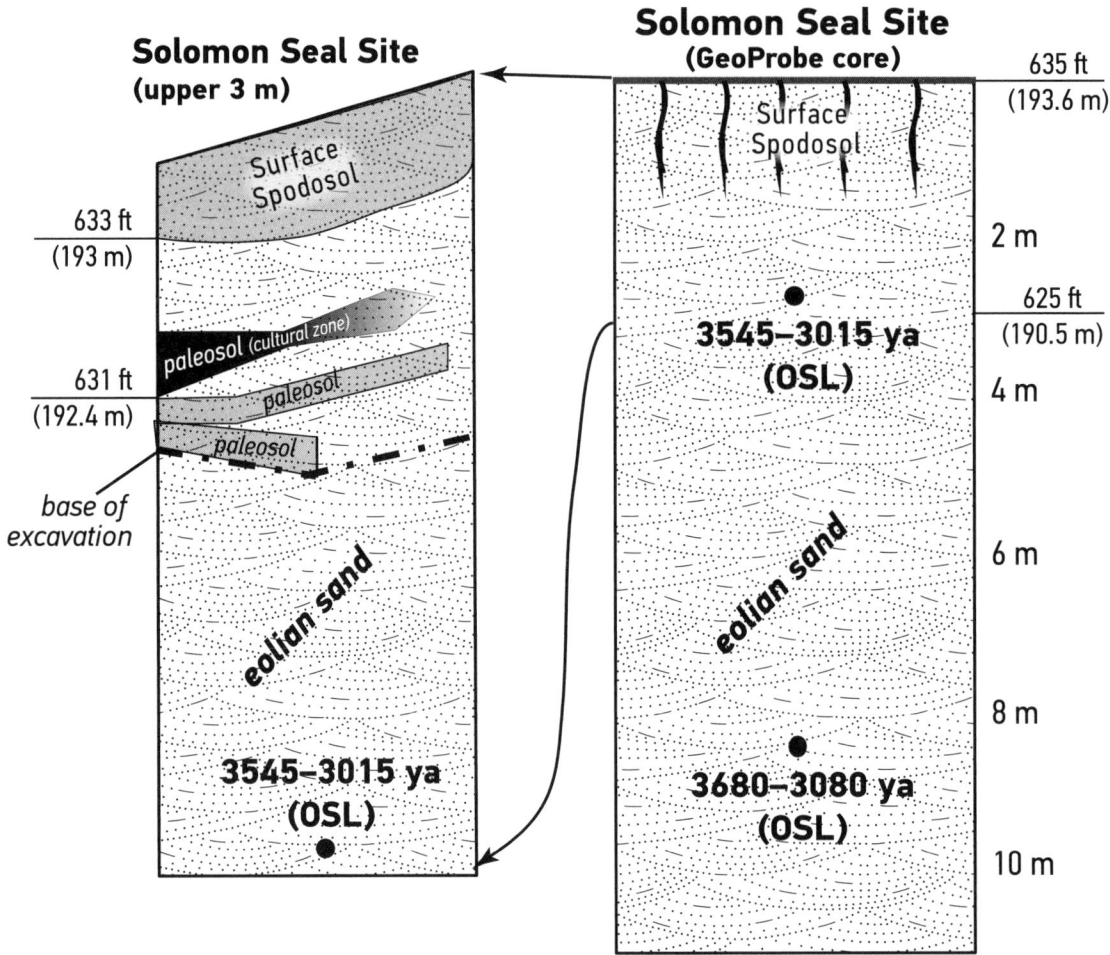

5-11. Generalized, composite stratigraphic column of the Solomon Seal site, Fisherman's Island State Park. Unit depths and boundaries are approximate.

from just below the basal cultural deposits encountered in 1976, and suggests that the Solomon Seal site is no more than circa 3,000 years old. The OSL ages obtained from the dune deposits associated with the Solomon Seal site suggest that eolian sand accumulated fairly quickly at the site circa 3,200 years ago. Given the overall lack of dunes between the ridge and the lakeshore, it appears that very little deposition of eolian sand occurred at the site after deposition of the ridge occurred.

Camp Miniwanca (20OA34), Oceana County

Location and Description

Camp Miniwanca is a an exposure of eolian sand west of Stony Lake in Clay Banks Township, Oceana County, in the SE ¼, NE ¼, Section 6, T13N, R18W (figure 5-12). This exposure faces to the southwest and is located on the northwestern side of a large blowout in a massive (> 50 m high) parabolic dune (figure 5-13). The site was reported to the project team by Dr. Edward Hansen of Hope College, who in turn had been informed of the site's location by the manager of the summer camp located immediately north of the dune. Dr. Hansen indicated that numerous pieces of chipped stone debitage had been seen at the site in association with a paleosol within a dune. Field reconnaissance verified the presence of the site and stratigraphic association (figure 5-13). It is the southernmost site in our sample. The Camp Miniwanca site is most likely the same site reported by Mr. Birt Darling and listed with an approximate location in the NE ¼ of Section 6 in the UMMA and OSA site files (B. Mead, personal communication).

As noted, the Camp Miniwanca archaeological site was brought to the attention of Alan Arbogast in 2007 by Dr. Edward Hansen. Hansen had observed multiple buried paleosols in the two large parabolic dunes situated on the camp property. Lovis and Arbogast subsequently visited the site on 26 October 2007, during which a small surface collection was made, and initial observations of the eolian stratigraphy relative to the cultural deposits was undertaken. This site visit

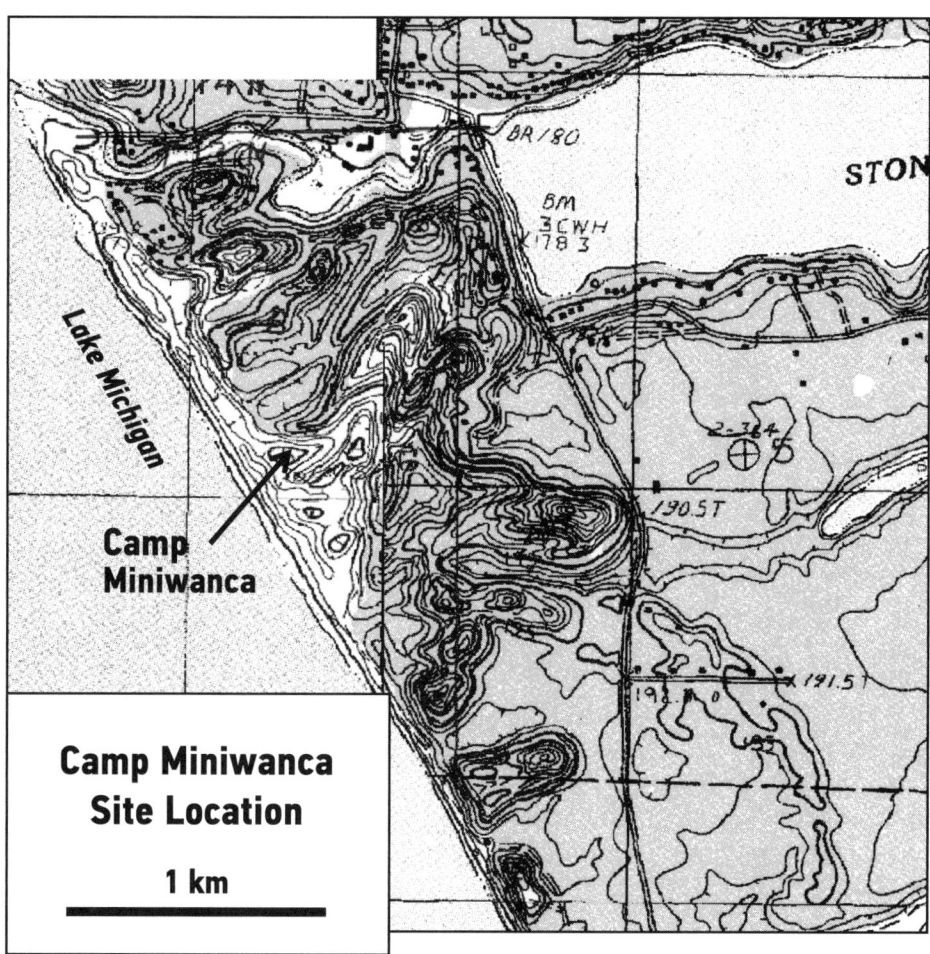

5-12. The Camp Miniwanca site location.

5-13. The Camp Miniwanca exposure showing buried hearth feature.

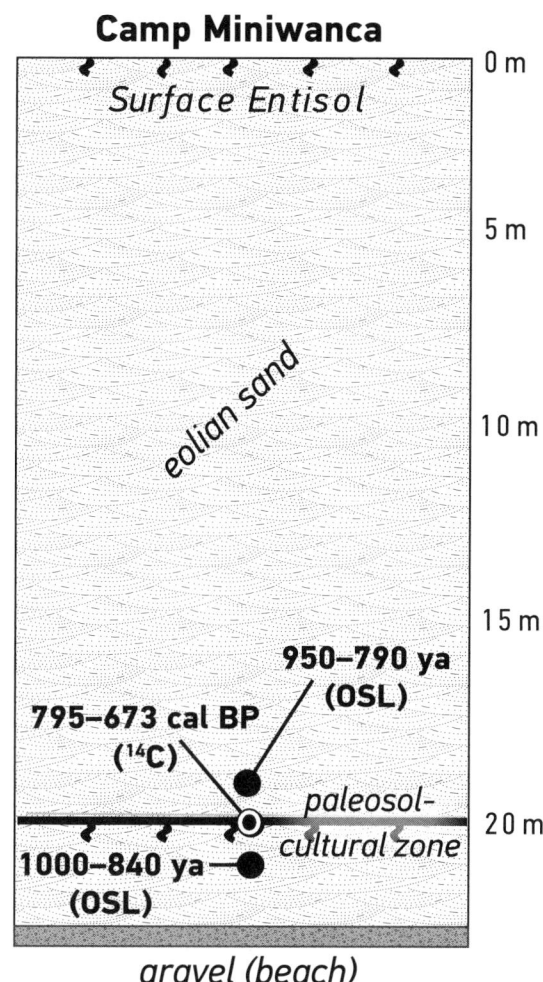

5-14. Generalized, composite stratigraphic column of the Camp Miniwanca exposure.

confirmed that cultural material was present in two locales in close proximity to one another; these were recorded as Locale A and Locale B. Locale A presented no visible intact stratigraphic context, appearing to be an eolian lag deposit of flint debitage typical of these deflated dune contexts. Locale B presented clear evidence of Late Woodland occupation, based on the presence of cord impressed potsherds eroding downslope from a deeply buried organic surface on the backslope, or slip face, of the first large dune crest behind the foredune. The large dune had been transected or cut by the parabolic dune deflation, thereby creating the eroding vertical exposure. Due to time constraints we decided not to attempt sampling, but to return to the site when more time was available.

The second visit took place on 16 November 2007 by a field team again consisting of Lovis and Arbogast, this time accompanied by graduate students Marieka Brouwer and Brad Blumer. During this visit a much larger area was inspected, but no other locales were recorded. The original two locales were relocated. Surface collection of Locale A on this visit revealed Late Woodland pottery sherds in addition to flint debitage, suggesting Locale A's general contemporaneity with

TABLE 5-2. Camp Miniwanca Artifact Assemblage, Locales A and B, 26 October and 16 November 2007

LOCALE	ARTIFACT	COUNT
Locale A	*Secondary thinning flakes*	
	Bayport chert	7
	Unidentified white pebble chert	5
	Unidentified tan pebble chert	1
	Total secondary thinning flakes	*13*
	Grit-tempered, exfoliated, and cordmarked sherds	3
Locale B	Rimsherd from hearth	1
	Grit-tempered, exfoliated, and cordmarked sherds	37
	Drill bit	1
	Blocky pebble flake with cortex, unidentifiable chert	1
	Secondary and retouch flakes	
	Unidentified mottled gray chert	14
	Unidentified mottled brown chert	2
	Total secondary and retouch flakes	*16*
	Animal bone fragments (*n* = 6 small bird long bone)	14

Locale B. While the evidence is not conclusive, it is likely that Locale A is derived from the same remnant paleosol at Locale B. During this visit the stratigraphic profile of the dune containing the buried paleosols was mapped and measured by Arbogast (figure 5-14). Cleaning of the profile of the occupied paleosol, subsequently labeled 4Ab as noted below, exposed a hearth in section (figure 5-13). The hearth contained abundant pieces of consolidated charcoal subsequently identified as spruce (Picea sp.; probably black spruce) by Drs. Frank Telewski (MSU Plant Botany and Horticultural Gardens) and Catherine Yansa (MSU Geography). It was observed that the tightly spaced growth rings revealed a stressed environment. In association with the charcoal was a large ceramic rim sherd bearing resemblances to Spring Creek Ware. Adjacent areas of the occupied surface produced additional artifacts itemized in table 5-2.

The Late Woodland occupation at Camp Miniwanca is ephemeral at best, and the assemblage is clearly reflective of short-term, transient use of the vicinity. There is no evidence of fire-cracked rock, or early stages of lithic reduction. It appears that either tools or partially finished items manufactured of Bayport chert, from the lower Saginaw River basin, were being brought to this location and further reduced or resharpened. Whether the Bayport chert is the result of exchange between regions, or transport by individuals using both areas, is speculative. The other raw material varieties appear to be local pebble cherts procured in the vicinity of the site. At least one ceramic vessel was present in the vicinity of the hearth, as well as a drill bit. The latter, while fire reddened and burned, suggesting it derived from the hearth, appears to be on a banded chert; this chert could be local Lesausky chert from Oceana County, Norwood chert from Charlevoix County, or even Charity Island chert from Saginaw Bay. Given the small exposed area, it may not necessarily be an accurate reflection of assemblage variability.

Stratigraphy and Dating

Camp Miniwanca posed an excellent control sample for our combined use of ^{14}C and OSL dating methods. The ability to date a large quantity of consolidated charcoal from the hearth could provide a clear control on the OSL ages we would obtain from the study area. We therefore placed priority on removing OSL samples from directly above the occupied paleosol, as well as from directly below it and adjacent to the hearth. Results of this controlled experiment are presented in the discussion of site stratigraphy.

The basal deposit at the site consists of beach gravels that occur at an elevation of circa 181 m (figures 5-13 and 5-14). Overlying the gravel is approximately 23 m of eolian sand in the northwestern part of the blowout. The lower part of the exposure contains a hearth laden with charcoal and a single potsherd. In an effort to determine the age of the cultural occupation and associated deposits, one conventional ^{14}C age and two OSL ages were obtained from the site. The radiocarbon age was acquired from charcoal in the hearth, whereas the OSL ages were acquired from the eolian sands immediately above and below the feature.

Dating results indicate that the eolian sand and associated archaeological remains in this portion of the dune are less than circa 1,000 years old. OSL ages indicate that the eolian sand in the lower part of the dune accumulated quickly. This conclusion is supported by an OSL age of 920 ± 80 ya (UIC2178) obtained from the basal eolian unit at the site, whereas the base of the upper sands provided an age of 870 ± 80 ya (UIC2179). The dip of visible eolian beds suggests that the portion of the dune was the slip face of (perhaps) a dune ridge at that time. Sometime during this period of early dune growth humans occupied the site. This conclusion is supported by the presence of the hearth within the sands, and the ^{14}C age of 820 ± 40 BP (Beta-238129; 2σ 795–673 cal BP; intercept 730) that was obtained from charcoal recovered within it. All ages overlap at 2 standard deviations and suggest that dune growth and occupation occurred between 810 and 710 years ago.

Discussion and Archaeological Implications

The Camp Miniwanca site is one of the few locales in our study that produced evidence of both unoccupied as well as occupied paleosols in eolian context. The closest parallel we have is at the Mt. McSauba locale north of Charlevoix, which dates to approximately the same time period. In this instance the soil labeled 4Ab lay near the base of the sequence, and was overlain by two other stabilization episodes prior to the development of the modern A horizon. From an archaeological discovery standpoint it should be observed that without the transecting exposure afforded by parabolic dune formation, the site's great depth of over 20 meters, and position on a dune backslope or slip face, would have made visibility of Locale B almost nil even with the use of deep coring devices. Locale A most likely would have been found, but its research utility is low since it lay on a deflated downslope surface. At best it is reflective of the presence of multiple short duration stays in the protective back dunes of the area.

Discussion

The combination of fieldwork and collections-based research detailed for each of the several archaeological sites we selected for further consideration has augmented existing information pertinent to the several project goals. In addition, it has created new data that allow the recognition of pattern at scales larger and more inclusive than the individual site—the historic scale of geoarchaeological interest. Of particular importance to us in this regard were observable regularities in site ages, the character of eolian deposits, the nature of the organic cultural deposits, and the spatial distributions of these variables at a subregional scale. From a macroregional perspective the issue of the effects of isostatic rebound was of particular interest. Much of the information on the nature of the eolian and organic deposits, as well as relative data on uplift, has already been presented in detail in the individual site or locale descriptions both here and in appendix C. While radiocarbon dates are an integral part of these site-specific discussions, the radiometric dates have been compiled in appendix B for uniform reference.

The nature of the organic occupation surfaces became of increasing interest as we compiled observations on the depositional sequences of individual sites. In particular, we noted buried soils at only a handful of sites in northern and northwestern Lake Michigan, in contrast to the many soils observed in the southern part of Lake Michigan (e.g., Arbogast et al., 2002). At first glance the few soils observed in this study appeared to be pedogenic, but it was quickly determined that they were very discontinuous and often confined to only the archaeological site itself, which is in stark contrast to paleosols in the southern end of the basin (e.g., Arbogast et al., 2002). In addition, these stratified organic zones appear to have formed much more quickly than would be expected. Of particular interest is the finding that Winter, Ekdahl-Goudreau, Scott Point, the basal stratum at Portage, all but the A horizon at O'Neil, the buried horizon at Antrim Creek, Porter Creek, and Camp Miniwanca display thin, densely organic paleosols *with cultural debris*. Winter, Scott Point, Portage, and O'Neil speak to rapid burial of such surfaces over relatively short periods of geomorphic time—200 to 400 years for the formation and burial of multiple occupation surfaces. Field observations by both the current research and original investigators speak of these surfaces being "greasy," "dense," "compact," and of course containing cultural materials. The consistency of such observations across multiple sites suggests that the occupation surfaces may in fact be anthropogenic rather than a consequence of ongoing soil formation processes, and that the surfaces were probably rapidly buried by eolian activation following occupation. Thus, some stable surfaces in dunes may be an accidental consequence of human activity in the dunes, rather than a proxy for climatic cycling per se.

The spatial aspects of chronology became quite evident with the additional dating—^{14}C, AMS, and OSL—conducted at multiple sites. Middle Woodland age stratified or buried components only occur north of a line between Petoskey and the Garden Peninsula. This group of sites includes Summer Island, Winter, Ekdahl-Goudreau, Wycamp Creek, and Portage. We also would predict that the Scott Point site, with a basal date of circa 1,180 years ago on a Mackinac Phase component, has additional deeper and earlier Middle Woodland strata not unlike those recorded at the Ekdahl-Goudreau site. The reasons for this prediction will become more evident when we discuss uplift and site depth further along in this discussion. The Portage site, despite its early radiocarbon dates, will add little beyond chrono-spatial information due to its position in a swale perched high in the Petoskey dunefield. It is not a coastal site, and is therefore not subject to the same agencies, particularly that of isostatic rebound, important in the formation of lakeshore dune and cultural deposits at lower elevations.

With the exception of the Portage site, all of the other sites in this northern sample of archaeological locales with earlier occupation components are stratified and buried by sand sheets deposited inland from the shoreline on uplifted beaches of varying age and distance inland. Such sand sheet deposition appears to be coupled with the degree to which the shore has been subjected to isostatic uplift. This relationship is systematic. Site depths increase with distance from the so-called hinge line on an axis generally trending east-northeast. The shallowest stratified coastal deposits with Middle Woodland components are Summer Island on the northwestern side of the basin and Wycamp Creek on the northeastern side. Importantly, there are archaeological sites further to the south in Wisconsin that have similar age Middle Woodland deposits, for example Rock Island. Site depths, and the number of discrete organic cultural strata, increase across the north coast of the Michigan basin on a west-to-east line from Winter, to Ekdahl-Goudreau, to Scott Point, the latter being the site with the greatest uplift rate among all the sites in our sample. These two sets of facts, site depth, and degree of isostatic rebound, suggest to us that Scott Point should have deeply buried Middle Woodland deposits that were not excavated by the Northern Michigan University excavations due to safety concerns.

From Petoskey, southward to about Muskegon, all but one of the dune-situated buried and/or stratified archaeological sites that have been either absolute-dated, or cross-dated with reference to diagnostic artifacts, are Late Woodland in age. More important, they almost all date more recently than circa 1,000 years ago. From north to south these sites include dated deposits at Mt. McSauba, O'Neil, and Camp Miniwanca. Sites with diagnostic Late Woodland ceramics in stratified context include South Manitou Light Site 1 and Porter Creek, neither of which was the subject of additional fieldwork. It should be noted that all of the sites north of Petoskey and the Garden Peninsula with earlier components also have later materials superimposed above the earlier occupations, or contain mixes of Middle and Late Woodland ceramics in their upper strata.

Just as site ages are different and more restricted to recent time, sites south of Petoskey on the eastern Lake Michigan shoreline are also found in different coastal settings than those in the northern part of the basin. Rather than being buried and stratified by sheet sand deposition, they are found either as constrained pockets in swales behind foredunes, such as at O'Neil or the buried occupation at Antrim Creek Natural Area, or as remnants on the slip faces or backslopes of larger dunes at highly variable elevations, such as Mt. McSauba, Solomon Seal, South Manitou Light 1 Site, Porter Creek, and Camp Miniwanca. Here organic occupation surfaces have been buried, at times quite deeply, either by the infilling of linear foredune swale features, or by sand blown and deposited on the downwind side or over the crest of a large coastal dune. The remaining occupation surfaces are of varying extent and are apparently dependent on the rate at which the dune is active and mobile. Thus, both the types of deposits and landscape settings, as well as the ages of these more southern sites, are different from those areas to the north. However, were it not for the fact that all of these sites also occur north of the hinge line, the continual shoreline exposure central to ongoing sand supply capable of activation would not be present, and the conditions for site burial would be minimal, nonexistent, or serendipitously idiosyncratic. This is precisely the case south of Muskegon, where buried dune occupations on intact organic surfaces have not been recorded. Importantly, though, in the subregion between Petoskey and Muskegon the depth of buried organic or cultural strata does not appear to be systematically related to the degree of uplift at a particular location, as it is in the northern part of the basin; sites do not necessarily trend to be deeper to the north, or to the south, but are contextually variable.

Summary

Based on the accumulated archaeological site data, the taphonomy of buried and/or stratified archaeological sites is subject to three subregional sets of coastal dune processes. Sites north of a line from the Garden Peninsula to Petoskey are situated further from the ambient coastal zone or water's edge. Such sites can preserve earlier, Middle Woodland, deposits. These site deposits are usually stratified and buried by thin sand sheets on gently sloping coastal dune surfaces. Sites in the northernmost part of the basin have the deepest deposits, at present attributable to the long-term effects of isostatic rebound. The eastern Lake Michigan coastal zone between Petoskey and Muskegon has the potential to preserve sites more recent than 1,000 years ago in coastal dune settings. This may be due to the effects of isostatic uplift on sand supply. These sites may be deeply buried in the inland aspects of foredune swales, or on the backslopes or slip faces of larger coastal dunes. Sites in the former landscape situation may be highly constrained in space. No stratified or buried archaeological sites in intact organic surfaces or cultural deposits have been observed on the eastern coastal zone south of Muskegon. Regardless of location, stable surfaces in dunes may be a consequence of both natural cycling and human occupation. Many occupation surfaces appear to be anthropogenic rather than pedogenic. There are long-term processes that most likely affected preservation potential that are not the purview of archaeology.

Appreciating the various ways in which human and natural factors interact within a dynamic coastal environment is necessary to develop an overall understanding of the factors that control the locations and burial of archaeological sites within dune settings. Several issues, however, remain to be addressed, including just how the regularities observed in the archaeological record of stratified and/or buried sites relate to or are consistent with the broader basin-wide processes of lake-level fluctuation, isostatic uplift, and dune activation. Placing these and other factors into their proper context is critical to formulation of a temporal and spatial framework and delineation of the natural and cultural processes that control the locations and burial of archaeological sites within coastal dunes settings.

CHAPTER SIX

Discussion and Synthesis of the Processes and Timing of Dune Formation and Archaeological Site Burial in Coastal Settings of Lake Michigan

Integrating the Natural Factors

Chronology of Dune Formation

Results from this study provide a comprehensive record of eolian sand mobilization and dune formation, particularly in the northeastern part of the Lake Michigan basin. In contrast to previous studies of dunes along the southeastern shoreline (e.g., Arbogast and Loope 1999; Arbogast et al. 2002; Hansen et al. 2003), which largely reconstructed dune histories through radiocarbon dating of charcoal in buried soils, this study systematically used OSL dating to establish chronological control to define intervals of eolian sediment deposition. Conversely, periods of dune stabilization were defined by the presence of paleosols using both ^{14}C ages of organic material within them and by the age and duration of archaeological deposits where they are included. As presented in appendix A, OSL dating estimates the most recent time that sand grains were exposed to sunlight; in other words, it estimates the most recent time the sediment was being blown by the wind and deposited to form dunes. As a consequence, we are able to more accurately assess the timing of dune construction because OSL provides an age estimate of the specific moment of final sand deposition. Conversely, as the age of the samples increases, for any given number of measurements, the confidence interval of the sample (i.e., the standard deviation, or σ) also becomes larger. Unless one obtains increasingly larger numbers of measurements as sample ages increase, measurements of older time periods are less precise than those of more recent time periods. Chronological interpretations derived from ^{14}C dating of buried soils, in contrast, may be erroneous due to uncertainties associated with the residence time of organic materials and record intervals of dune stability, not construction.

Despite our intentional focus on sampling and dating middle Holocene dunes, the outcome of our research revealed that few such dunes actually exist. Moreover,

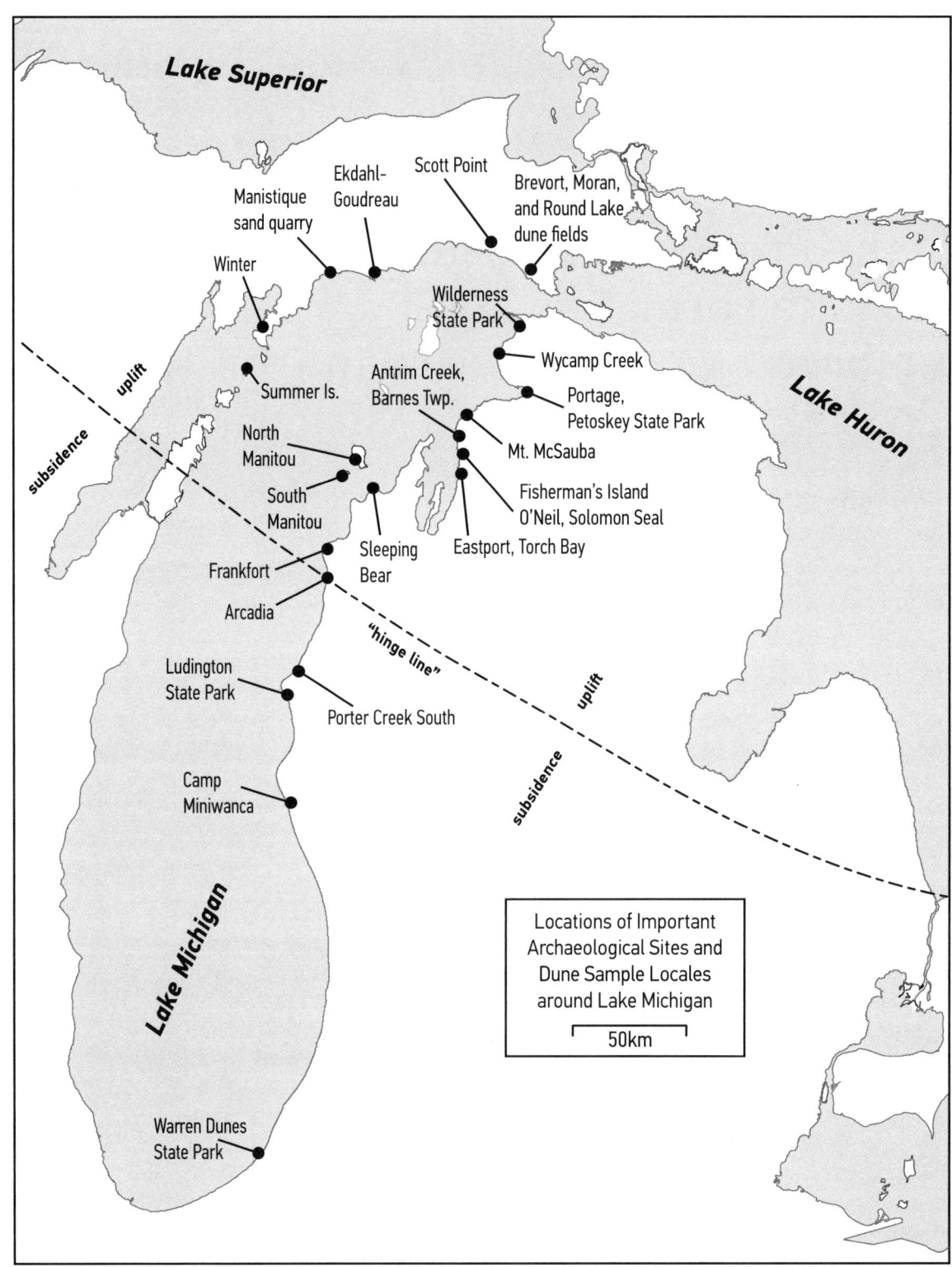

6-1. Locations of archaeological sites and dune-sampling locales around Lake Michigan. Dashed line indicates the approximate position of the Algonquin and Nipissing hinge line; after Leverett and Taylor (1915). Areas of presumed subsidence and uplift labeled on either side of "hinge line."

the few middle Holocene dunes identified are also not the largest dunes within any given dune series. The oldest dunes recognized in this study occur near the village of Eastport and date to circa 5,300 years ago. This pulse of dune formation correlates very well in time and altitude with Lake Nipissing (Lewis 1969; Monaghan and Lovis 2005). Similar age dunes have also been generally recorded in nearby high-perched dunefields at Sleeping Bear Dunes National Lakeshore (Snyder 1985) and at the Arcadia dunes (Blumer 2008). A significant finding of this study is that construction of coastal dunes on topographically low lacustrine surfaces in the northeastern part of the basin during the middle Holocene was apparently rare. The dune at Eastport was the only location sampled during this research that provided an age associated with the Nipissing phase. Although a similar age of circa 4,800 years ago was acquired at Petoskey State Park (Cordoba-Lepczyk and Arbogast 2005), it is the only other known sample dating to this period in this part of the basin. This chronological pattern stands in stark contrast to the record of dune construction along the southeastern shore of Lake Michigan, where significant mid-Holocene dune deposits have been recognized near Muskegon (Arbogast and Loope 1999), Holland (Arbogast et al. 2002), and Van Buren State Park (Van Oort et al. 2001). In this context, backdunes dating to this period of time near Holland are built up to 10 m high.

The first extensive period of dune construction recognized on topographically low lacustrine surfaces in the northeastern part of the Lake Michigan basin began circa 3,500 years ago. The onset of this period of dune growth was identified at four locales: the Solomon Seal site, the Fisherman's Island transect, the Torch Bay transect, and Ludington State Park (figure 6-1). At each of these locations, deposition of eolian sand built a dune ridge that now lies about 0.5 km east of the modern lakeshore. Deposition of eolian sand at this time was also recognized in a high-perched dunefield near Arcadia (Blumer 2008). Although dune growth ceased at Arcadia after this episode, accumulation of eolian sand continued on topographically lower lacustrine surfaces elsewhere in the northeastern part of Lake Michigan until soon after 2,000 years ago. Dunes dating to this period of time were recognized as far west as Manistique and were also identified at the Moran dunefield, Wilderness State Park, the Antrim Creek Natural Area, and at Ludington State Park (figure 6-1). Extensive deposition of eolian sand during between circa 3,500 and 2,000 years ago was previously recognized at Petoskey State Park (Cordoba-Lepczyk and Arbogast 2005). In general, dunes dating to this period are also the most inland along the coast of Lake Michigan in the Upper Peninsula, whereas they tend to lie within the central part of the dunefields in the northwestern part of Lower Michigan.

Consistent with these observations, the deposition of eolian sand and dune growth between circa 3,500 and 2,000 years ago has been well documented along the southeastern part of the Lake Michigan shore (Van Oort et al. 2001; Arbogast et al. 2002). Dunes in the southeastern end of Lake Michigan grew extensively in a vertical fashion during this period and about 50 percent of the total volume of eolian sand in this region was emplaced. Although dunes grew rapidly during this period along the southeastern coast, deposition of eolian sand was episodic, with several brief periods of stability that resulted in the formation of weakly developed soils (A/C horizonation) that were subsequently buried by eolian activity. This deposition pattern apparently did not occur on topographically low lacustrine surfaces in the northern part of the basin, however, as no buried soils were recognized in dunes on this landscape position in this study.

Following the major pulse of dune growth between circa 3,500 and 2,000 years ago, deposition of eolian sand slowed considerably in the northern part of the Lake Michigan basin for several hundred to about 1,000 years. This hiatus correlates very well with an extended period of reduced sand supply at Petoskey State Park (Cordoba-Lepczyk and Arbogast 2005) and in high-perched dunes near Arcadia (Blumer 2008). Reduced sand supply and landscape stability during this period of time has also been widely recognized in the southeastern part of the coast (Van Oort et al. 2001; Arbogast et al. 2002, Hansen et al. 2004, 2010), resulting in the formation of a weakly developed Spodosol (A/E/Bs/C horizonation) across most dunes in that part of the shoreline. Arbogast et al. (2004) informally named this paleosol the Holland Paleosol and argued that it is a major stratigraphic marker along the southeastern coast of Lake Michigan. Although this soil was identified in the southern end of Ludington State Park in this study (figure 6-1), it was not recognized elsewhere.

Our evidence indicates that a second major interval of eolian sand deposition began again along the northeastern coast of Lake Michigan circa 1,000 years ago. A number of sites provide OSL ages dating to this period, including the southwestern side of the Moran dunefield in the Upper Peninsula, the Sturgeon Bay Point exposure at Wilderness State Park, and a dune in the Torch Bay transect (figure 6-1). Increased sand supply at this time was also reported in high-perched dunes near Arcadia (figure 6-1) by Blumer (2008). Still younger deposits of eolian sand were identified at Mt. McSauba and the Torch Bay transect (figure 6-1). This most recent period of dune growth was recognized in a similar setting in the region at Petoskey State Park (Cordoba-Lepczyk and Arbogast 2005), as well as in high-perched dunes near Arcadia (Blumer 2008). It is very well documented in the southeastern part of the basin (Van Oort et al. 2001; Arbogast et al. 2002; Hansen et al. 2004, 2010), where it resulted in the burial of the Holland Paleosol (Arbogast et al. 2004) as dunes enlarged vertically. In the northeastern part of the basin, eolian sand deposits younger than 1,000 years in age stand alone and tend to be located in positions relatively close to the lakeshore.

Factors Responsible for Dune Formation in Coastal Settings of Lake Michigan

Our research has provided a temporal and spatial framework that describes and clarifies the chronology for episodes of coastal dune formation and stabilization, as well as the spatial variability of these episodes around Lake Michigan. More critically, however, we also aim to explain the processes that control the temporal and spatial patterns of dune formation across the study area. The controls and influences for dune formation, growth, and preservation within coastal settings are numerous, complicated, and often not independent of each other. At the most basic level, three major factors must occur to begin and sustain dune formation and growth. First, an extensive source of erodible sand must be present. Second, sufficient wind must be present to erode and transport the sand. Finally, in order to build large dunes, the sand supply must be replenished. Importantly, these three factors must occur concurrently to actually build sand dunes. Our goal in this research is to describe the environmental settings and temporal circumstances that promote the simultaneous occurrence of these three conditions along the Lake Michigan shoreline during the middle and late Holocene.

Many factors influence or control dune formation and preservation around Lake Michigan and have been discussed in detail in chapters 3 and 4. We believe, however, that the most important of these factors relate to two major processes: geographical differences in isostatic uplift rates and the multiscalar transgressions and regressions of Lake Michigan during the middle and late Holocene. The variation in the pattern of these processes through time and space provides the comparative basis to delineate the controlling events that initiate the formation, reactivation, and preservation of coastal dunes. If we compare the temporal and spatial differences in the effects of differential rebound and secular variations in transgressions and regressions of Lake Michigan, with the episodes of coastal dune construction and stabilization across the study area, the critical factors that control the distribution, stratification, and preservation of archaeological sites within eolian contexts become apparent.

Temporal Variations in Middle and Late Holocene Lake Level

Lake level has an important influence on dune formation, and several reconstructions of the lake-level history have been presented (see figure 4-3). Although the lake-level curves developed for these reconstructions are based on different types of data, and derive from different areas in the Lakes Michigan and Huron basins, they share some general commonalities that probably reflect significant basin-wide phenomena.

The middle Holocene (ca. 6,000 to 4,500 years ago) is historically characterized by the Nipissing phase, which includes the most significant transgressive lake sequence within the upper Great Lakes. This transgression culminated in Lake Nipissing, the highest Holocene lake phase in the Lake Michigan basin. A significant regression, which probably dropped to at least modern lake level in the southern parts of the Lake Michigan and Huron basin, generally follows that Nipissing phase. The fall from Nipissing to modern levels occurred between circa 4,000 and 3,500 years ago and is mainly a consequence of erosion of the outlet channel at Port Huron. Some relatively minor lake-level variations that are linked to climate cycles, however, may also have occurred as the outlet eroded (i.e., what has variably been referred to as post-Nipissing regression, the Nipissing II and Algoma stages).

Immediately after the post-Nipissing regression, the water level near the Port Huron outlet in southern Lake Huron stabilized at approximately that of modern Lake Huron. Similarly, the level of southern Lake Michigan was also stabilized as that of modern Lake Michigan. Even though modern levels of Lakes Michigan and Huron were established in the southern end of the basins, however, the upper Great Lakes still formed a single lake in the Lakes Michigan, Huron, and Superior basins. The continued connection between these lakes was a consequence of the relatively greater amount of isostatic depression in the northern regions of these lake basins at this time. The combined lake drained through Port Huron and apparently had a relatively stable level from circa 3,500 to 4,000 years ago until circa 2,000 years ago (Monaghan and Lovis 2005). By 2,000 years ago, isostatic rebound had raised the rapids at Sault Ste. Marie above Port Huron and separated

Lake Superior from Michigan-Huron. This event essentially established the modern configurations of the upper Great Lakes.

Once Lake Superior became separated, the hydrology of Lakes Michigan and Huron seemed to change and become more responsive to variations in climate. This is reflected by a series of climate-induced high- and low-water-level (i.e., transgressions and regressions) cycles after circa 2,000 years ago. These cycles were, individually, several hundred years in duration (e.g., Larsen 1985a; Monaghan and Lovis 2005). High-water (transgressive) phases are believed to be related to cool/wet intervals, while low-water (regressive) phases are believed to relate to warm/dry conditions (e.g., Little Ice Age and Medieval Warm intervals, respectively; Monaghan and Hayes 1998, 2001; Monaghan and Lovis 2005).

Spatial and Temporal Variations in Middle and Late Holocene Uplift and Subsidence

Differential uplift around Lake Michigan also influenced the timing and character of dune formation and preservation. The dominant trend is north to south; the lowest rates of uplift in the southwest part of Lake Michigan (i.e., north of Chicago) to the highest rates in the northeast (i.e., near the Straits of Mackinac; figures 4-1 and 4-4). Direct measurements of isostatic rebound recently collected in the upper Midwest and Great Lakes regions suggest southern Lake Michigan is actually subsiding, and that the hinge-line concept (Goldthwait 1908; Leverett and Taylor 1915) may actually mark the inflection separating regional uplift (north) from subsidence (see figures 4-4, 4-6, and 4-7).

The magnitude, spatial distribution, and chronological change evident in differential rebound collectively have some important implications for modeling the timing and character of dune formation. Two salient facts are vital to modeling the relationship between rebound and dune formation: the southern end of Lake Michigan is currently undergoing relative subsidence, while the northern end of the basin is uplifting, and the rate of differential isostatic recovery has lessened through time. This latter fact means that the relative effects of rebound around the basin were greater during the middle Holocene than now and, therefore, implies that the effects of uplift or subsidence on dune formation (discussed below) were accentuated between 2,000 and 4,000 years ago and then become progressively less significant after circa 2,000 years ago.

From the standpoint of coastal dune processes, beaches within subsiding portions of the basin will effectively undergo long-term transgression, which will promote shoreline and bluff erosion. In fact, as much as 800 m to 1,500 m of shoreline may have been eroded in southeastern Lake Michigan since the middle Holocene (Buckler and Winters 1983), which suggests that in the areas of subsidence in southern Lake Michigan older coastal dunes will tend to be eroded and destroyed, or at best be only partly preserved. Conversely, beaches that form along the northeastern Lake Michigan coast, an area of relatively high differential uplift, will effectively undergo long-term regression. We know that over time, this pattern will result in a series of raised beaches and that each older beach will be positioned further inland from the younger beach. Unlike the case in areas of subsidence, such a pattern will also tend to preserve former coastal dune sets that, like the beaches, are offset progressively more inland as that shoreline is raised.

Spatial-Temporal Model for Coastal Dune Formation around Lake Michigan

OSL dating of eolian sand deposits clearly demonstrates a distinct history of eolian sand deposition in the northern and northeastern part of the Lake Michigan basin. Overall, our results indicate that Nipissing dunes are rare and that the first major pulse of dune growth occurred between circa 3,500 and 2,000 years ago. In many places, such as at the Solomon Seal site, the Fisherman's Island transect, and the Torch Bay transect, the largest dunes formed during this interval. A similar history was reported at Petoskey State Park (Cordoba-Lepczyk and Arbogast 2005). The supply of eolian sand to dunes subsequently slowed considerably until circa 1,000 years ago, when it again increased. Dunes have continued to form over the past 1,000 years, particularly between 1,000 and 500 years ago. It appears that the supply of sand may have decreased in the past 500 years, although this currently remains uncertain.

Given these results, important questions associated with the causes of eolian sand deposition through space and time remain to be addressed. Our research demonstrates that the vast majority of dunefields in the northeastern part of the Lake Michigan basin are associated with embayments, including contexts at Manistique, the Moran dunefield, Wilderness State Park, Mt. McSauba, the Solomon Seal site, the Fisherman's Island transect, and the Torch Bay transect. Another large dunefield occurs adjacent to Little Traverse Bay at Petoskey State Park (Cordoba-Lepczyk and Arbogast 2005). These sites are natural places for littoral sand to collect because the longshore current tends to slow due to changes in shore angle (e.g., Hesp 1999). In addition these sites tend to be associated with dissipative beaches (Short and Hesp 1982) where the offshore bathymetry indicates a shallow gradient (figure 6-2). The shallow offshore slope gradient reduces wave energy as waves approach the shore (Short and Hesp 1982). As a result, littoral sand has a tendency to be nudged up into the beach zone, where it can subsequently be reworked by the wind to form dunes (Short and Hesp 1982; Sherman and Bauer 1993).

Aside from the basic patterns linking the location of dunefields and specific coastal configurations, our research also demonstrates a distinct space-time relationship within individual dunefields. In general, the distribution of eolian sand within individual dunefields is time-transgressive, with older deposits the most inland and progressively younger sediments closer to the shore. The most likely explanation for this geographical pattern is that dunefields in this part of the Lake Michigan basin occur within an area of relatively rapid uplift. Consequently, as noted above, persistent rebound throughout the late Holocene has caused progressively younger surfaces to emerge and become the focal point for eolian sand deposition. These findings support Scott's (1942) observation that dunes north of the hinge line consist of well-defined ridges that are spaced far apart due to isostatic uplift. Our findings are also supported by work on beach-ridge complexes at Wilderness State Park that indicate a similar time-transgressive pattern (Lichter 1995). A similar pattern has also been reported at Petoskey State Park (Cordoba-Lepczyk and Arbogast 2005), which lies within the core of our study area.

The most difficult question to answer, however, relates to the specific cause(s) for variations in the supply of eolian sand over time in the northeastern part of

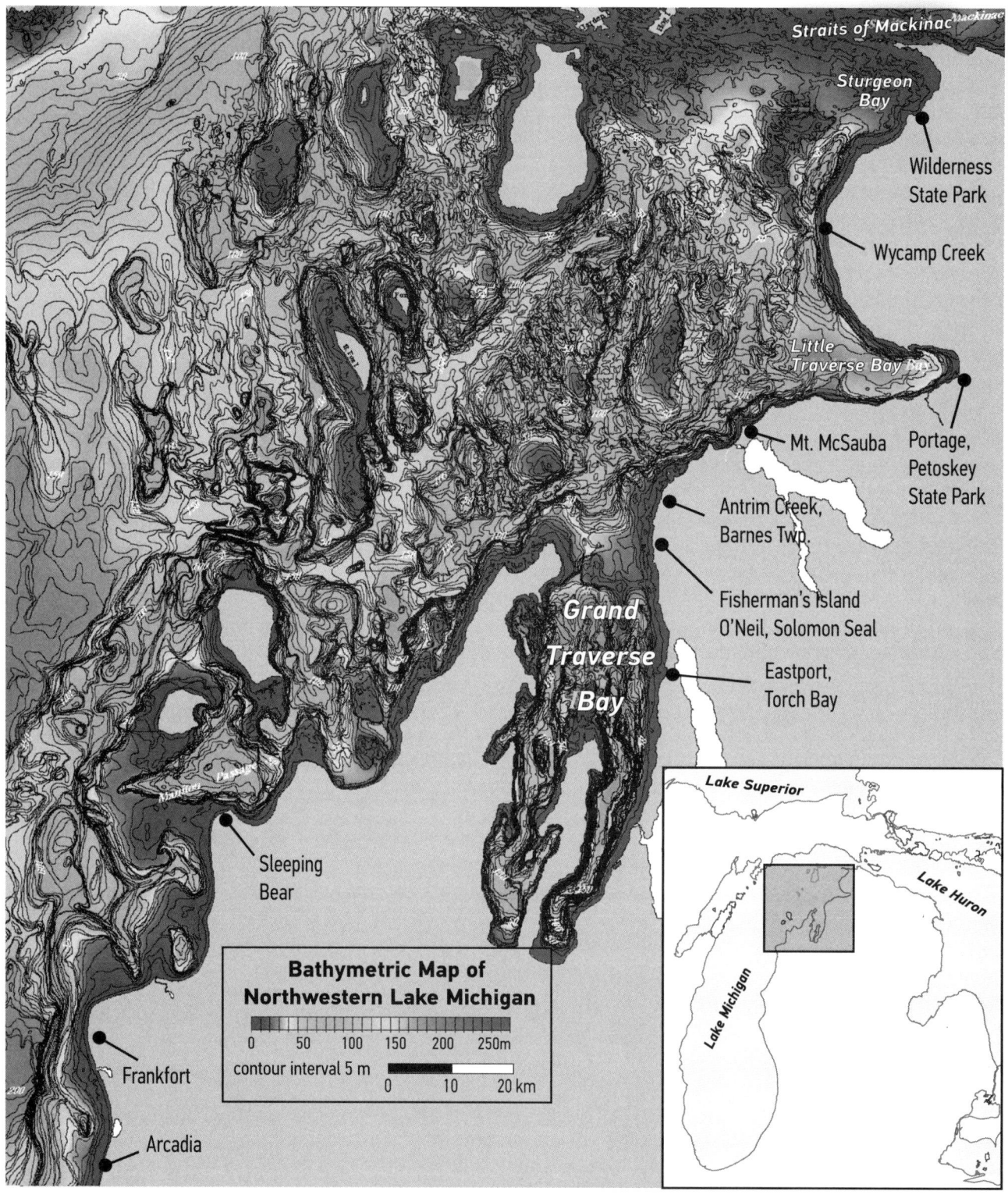

6-2. Bathymetric map of Lake Michigan in northwestern Lower Michigan. Dunefields are associated with arcuate embayments that trap littoral sediments. Because the offshore slope is gradual, beaches in these areas tend to be dissipative and, thus, are zones of littoral sand accumulation.

the Lake Michigan basin. Several studies (Arbogast and Loope 1999; Van Oort et al. 2001; Arbogast et al. 2002; Hansen et al. 2003; Fisher and Loope 2005) on low-perched dune complexes along the southeastern shore of Lake Michigan linked increased sand supply to high-water lake phases. These studies generally linked the model governing sand supply in high-perched dune systems in the northern part of the Lake Michigan basin (Dow 1937; Snyder 1985) and along the southern coast of Lake Superior (Anderton and Loope 1995) with lake-level fluctuations. In general, these studies indicate that increases in sand supply occur during high lake phases, when waves aggressively erode high bluffs below the dunes. This erosion increases bluff destabilization in a way that promotes mobilization of eolian sand liberated from lake-facing exposures. Sand supply to high-perched dunes subsequently decreases when lake level falls and bluffs stabilize. According to Loope and Arbogast (2000) increases in eolian sand supply in Lake Michigan coastal dunes may be related to lake-level cycles with durations of about 150 years.

Although previous research on low-perched dunes has supported the hypothesis linking increased sand supply with high lake phases (e.g., Arbogast and Loope 1999; Loope and Arbogast 2000; Arbogast et al. 2002; Fisher and Loope 2005), this association is based entirely on the ^{14}C ages of buried soils. In addition, it is assumed that the dates derived from these soils estimate the time of soil burial by younger deposits of eolian sand. This assumption may be erroneous because the residence time of charcoal in soils is unknown. Other sources of error may lie in the uncertainties associated specifically with the dating method. Given these issues, an initial goal of this study was to more accurately test the linkage between sand supply to dunes and lake-level fluctuations by directly estimating the age of dune deposits via OSL dating. Unfortunately, given the age dependency of statistical confidence, the earliest OSL ages produced larger standard deviations (e.g., ca. 300 years), which makes a detailed correlation of dune activation and lake-level fluctuations impossible. With increased numbers of samples, derived from close-interval vertical sampling techniques, it may be possible to reduce the statistical error.

Nevertheless, a more general analysis of dune behavior and lake-level history reveals some interesting and potentially significant relationships. A major finding in this study is that dunes relating to the Nipissing phase high stand are quite rare in the northern part of the Lake Michigan basin. This conclusion is significant because it directly counters a long-standing assumption (e.g., Scott 1942; Dorr and Eschman 1970; Buckler 1979) that the most active dune-forming episode along the eastern shore of Lake Michigan occurred during the Nipissing high stand. The default hypothesis associated with this assumption was that the supply of eolian sand was high during this lake phase due to extensive coastal erosion. Although several studies (Loope and Arbogast 2000; Arbogast et al. 2002; Fisher and Loope 2005) along the southeastern shore of the lake indicate that dune growth occurred throughout much of the late Holocene, significant deposits of eolian sand also accumulated during the Nipissing stage (Hansen et al. 2003). Eolian deposits of similar age are unusual in the northern part of the basin, which suggests that dune formation during the highest Holocene lake phase in this part of the system was limited.

Among the additional significant findings of this study is that the first major, and probably volumetrically most significant, pulse of eolian sand in the northern half of the Lake Michigan basin occurred between circa 3,500 to 2,000 years ago.

This period of time fundamentally coincides with a significant lake regression following the Nipissing high stand (figure 6-3). In fact, many of the largest dunes in the study area formed during this period of time, suggesting that the supply of eolian sand was high. This finding runs counter to the conclusions of previous studies (Loope and Arbogast 2000; Van Oort et al. 2001; Arbogast et al. 2002; Fisher and Loope 2005) that related high sand supply in low-perched dunes with high lake phases. Instead, our findings imply that dunes formed in association with Olson's (1958b) foredune model, and that the beach was a significant supply of eolian sand. Conceivably, many of the dune ridges we observed actually may be very large foredunes.

A more important question relates to sand supply during this initial growth phase and why it was so high. An emerging hypothesis from this study is that this high sand supply is actually related to extensive coastal erosion during the Nipissing transgression and high stand. Prominent coastal bluffs of Nipissing age are preserved throughout northern Michigan and attest to extensive wave erosion during this period of time. We propose that the eroded sediments were subsequently stored in the offshore, or lake-bottom, environment for a period of time and gradually were nudged toward the shore in this dissipative environment. In this fashion, they accumulated on the shoreline in embayments, where they became active and were eventually reworked to form dunes.

If it is accurate, such a scenario is highly significant because it means that a substantial temporal lag occurred between the erosion of coastal sediments during the Nipissing phase and their ultimate movement into the coastal dune system. This lag model is supported by the fact that the supply of eolian sand apparently ended sometime just after 2,000 years ago, with few dunes apparently forming during the ensuing 1,000 years. In other words, a finite supply of eolian sand was created during a particular point in time (i.e., the Nipissing high stand). The following hiatus in dune growth occurred during a transgression to a high lake stage circa 1,700 years ago and a subsequent regression. This hiatus in the northern part of the Lake Michigan basin correlates with the formation of the Holland Paleosol, which is a widespread, important stratigraphic marker within dunes in the southeastern part of the basin (Arbogast et al. 2004).

Approximately 1,000 years ago the supply of eolian sand apparently increased again in the northern part of the Lake Michigan basin, as a distinct period of dune formation or growth dating to this period of time was identified in this study. This period of dune growth correlates reasonably well with burial of the Holland Paleosol and the enlargement of dunes along the southeastern part of the shore at that time (Arbogast et al. 2004). We propose that the increased supply of eolian sand at this time may be associated with a similar temporal lag that

6-3 (OPPOSITE). Dune ages and magnitude of lake-level changes with dune activation during the middle and late Holocene. A. (*upper diagram*) Models of lake-level changes developed by various researchers (corrected for uplift at Port Huron; see figure 4-2 for citations). Range of cultural periods shown above time axis. B. Details of the post-2,500 BP lake hydrographs. C. Probability Density Diagram (PDD) of OSL ages from dunes around northern Lake Michigan. D. PDD of ages of archaeological sites and paleosols buried within coastal dunes around northern Lake Michigan.

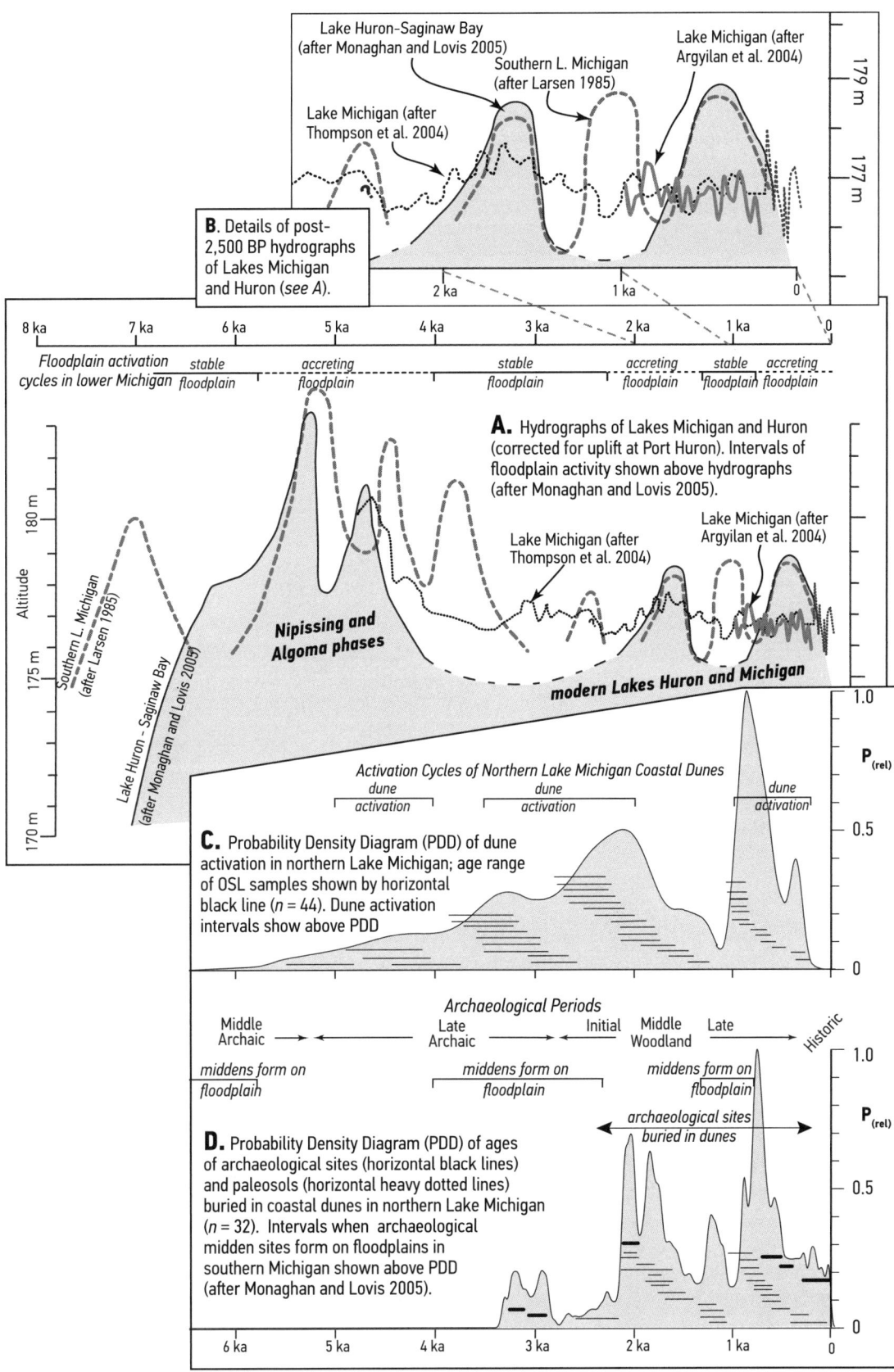

followed the Nipissing lake phase. In the more recent case, however, increased sand supply hypothetically relates to coastal sediments eroded during the high lake phase that occurred circa 1,700 years ago. As in the earlier lag, these sediments were stored for a period of time in the offshore environment before they ultimately became part of the dune system. This hypothesis is supported by the fact that this most recent dune-forming interval was apparently less intense than the earlier episode, which is logical because the amount of coastal erosion circa 1,700 years ago would have been less than during the earlier (Nipissing) stage. This is most likely due to the fact that the 1700 BP transgressive event achieved elevations 5 m less than that of the Nipissing phase maximum, resulting in less coastal erosion. Our results also suggest that the supply of eolian sand has gradually reduced during the past few hundred years, which further supports the lag hypothesis. The differential effects of isostatic rebound on coastal dunes apparently became an important factor in the past few hundred years. Although dunes are well preserved in a time-transgressive fashion in the northern part of the basin, extensive erosion of dunes has occurred in the south (Hansen et al. 2004).

Although our findings suggest that the primary variable associated with dune construction in the northern part of the Lake Michigan basin may be the lag response to coastal erosion, it is possible that secondary variables are also important. A very important factor that influences coastal dune formation is the frequency and intensity of storms. In northern Europe, for example, formation of dunes has been linked to increased storminess at several times during the Holocene (e.g., Jelgersma et al. 1970; Lamb 1995), as well as during the Little Ice Age (Gilbertson et al. 1999). Dune activity related to increased storminess has also been observed along the Atlantic Coast of North America at Cape Cod (Forman et al. 2008). Another variable that may be related to coastal dune formation is the impact of drought. According to Booth and Jackson (2003), a major drought occurred in the Great Lakes region circa 1,000 years ago, which correlates very well with the onset of the most recent episode of dune growth in the northern part of the Lake Michigan basin. Conceivably, stabilizing vegetation cover on dunes may also have been reduced during the dry phase, which allowed more mobilization of eolian sand to occur. A similar linkage was identified in dunes on Cape Cod (Forman et al. 2008). Yet another factor that may have contributed to the growth of coastal dunes in the northeastern part of Lake Michigan is subtle fluctuations in lake level. A series of small transgressions and regressions have occurred in Lake Michigan throughout the past 1,000 years (e.g., Lichter 1995; Thompson and Baedke 1999). Like the potential impact of drought and storminess, these small fluctuations in lake level in the latest part of the Holocene had the effect of producing a consistent supply of littoral sand that could be subsequently reworked into dunes. At this point in time, it is impossible to prove that such a linkage, or for that matter any other, actually exists.

Archaeological Site Formation within Coastal Dune Contexts in Northern Lake Michigan

The interrelated processes of sand supply, dune activation, lake-level variation, and uplift act in different ways to determine how coastal dune archaeological sites are buried, may perhaps become stratified, and/or are preserved at discoverable depths or in exposed landforms. Understanding how these processes interact at individual sites and groups of sites requires a multivariate perspective, including a consideration of site ages and their positions on coastal dune landforms, the evolution of organic soils containing cultural materials, and the role of humans in fostering site formation and preservation.

Stabilization and Burial of Occupied Surfaces

The interdisciplinary collaboration undertaken for this study provides some significant observations about and insights into the nature of organic horizons on buried and stratified dune archaeological sites. Under normal conditions the development of a soil, with or without cultural materials embedded in the soil matrix, is a relatively long-term phenomenon measured in multiple decades or centuries. Soils result from the complex interaction of climate, organisms, topographic relief, parent material, and time to produce pedogenic diagnostic horizons. In Michigan, well-developed soils in sandy sediments typically consist of an organic-rich A horizon, leached (light gray to white) E horizon, and spodic (reddish) B horizon enriched with sesquioxides and organic acids from above. Almost without exception, our observations of organic horizons bearing cultural materials are at odds with or differ from this categorization, begging the question whether stratified archaeological sites in the Lake Michigan coastal dunes are the result of occupations on short-duration, stable soil surfaces, or whether the occupation themselves create organic-rich, anthropogenic surfaces.

The majority of the stratified and buried dune archaeological sites in our study (i.e., those associated with buried paleosols) display minimal or no evidence for soil development, lack E or B horizons, and thus display A/C soil horizonation. The inventory of such sites is numerous, including Winter, the basal horizons at Ekdahl-Goudreau, Scott Point, Mt. McSauba, O'Neil, Antrim Creek Natural Area, Camp Miniwanca, and Porter Creek South (figure 6-1). Other candidates include Portage and South Manitou Light Site 1 (figure 6-1). Almost uniformly the occupation-bearing organic strata at these sites are thin, densely organic, and display no evidence of organic leaching. Other than the organic-rich A horizon that typically corresponds with the occupation strata at these sites, no evidence of even incipient soil horizonation is apparent. In fact, only two sites employed in our study display classic A/(E)/B/C horizonation sequence, the Summer Island and Wycamp Creek sites (figure 6-1). Accounting for this observation among a large number of sites, and an even larger number of discrete cultural strata within sites, is no easy matter. Here, we take the position that the most parsimonious fashion by which to explain the majority of our observations is by exploring the coupling of human behavioral processes with dune formation dynamics.

The organic horizons at the aforementioned sites lacking or displaying minimal soil development are thin but contain abundant organic matter to the degree that some possess a "greasy" texture (M. Buckmaster personal communication, observations at the Scott Point site; Lovis field observations at O'Neil, Portage, and Wycamp Creek sites; see also the comments of Brose [1970a] at Summer Island). The soils also tend to be highly localized and discontinuous, which contrasts with the numerous relatively continuous buried soils in dunes in the southeastern part of the Lake Michigan basin (Van Oort et al. 2001; Arbogast et al. 2002; Hansen et al. 2004). Finally, some sites such as Scott Point contain several (stratified) organic horizons that lie within a matrix that is only a few hundred years old. Given that a minimum of circa 150 years is thought necessary to form a weakly developed A horizon in Lake Michigan coastal dunes (Loope and Arbogast 2000), it appears that insufficient time transpired for a series of organic horizons to develop "naturally" and completely independent of cultural processes. These combined factors suggest that organic enrichment may well be a consequence of the deposition of fats, oils, charcoal, and other organic debris related to the cultural occupation and other human usage of the site.

The topographic position of these deposits within the dune complex is also significant. These sites are almost exclusively preserved in shallow depressions, hollows, or swales behind foredunes, which presumably provide protection for the inhabitants, as well as helping to preserve the resulting organic matrix embedded with the cultural debris. However, a general lack of formal soil development characteristics, in addition to the typical short-duration cultural chronology noted above, suggests that these organic-rich horizons are rapidly buried, rather than exposed for the decades or few centuries necessary for even weak soil horizons to develop. We interpret these observations as an indicator that human occupation itself may be the primary agent that creates, or at least fosters, the conditions for short-term stabilization of these surfaces. While the fact that they occur within protected locations such as dune backslopes or inland swales behind foredunes is important to their preservation, the organic enrichment by humans, with the associated low-level cementing or aggregation of the sand matrix with the fats, oils, charcoal, and finely fragmented shell or bone related to occupations, is the most important process for development of these horizons. Such horizons are fundamentally anthrosols, but are also commonly referred to by archaeologists as "middens." The formation of these horizons is also not wholly related to human occupation. The organic enrichment by humans, for example, may also result in short-term vegetation growth on the resulting surface. Under these conditions thin, hard, organic-rich surfaces will be formed such as are observable across the northern and northeastern parts of the Lake Michigan basin.

The fact that little or no soil development has been observed for these horizons suggests short-term longevity and relatively rapid burial of the organic-rich surfaces. To explain this phenomenon, we propose the hypothesis that a combination of human traffic and natural destabilization processes, including removal of vegetation, over the crests of the lower lakeward dunes regularly destabilized them and initiated or exacerbated eolian activity within the once stable foredune complex. As a result of such destabilization, sand within the dunes was eroded from dune crests and redeposited onto lower elevation and occupied inland surfaces within backdune swales. This pattern of human-induced mobilization of eolian sand has been recognized at the present time at Petoskey State Park (Cordoba-Lepczyk and Arbogast 2005), which lies within our study area (figure 6-1). Given current knowledge of rates of soil development and formation we cannot invoke natural processes, for example, to explain a series of four or more undeveloped organic surfaces buried during a period of 400 years at the Scott Point site, or three similar surfaces buried in a span of 400 years at the O'Neil site, or three or more surfaces at the Winter site over a similar span of time, and so forth.

In summary, we interpret the poorly developed, organic-rich A/C horizons that are characteristic of some eolian sites along the northern and northeastern shore of Lake Michigan as fundamentally a consequence of human occupation. As such, they are very short-lived Anthrosols that underwent distinctive formation processes that are somewhat different from the pedogenesis of other soil units such as the Holland paleosol. Such anthropogenic A/C horizons initially formed by stabilizing the surface of topographically low swales through the deposition of culturally derived organic materials and debris to create a relatively firm matrix. The crests of lakeward dunes adjacent to the site were subsequently destabilized and reactivated by human activity, which enhanced eolian redeposition of sand over these backdune environments to bury the horizons in these swales. The number of such surfaces buried in stratigraphic superposition is probably related to the regularity of cycling of human reoccupation at any given location. Sites with greater numbers of buried organic strata in a restricted area, such as Summer Island, Winter, Ekdahl-Goudreau, Scott Point, and O'Neil, were either utilized more regularly or possessed more constrained areas that local populations found desirable for occupation. With all of this in mind, however, we do not view the local occupants as active, premeditated agents in the stabilization or burial of surfaces, but rather that stabilization, organic enrichment, and burial are unintentional consequences of human activity on specific landforms.

Cycling and Chronology of Archaeological Site Burial

Three chronologically distinct sets of buried or stratified dune archaeological sites occur in the Lake Michigan basin, increasing in frequency through time. The Eastport site, which is a buried Archaic occupation surface with depleted organics, is the earliest and sole representative of the first set (figures 6-1, 6-3 and 6-4). Although not directly determined, the age of the Eastport site occupation is assumed to be as old as circa 5,300 years ago based on the OSL date of the dune in which the site is buried. A hiatus of about 2,500–3,000 years follows Eastport. We cannot confidently attribute any preserved buried sites within dunes during this interval. Between circa 2,500 to 1,700 years ago, during the Laurel Middle or Initial Woodland period, there was, it appears, another episode during which sites were buried and preserved, observable at Summer Island, Winter, probably Ekdahl-Goudreau, and Portage (figures 6-1 and 6-3). The Solomon Seal and Antrim Creek Natural Area sites probably also date to this period, although precisely when within this span of time is not possible to assess.

The hiatus in buried coastal dune sites after circa 1,700 years ago is of particular interest. Almost no buried later Laurel Middle Woodland period dune sites have been noted in the coastal zone, although dated terminal Laurel Middle Woodland sites are known, such as Wycamp Creek and perhaps, but not confidently, the Ekdahl-Goudreau site. The Middle Woodland deposits at the Rock Island site and other stratified sites on the Door Peninsula of Wisconsin can also be invoked (Mason 1966, 1967). However, a spate of buried and preserved dune sites occurs between circa 1,000 and 700 years ago. These sites include certain components at Summer Island, the upper occupation at Ekdahl-Goudreau, Scott Point, the lower beach at Wycamp Creek, Mt. McSauba, O'Neil, South Manitou Light Site 1, Camp Miniwanca, and Porter Creek South. Some sites, such as O'Neil and

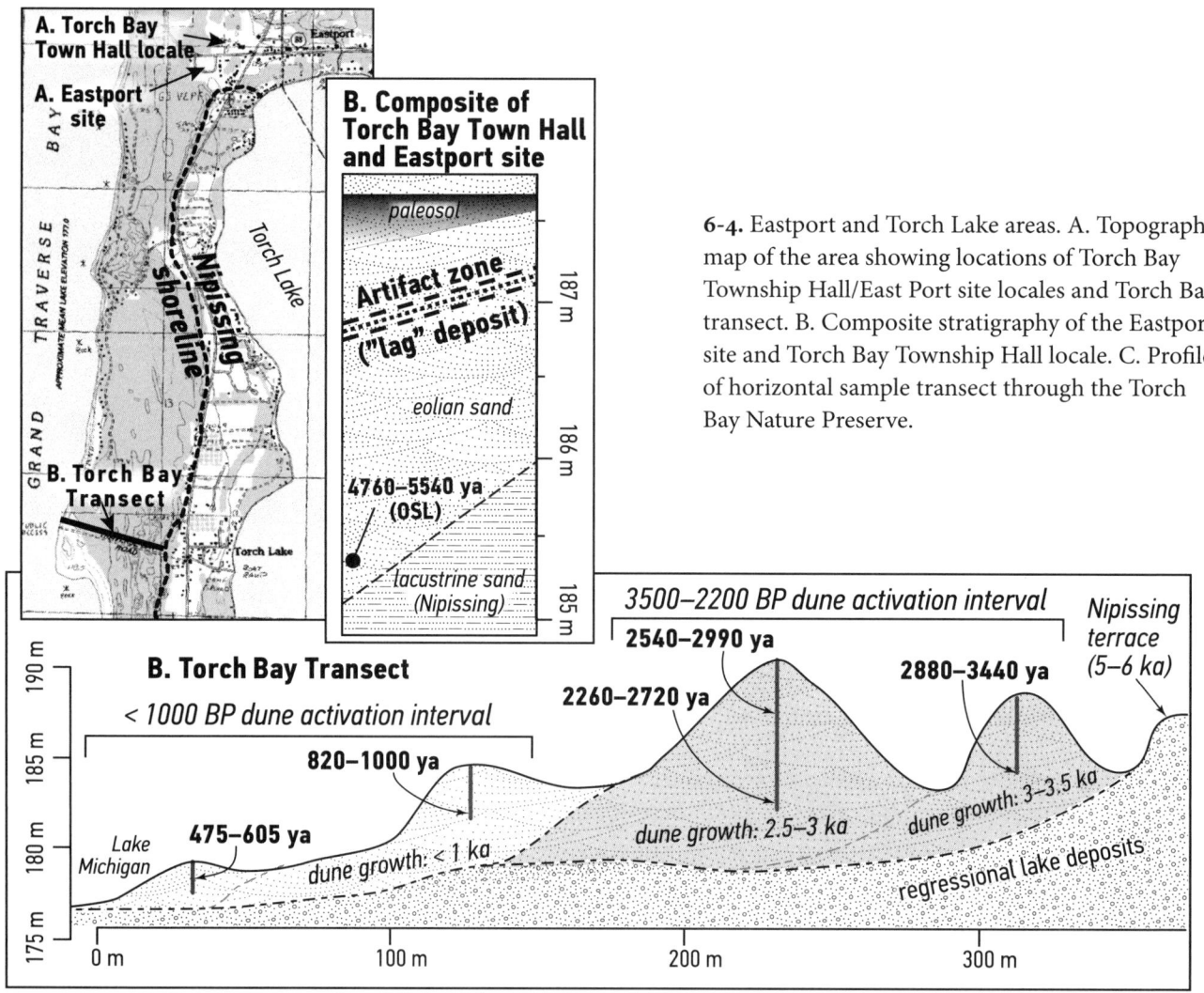

6-4. Eastport and Torch Lake areas. A. Topographic map of the area showing locations of Torch Bay Township Hall/East Port site locales and Torch Bay transect. B. Composite stratigraphy of the Eastport site and Torch Bay Township Hall locale. C. Profile of horizontal sample transect through the Torch Bay Nature Preserve.

Wycamp Creek, even include postcontact components buried by eolian activity (appendix C; Lovis 1973, 1990a).

The evident question, then, is: Why are there hiatuses in site preservation between circa 5,300 and 2,500 years ago, and between circa 1,700 and 1,000 years ago? Two possible explanations emerge: (1) the hiatuses mainly relate to diminished use of coastal locales and resources by local indigenous populations, or (2) they are a consequence of larger-scale, natural regional processes of dune formation and site taphonomy and are unrelated to cultural behavior patterns. Alternatively, they might also derive from a coupling of these two explanations. With respect to future research, these two explanations have important but different implications for our understanding of human settlement and subsistence patterns along the Lake Michigan shoreline.

Current evidence reveals that the seasonality of coastal occupations is clearly late spring through early or mid autumn, as the faunal and floral analyses summarized for our site sample in chapter 5 and appendix C clearly reveal. That said, it is almost impossible to identify convincing, clearly cold season or winter indicators of occupation for any site. To invoke coastal zone abandonment, or diminished use of coastal zones for occupations attached to local resource procurement, is difficult for us to accept, but not an improbable hypothesis. Is it possible, for example, that there is an economic transformation catalyzing settlement pattern change favoring more limited use of the coastal zone in the later Middle Woodland period. Although there is currently little evidence to suggest this is the case, such a scenario might relate to a time lag between maturation of coastal resources following the post-Nipissing regression and the incorporation of those new resources into human seasonal exploitation patterns.

The relationship between site preservation and age is of particular interest primarily during two episodes starting circa 2,500 years ago, as is the timing and cycling of dune formation across the basin generally. Sites are preserved at the later end of the major episode of dune formation circa 3,500 to 2,000 years ago. It appears that sites formed during earlier parts of this dune-building process may be deeply buried and not discoverable using standard archaeological exploration methods. Alternatively, they may also have been sufficiently reworked that they were destroyed and only those sites formed later in the dune-building episode, which were subjected to less reworking, survive within more shallowly buried contexts. As noted elsewhere, the processes responsible for earlier Archaic/Middle/Initial Woodland and later Late Woodland episodes of dune formation were different. Late Woodland age sites occur during a period when persistent pulses of coastal dune cycling took place, which resulted in a greater propensity for them to be rapidly buried and preserved throughout the entire duration of dune building. Moreover, because these occupations are younger, less time has elapsed during which they might have been altered or destroyed.

If a relationship between site burial and preservation and the processes associated with major cycles of dune activation exists, as we suspect, then the taphonomy of sites between 1,700 and 1,000 years ago might have been due to conditions that did not promote rapid site burial, the sealing of occupation surfaces, and consequent preservation. For example, this interval is apparently associated with a relative stabilization of the dunes (figure 6-3). Assuming that the coastal zone was continuously used during this interval and that site preservation is in fact linked to cycles of dune activation, then the absence of sites might be explained by taphonomic processes. Without rapid burial, several site taphonomic factors would act to promote site destruction rather than preservation. First, prolonged surface exposure would result in the rapid degradation of organics at both the macro- and micro-levels. The more deeply buried a site is, the more isolated it becomes from the physical, bio-, and cultural turbation or weathering processes that act to destroy site context and content (Monaghan and Hayes 1998, 2001; Monaghan and Lovis 2005). Second, relatively reliable but fragile time-diagnostic artifacts, such as low-fired, poorly tempered, relatively thin-walled Laurel Middle/Initial Woodland ceramics, would rapidly weather and decompose, which would result in few if any preserved ceramics. These simply do not survive the rapid changes in moisture and thermal ranges common in near surface contexts. Finally, eolian surfaces exposed to prevailing winds for extended periods would slowly deflate and create a lag of undifferentiated artifact pavements.

The cumulative effects of these multiple taphonomic processes would be to create lithic scatters composed of flint, fire cracked rock or FCR, and possibly tempering materials on deflated surfaces. In the absence of diagnostic lithic tools in the pavement debris, dating such scatters becomes difficult, if not impossible. Thus, it is possible that even continued use of the coastal dunes for occupation between circa 1,700 and 1,000 years ago might result in a relatively indistinguishable archaeological signature. In fact, such pavements might well be incorporated into a basal stratum of later, Late Woodland age, stratified deposits.

Summary

The outcome of this multiyear research into the taphonomy and preservation of coastal dune archaeological sites in some respects bodes both well and ill for archaeologists. The perspective that we take on our results begins with notions of archaeological site populations in multiple dimensions, both temporal and spatial, and at multiple scales. As archaeologists, we know that there are a specific number of human occupations that took place in any bounded space over a given span of time, although we really can't provide a precise number; it is an unknown population. This unknown number, however, is the population of occupation sites that were initially present in that time and space. As we also know, subsequent to the formation of sites by human occupation they become subject to postdepositional processes that may preserve, alter, or destroy them. Those sites that are not destroyed form the population of preserved sites, all of which have been altered to some degree. Very few, if any, sites have been left precisely as they were when the occupants abandoned them (the Pompeii Effect).

Additionally, not all of the preserved sites are discoverable using the field techniques and equipment that are available for archaeological fieldwork. For example, sites that are deeply buried in dunes beneath tens of meters of sand are not likely to be found using standard, primarily relatively near-surface field discovery techniques. So the archaeological record that we employ for research, analysis, and inference is actually a subset of the original population, and individual members of the subset are of variable integrity based on postdepositional alteration. Of these we assume that all of the remaining intact sites are potentially discoverable. With this in mind we make strategic decisions about how well this remaining subset, and particularly those sites that are discoverable, represents the original population. Archaeologists often assume that the sites we see and have discovered are representative of the whole population of original sites, whether or not such assumptions are justifiable. Research such as that conducted here provides significant correctives to this latter perspective. Correctives in this context can, unfortunately, often be viewed as "spoilers"; that is, they can be viewed as perspectives that make the process of archaeological research meaningless. This is far from the case.

The insights derived from this research have the ability to constrain strong inferences to those sets of bounded spaces and time periods for which we have the greatest confidence, and that is what the outcome of our research has helped to define, both in greater detail and at a larger spatial scale than has been done in the past. Of particular importance is the recognition that not all coastal subregions of the Lake Michigan basin in Michigan were subject to the same coastal dune formation processes, and the same temporal cycling of formation,

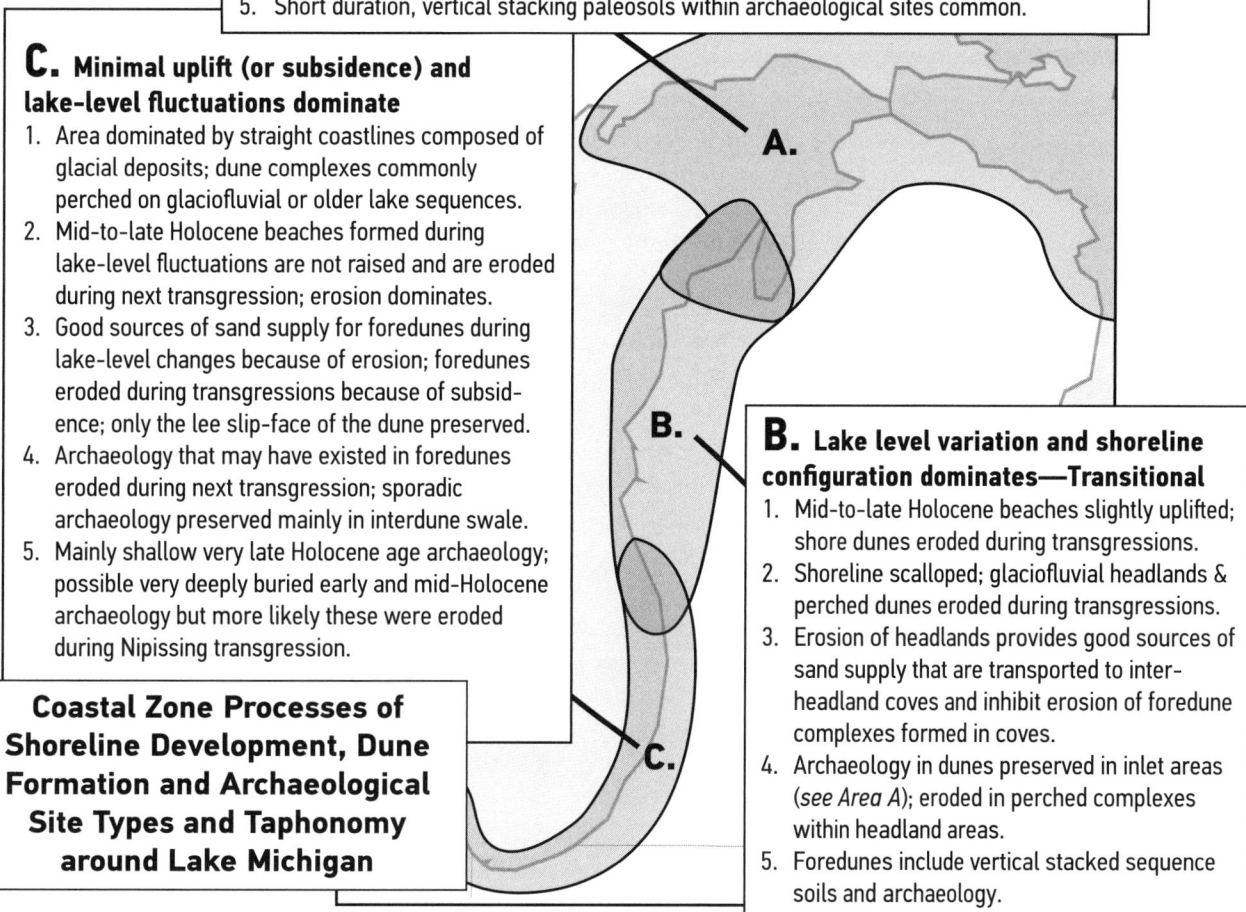

6-5. Zones on the Lake Michigan shoreline where groups of processes dominate to create specific types of beaches, dunes, and archaeological sites.

activation, and stabilization cycles. The most overt result of this recognition is that different subregions have different potentials for site burial and preservation, and that specific cultural time periods also have different potentials to be preserved. Thus, not all subregions have remaining preserved site populations that are even marginally representative of all time periods (figure 6-5). Moreover, depositional and site formational processes within the subregions may also result in selective preservation of sites of specific periods more than those from another period.

As one example, to our knowledge, there are no known dune-based buried or stratified archaeological sites that contain Hopewell Middle Woodland materials in the southeast part of the basin, where one might expect them, whereas there are multiple Laurel Middle Woodland sites preserved and buried in the northern third of the lake basin. This begs the question of whether one can make confident statements about Hopewell Middle Woodland use of coastal dune environments specifically, or coastal zones generally. To a degree this phenomenon is related to larger basin-wide coastal zone formation processes. Areas subject to isostatic rebound are more likely to have intact coastal deposits, whereas those coastal areas of the Lake Michigan basin south of the zone of rebound or uplift have undergone subsidence and been heavily eroded (figures 6-6 and 6-7). By some estimates more than a kilometer of shoreline may be gone (figure 6-6). In sum, coastal zone and coastal dune preserved site populations are decidedly not representative of the initial population of sites that was once present.

Time also figures prominently in the correctives our research can provide. Within the coastal dunes north of the hinge line (i.e., line of zero uplift; figure 6-1), where we have the largest number of buried and preserved stratified archaeological sites, there is a clear temporal trend to site preservation. There are far more Late Woodland archaeological sites than Middle/Initial Woodland archaeological sites, and more Middle/Initial Woodland archaeological sites than Archaic archaeological sites—at least in the coastal dunes. In other words, our confidence that the population of preserved sites is more representative of the original site population is greater for more recent time periods, and less for earlier time periods. Although this phenomenon is general and results from the fact that the older a site is, the more likely that taphonomic processes will act to destroy it, it also relates to some of the larger basin-wide temporal and spatial variations occurring around Lake Michigan.

Coastal and coastal dune sites occupied prior to, during, or immediately following the Nipissing stage transgression would have been subjected to massive postdepositional modification and most likely destruction during the major episode of dune building that took place circa 3,500 to 2,000 years ago. It appears that only parts of very few Archaic age sites are preserved in the coastal dunes, and they most likely date to the end of the dune-building period when eolian activity was dissipating; Archaic age examples include sites such as Solomon Seal and the horizon at Antrim Creek Natural Area buried in a dune OSL-dated to the Archaic. One can also readily view the preserved Laurel Middle Woodland sites in the northern end of the basin as part of this continuum. However, a lack of sites with mean ages between circa 1,700 to 1,000 years ago suggests that the latter part of the Laurel Middle/Initial Woodland period is not well represented in the archaeological record. Although we pose alternative hypotheses about this being a consequence of diminished use of coastal dune environments or an outcome of multiple taphonomic processes, it suggests that one cannot *ipso facto* assume representativeness of the sites that we use to build and test these alternative hypotheses.

The issue of selective site preservation in time and space aside, we can more

6-6 (OPPOSITE). Descriptive model of middle and late Holocene shoreline and dune development in areas of subsidence or stability along southern and southeastern shore of Lake Michigan.

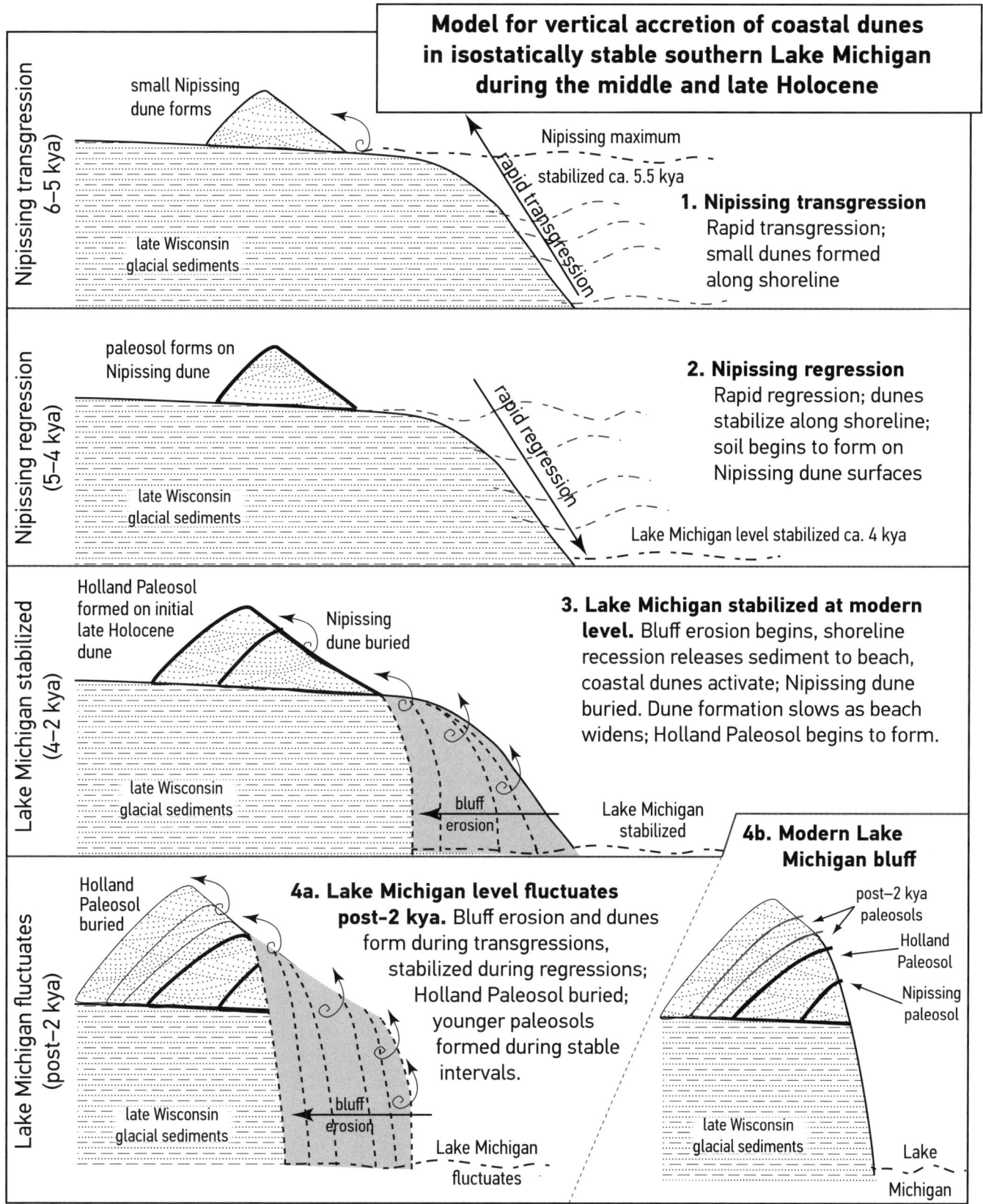

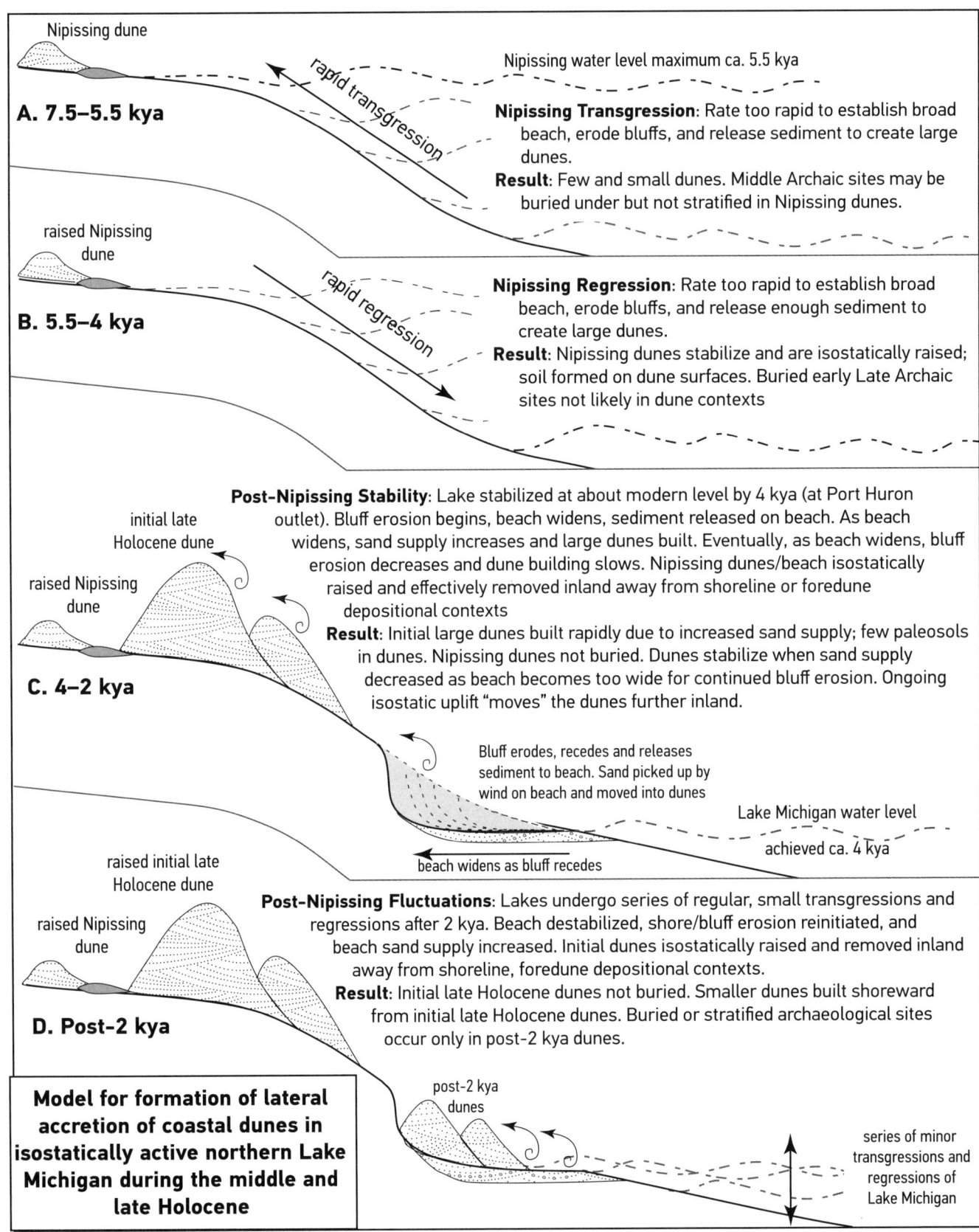

confidently assess *where* coastal dune sites will be preserved. In terms of basin-wide processes, dunefields are present where the coastal zone consists of a point of land, or an embayment of varying magnitude with an aspect to the northwest or the southwest, the latter not being a particularly difficult criterion to meet on the eastern side of the lake basin. These characteristics lend themselves to a position that maximizes the potential sand supply. Beyond coastal landforms, topography plays a significant role in determining where preservation of sites is most likely to occur. Depressions, swales behind foredunes, and the slip faces or backslopes of dunes are the topographic positions where all known buried or stratified dune sites occur (figures 6-6, 6-7, and 6-8). Often, such depressions or swales may be minimal but it is precisely these situations that promote eolian infilling during periods of activation. Occupations on the exposed foreslopes of dunes will rapidly deflate to lag pavements and be subjected to reworking and redeposition. Thus, coastal dune archaeological sites may only represent sites situated somewhat inland of the coast rather than representative of the entire population. Regardless of whether we would like to attribute protected locations to intentional selection, and attribute the majority of sites from any given time period to such locations, taphonomic considerations do not allow us to know what relative proportion of all coastal sites were actually placed in such positions.

These same phenomena, however, may provide us with an as yet undiscovered, and perhaps currently undiscoverable, set of preserved dune archaeological sites. In the north, there may well be intact Nipissing age beach terrace surfaces with occupations that were deeply buried by eolian activity during the major period of dune growth starting 3,500 years ago. In those areas south of the Muskegon/Ludington area, where steep, eroded, but fundamentally stable low-perched dunes are present, for example in Warren Dunes and Van Buren state parks, other processes might act to preserve sites. This is the region where kilometers of the coastal zone have probably been eroded, so that the remaining dunes, regardless of their current proximity to the coast, may actually have been relatively far inland during the Nipissing high-water stand circa 5,000 years ago. Here the expectation is that preserved organic surfaces or paleosols on dune backslopes and in swales and depressions have been regularly buried by eolian activity. However, burial may be at great depths, suggesting that standard, relatively near-surface, field techniques may be insufficient for their discovery. Premonitions of such deep burial are evident at sites such as Camp Miniwanca, which includes a Late Woodland hearth buried under 20 meters of eolian sand. The fact that it was discovered within an eroding dune face indicates the fragility of such sites from the perspective of long-term preservation.

In sum, the results of the site taphonomy and coastal dune cycling research have produced significant new insights into the nature of a specific set of coastal archaeological sites, their locations, and the processes that tend to preserve them. The outcome provides clarification for future researchers in terms of both

6-7 (OPPOSITE). Descriptive model of middle and late Holocene shoreline and dune development in areas of rapid uplift on northern and northeastern shore of Lake Michigan.

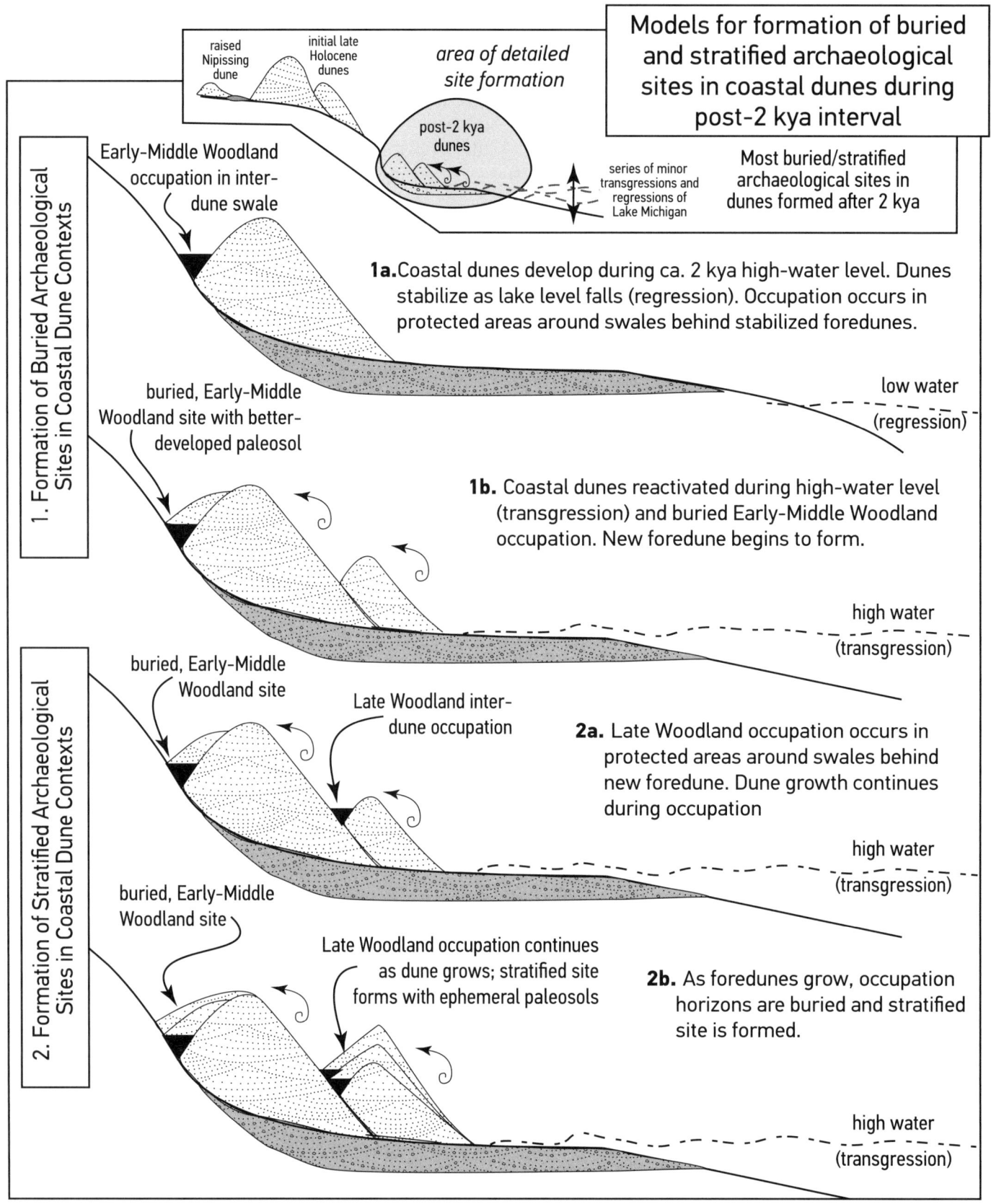

expectations and assumptions that can be applied to coastal dune site populations. Moreover, it directs future research to time periods and subregions that allow strong inferences to be generated about the processes responsible for the timing of archaeological sites' formation, their potential for preservation, and the potential for their discovery.

6-8 (OPPOSITE). Diagrammatic model of the formation, burial, and stratification of archaeological sites in areas of rapid uplift on the northern and northeastern shore of Lake Michigan (see figure 6-7 for dune formation processes in similar areas).

CHAPTER SEVEN

The U.S. 31 Case Study

Torch Bay to South Point

An oft-asked question at the conclusion of a major study such as the one presented here is, "So what practical value does all of this technical scientific research have?" Recall that at the outset of our work we posed a series of three questions regarding (1) the age of Lake Michigan coastal sand dunes, (2) their cycles of activation and stabilization over a long span of time, and (3) the relationship between their age and cycling to lake-level fluctuation, postglacial uplift, and the potential for the preservation or burial of intact archaeological sites. To varying degree, and for different subregions of the Lake Michigan coastline, we are now able to answer all three of these questions.

Synoptically, we now know that very few dunes on the Lake Michigan coastal zone date as early as the middle Holocene, or the Nipissing stage, of the Great Lakes. These are also not necessarily the largest of the coastal dunes. Such dunes occur in the lee of larger dunes in the southeastern part of the lake basin and more randomly (or at least less predictably) in the northern end of the basin. The first major sustained episode of post-Nipissing dune growth began circa 3,500 years ago, which is about 800 years after Lake Nipissing fell to the modern level of Lake Michigan. Regional dune growth and formation continued until circa 2,500 years ago. This resulted in the deposition of most of the eolian sand present along the coast. The dunes associated with this building interval are the most prominent along the coastal zone and often exceed 20 to 30 m in height. Following a distinct period of stability, a second and final regional episode of dune growth started circa 1,000 years ago and continued sporadically until historic times. This interval also apparently followed a resurgent late high-water stage in the Michigan basin that occurred 500–1,000 years earlier, circa 1,700 years ago. These data suggest that on a regional level or scale that dune growth is episodic. In addition, major intervals of dune growth are clearly not an immediate response to large-scale transgressive sequences, but rather occur as time-lag response several hundred years after higher water levels (i.e., transgressions) of Lake Michigan. This observation is important because it means that eolian site burial or stratification potential is not uniform throughout prehistory.

The burial of stable surfaces, many with cultural materials associated, appears to be more common in the northern end of the basin (i.e., north of the "hinge line"), although this may be a product of the relative depth and discoverability of sites with standard exploratory methods. While relatively rare, some Archaic-age sites are preserved, at least one as a potential lag deposit. The majority of buried and stratified sites, however, date more recently than circa 2,500 years ago. On the northern coastal zone of Lake Michigan (i.e., Upper Peninsula shoreline) archaeological sites are most often buried and stratified by relatively thin eolian sheet sand deposits. On the northeastern side of the Lake Michigan, on the other hand, sites are stratified and/or buried in swales behind foredunes, and may be variably preserved on the inland aspects, or the slip faces, of dunes. The southeastern Lake Michigan basin may, in fact, be heavily eroded, and most coastal sites largely destroyed. In sum, the results of this research have some clear and substantial predictive potential related to both the geographic and the temporal configurations of sites.

With this basic model in mind, it is useful to think of an application that speaks to the management of both known and as yet unknown sites at a subregional planning level. To this end, we have chosen a well-sampled area that allows more fine-grained statements about land use and management, the area of the Lake Michigan coastal zone between the towns of Torch Lake and South Point just south of Charlevoix, a linear distance of approximately 16 miles. Perhaps more importantly, this is an area with a major U.S. highway and a variety of public land parcels, many with coastal dunes, held or managed by township, county, and state agencies. The U.S. 31 right-of-way follows the Lake Michigan coastline in this area at varying distances from the shoreline (figure 7-1). There are major public landholdings, from south to north including the Torch Bay Nature Preserve, Banks Township Park, Antrim Creek Natural Area, and parts of Fisherman's Island/Young State Park, all of which are on the coastal zone, and all of which contain coastal dunes of varying size and age.

This portion of the coastal zone also has two further characteristics making it suitable for use as a case study. First, it contains a large number of archaeological sites that have been well reported in the literature. Second, it has been well sampled and dated as part of the current project. We also have a major transect and multiple OSL dates from across the dune sequence at Torch Bay Nature Preserve at the southernmost end of the case study area. These data are supplemented by an individual date on a Nipissing age dune situated behind Torch Bay Township Hall to the north, and also in close proximity to the Eastport archaeological site. Another major transect, the Antrim Creek Nature Preserve, occurs between Atwood and Norwood and is associated with two archaeological sites as well as a series of three OSL dates. A string of archaeological sites, many of which have been intensively excavated or test excavated and reported, occurs along the coastal zone from Norwood to Charlevoix. One of these, the O'Neil site on the north bank of Inwood Creek, is stratified, and well dated by ^{14}C. Importantly, an OSL-dated dune transect is also available from the Fisherman's Island Campground just north of Inwood Creek and its associated archaeological site. The northernmost end of Fisherman's Island Campground

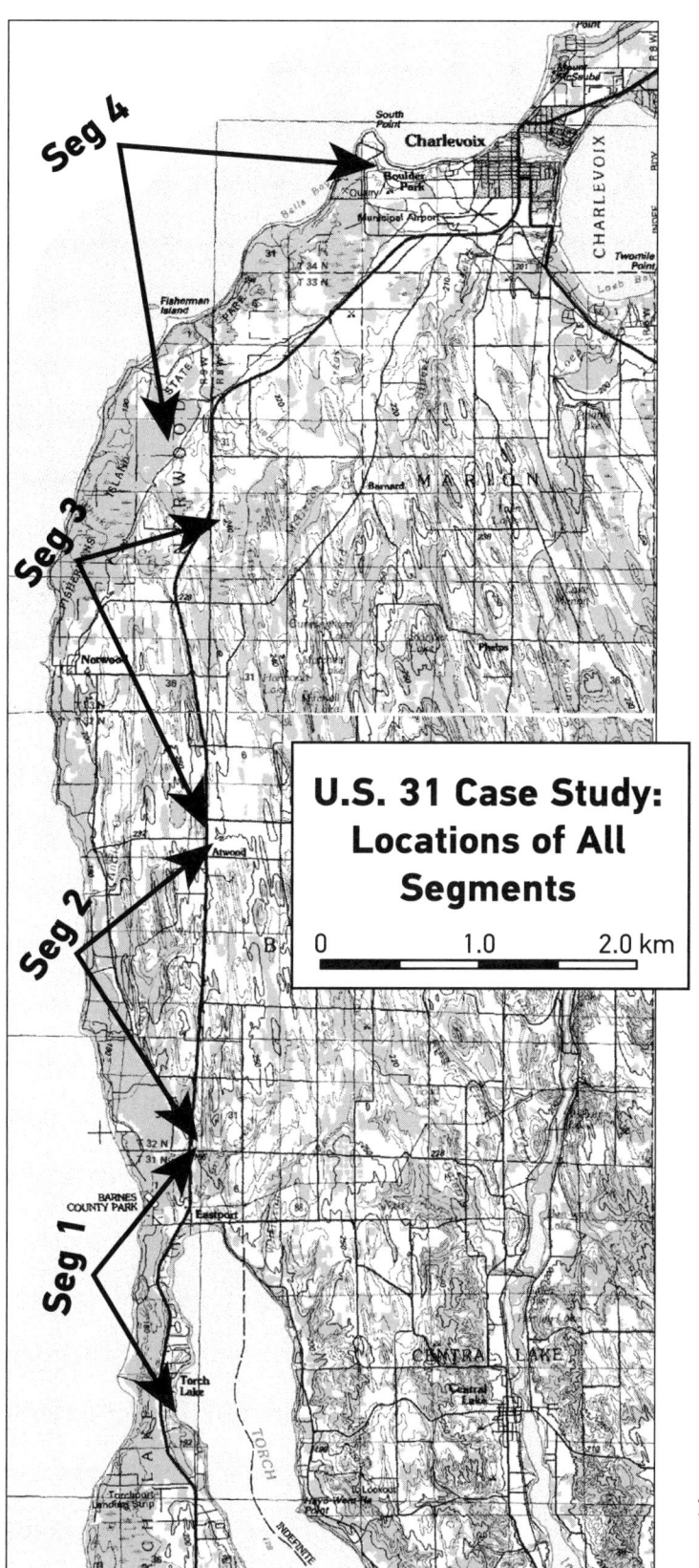

7-1. Locations of the four U.S. 31 case study segments.

has a pair of OSL samples extracted from a dune housing an archaeological site, the Solomon Seal site.

Within the area of interest in this case study, the prominent landform features to be considered for management, or various kinds of public interpretation, include:

- Exposed near-shore and shoreline features of the Nipissing stage
- Dunes remnant on this Nipissing and immediate post-Nipissing age surface
- Large and prominent post-Nipissing dunes that formed circa 3,200 to 2,500 years ago.
- In combination, foredune swales of post-Nipissing dunes circa 2,000 to 1,500 years ago, and foredune swales more recent than circa 1,000 years ago

These features are variably preserved and visible across the case study area. For ease of presentation the case study transect has been subdivided into four segments, which we discuss individually.

Segment 1: Torch Lake to T32N

The southernmost part of this coastal transect, approximately four linear miles from just south of Torch Lake to about one mile north of the town of Eastport (figure 7-2), was at one time an open postglacial spillway or channel between Torch Lake and the Michigan basin. It was submerged during glacial lake stages, but became exposed when the water level fell during the Chippewa-Stanley low-water period. It was partially inundated again during the rise to the Nipissing stage, and reexposed to approximately its current configuration with the post-Nipissing recession and subsequent stabilization at modern water level. Pre-Nipissing bottom sediments, Nipissing terraces, Nipissing eolian features, and post-Nipissing eolian and wetland features are present in this area. Notably, the U.S. 31 corridor closely follows the edge of the Lake Nipissing terrace to approximately Eastport, where its right-of-way swings further inland and crosses relatively level, exposed lake bottom. This entire distance also has small dunes, less than 3–5 m in height, adjacent to the right-of-way and above the Lake Nipissing terrace. Post-Nipissing dunes of various ages are present lakeward from the Nipissing terrace. Recent borrowing activity has occurred and continues to occur in the sand dunes situated at elevations above the Nipissing terrace.

Archaeological management, planning, and interpretation in this vicinity can take several significant directions. First, because it generally follows the crest of the Nipissing terrace, the U.S. 31 corridor has a high potential of encountering archaeological sites of widely varying age, including both those of pre- and post-Nipissing age. This notion is not new, and is evident from regional reconstructions of site locations and elevations. However, what is new is the reinterpretation of the contexts within which such sites might occur. Our regional data on dune formation processes, ages, and the potential for preservation of preserved surfaces reveals that they are most often buried and partially preserved on the interior or slip faces of eolian features. The southernmost four miles of our case study transect has intact as well as partially or completely destroyed dune features. Our data also reveal that the initial formation of such low-lying dunes occurred circa 5,300 years ago. Sites such as the Archaic Eastport site, in an area of dune

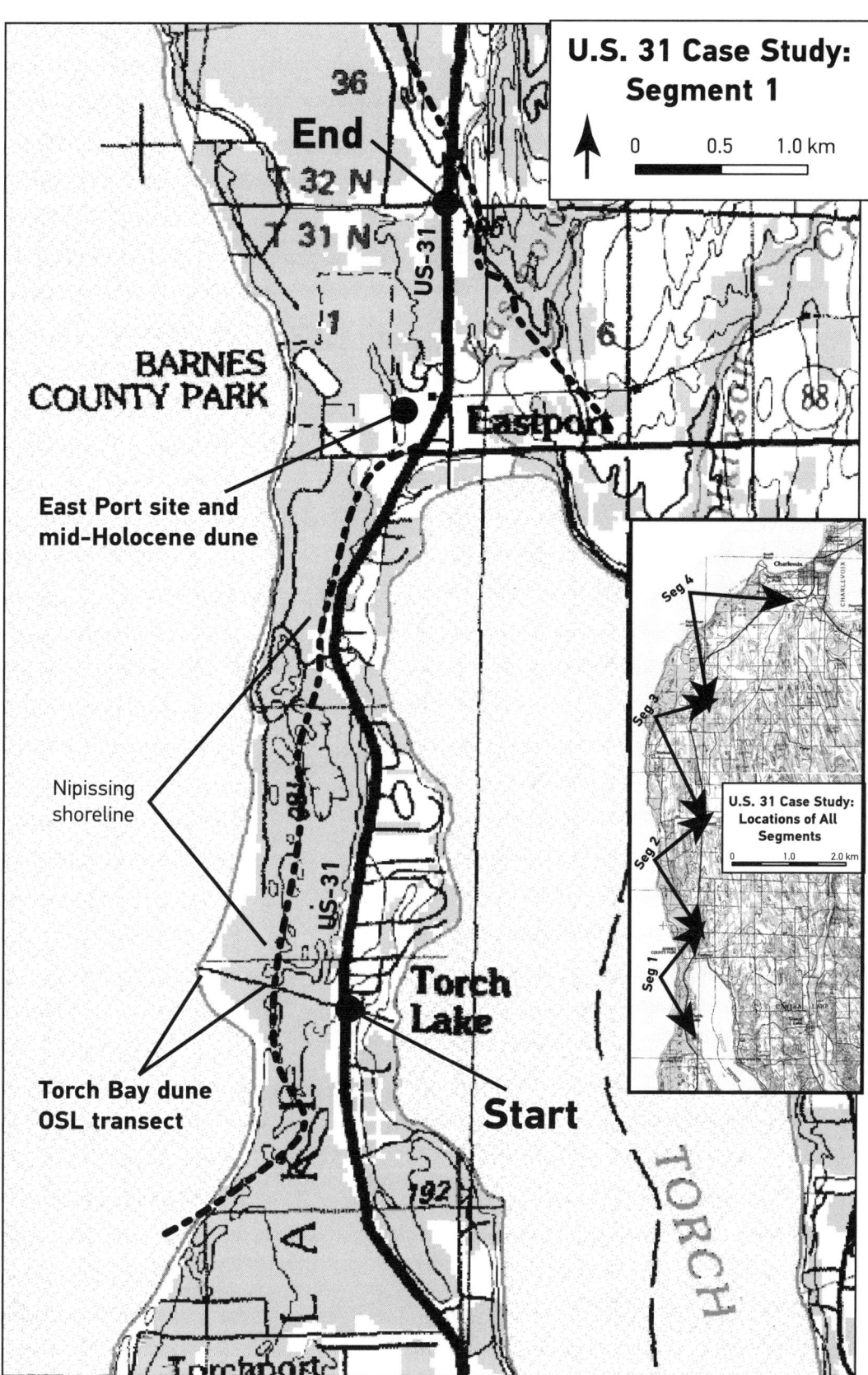

7-2. Geological and cultural features and locales of interest within U.S. 31 case study Segment 1.

borrowing, *underlie* these surfaces. The deposits at the Eastport site may occur on an intact, unmodified paleo-ground surface or alternatively as an erosional lag deposit; no definitive interpretation can currently be offered based on the current data. This means that areas in the vicinity of remnant dunes on this surface have high potential for buried and preserved archaeological sites of middle or early Holocene age. The archaeological site at Barnes Township Park, which lies on the crest of the Nipissing dune, is Late Woodland in age—Woodland sites can readily occur on Nipissing age features, particularly on eroded shorelines, an observation replicated elsewhere in the case study.

As a consequence of lake-level recession, the post-Nipissing sequence in this area predictably occurs lakeward of the Nipissing terrace. Our regional data suggest that, if preserved *in situ,* buried archaeological sites and surfaces occur mainly within the inland troughs behind foredunes. If such sites are not preserved unmodified, they will probably occur as deflated lag surfaces within either buried or surface contexts. This suggests that such areas have at least a moderate potential for preservation of deeply buried sites associated with the swales of any dune sequence. Further, the sites preserved should be more recent than the age of the lakeward dune. Well-dated dune sequences, therefore, such as those at the Torch Bay Nature Preserve, may well have preserved sites despite the fact that our coring did not encounter early cultural material. In part, this lack of discovery may reflect the fact that because of the nature of our work, coring focused on the crest of dunes rather than the interdune swales. The implication of our work indicates that subsurface work should, particularly related to backdune and interdune swales, be carefully planned and monitored.

Segment 2: T32N to North of Antrim Creek

North of Segment 1 (Eastport vicinity of U.S. 31; figure 7-3) the corridor trends more inland of the coastal features and traverses an extensive drumlin field that includes morainal features. A strongly developed Algonquin terrace parallels the Lake Michigan shoreline at elevations of circa 710 ft (200 m) asl. Weakly developed Nipissing features are apparent below the Algonquin shoreline, but eolian dunes are absent, being replaced by poorly drained and thickly vegetated wetlands. This situation changes markedly five miles north of Eastport, on the Lake Michigan coastline west of Atwood on U.S. 31, where the Nipissing terrace reemerges as a strong, wave-cut coastline feature diverging visibly from the topographically higher Algonquin shoreline. Here, Antrim Creek cuts through the Algonquin and Nipissing terraces and ultimately discharges into Lake Michigan. Low-lying, largely linear, coastal dunes are prominent for a mile north of Antrim Creek, within the Antrim Creek Natural Area. These eolian deposits are more active on the north bank of the creek, which has been heavily disturbed by both vehicle and foot traffic during the past decades. Similar to observations at other creeks where they discharge into Lake Michigan, archaeological sites are present at the mouth of Antrim Creek, where they occur at multiple elevations on the north bank. Deflated surfaces include lag deposits of Late Woodland period artifacts, this being expectable in such contexts.

Of particular interest in this county-managed parcel, however, is the presence of a deeply buried organic surface or paleosol that includes an archaeological

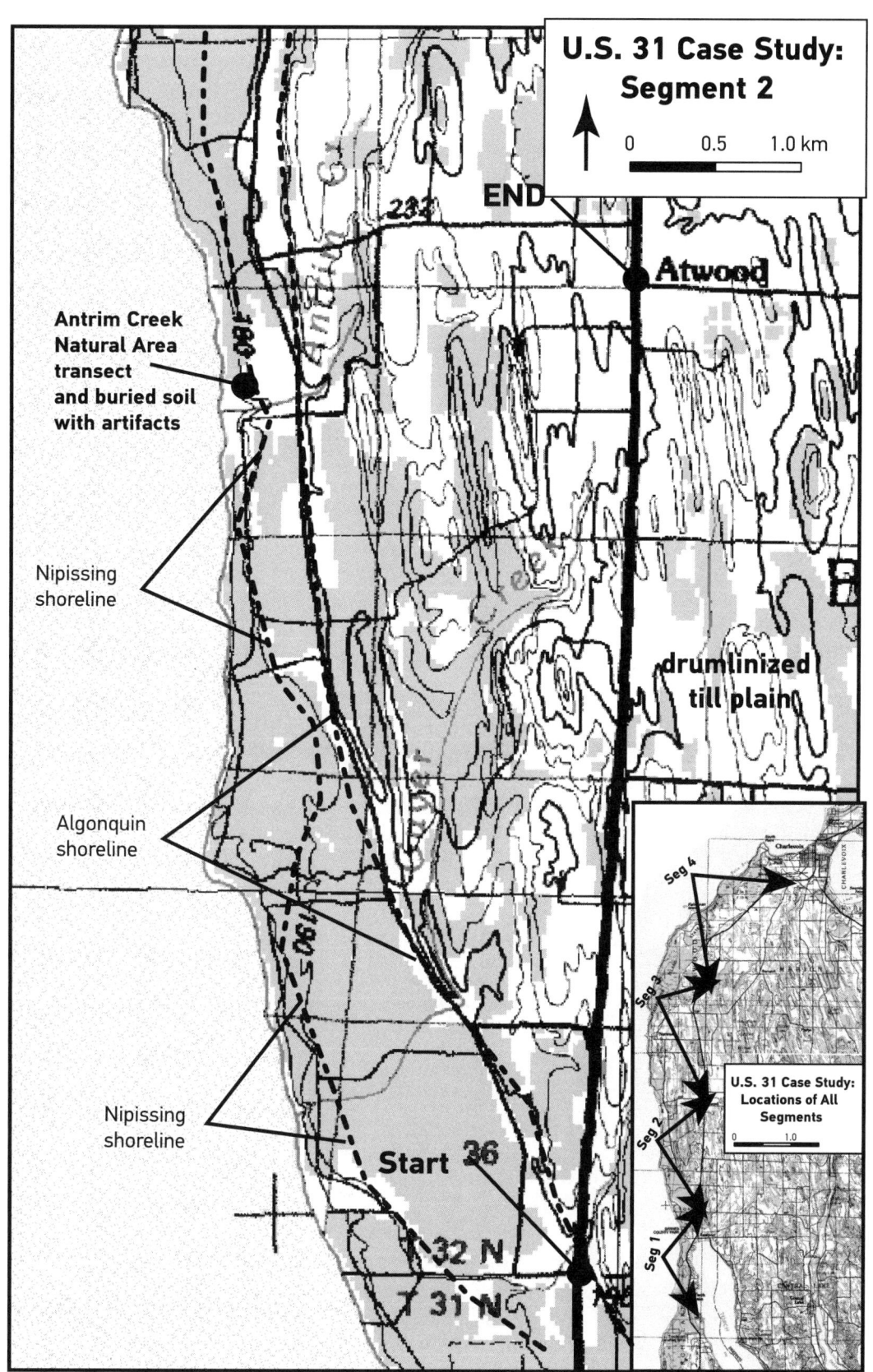

7-3. Geological and cultural features and locales of interest within U.S. 31 case study Segment 2.

assemblage that lacks ceramic artifacts, which suggests, albeit inconclusively, that it may be Archaic in age. The paleosol and associated artifacts occur within a foredune swale behind or inland of an eolian feature (Raviele 2006) that has been OSL-dated to circa 2,000 years ago (Holmstadt 2008). This age suggests that the dune was active circa 2,000 years ago and that, assuming an Archaic age for the cultural material, the buried surface (paleosol) predates the dune. The buried surface is discontinuous across the site, as both other test excavation (Raviele 2006) and subsequent sampling by Holmstadt (2008) reveal. Regardless of the age or association of the swale to the present dune, the fact that a buried site occurs in the predicted swale location is of importance. This observation indicates that any activities involving subsurface disturbance in dune swales north of Antrim Creek (or within similar geomorphological contexts) must consider the potential for early, buried surfaces to be present.

Segment 3: North of Antrim Creek to North of Whisky Creek

Our study lends little to this segment (figure 7-4) of the coastline, large parts of which are state land within the Pigeon River State Forest. Much like the area intervening between Eastport and Antrim Creek, there are few if any eolian landforms present. This lack of eolian landforms probably relates to the fact that bedrock is close to the surface in the northern part of this segment, and it is poorly drained. As discussed previously, both of these factors inhibit sustained dune growth. Our predictive models have little value here, although it should be recognized that the surface bedrock is home to two major quarry sites; the Pi-wan-go-ning Quarry site, and the Fritz Trail site, both locales being the sources of varieties of Norwood chert from Late Paleoindian times through Euro-American contact.

Eolian activity is evident once again, however, north of this area where Whisky Creek empties into Lake Michigan. The Whisky Creek site occurs on the south bank of Whisky Creek. As noted in other locations, the Whisky Creek site is deep, but heavily reworked; it does not have preserved organic horizons and is not stratified. Of some surprise is the fact that no site is recorded on the north bank of the creek mouth, as we might expect given our regional data. However, there is neither eolian activity nor evident dunal landforms on the north bank of the creek, suggesting a topographic situation unsuitable for preserved occupation surfaces. Future planning should be aware of the sensitivity of the area north of Whisky Creek, despite the lack of prominent dunes.

Segment 4: North of Whisky Creek to South Point

Except for about 2.5 square miles on the south, and South Point proper on the north, large parts of this segment (figure 7-5) are in State of Michigan ownership, lying within Fisherman's Island State Park Campground. The area has been well studied archaeologically, and has numerous known archaeological sites presented or discussed elsewhere in this book. Except for the zone directly along the Lake Michigan beach, few eolian features occur within an approximate two-mile stretch of shoreline north of Whisky Creek. Remnants of nineteenth- and twentieth-century limestone and gravel quarry activity do occur in this vicinity, primarily between Algonquin beach/bluff at circa 710 ft (200 m) asl and Nipissing shore features at

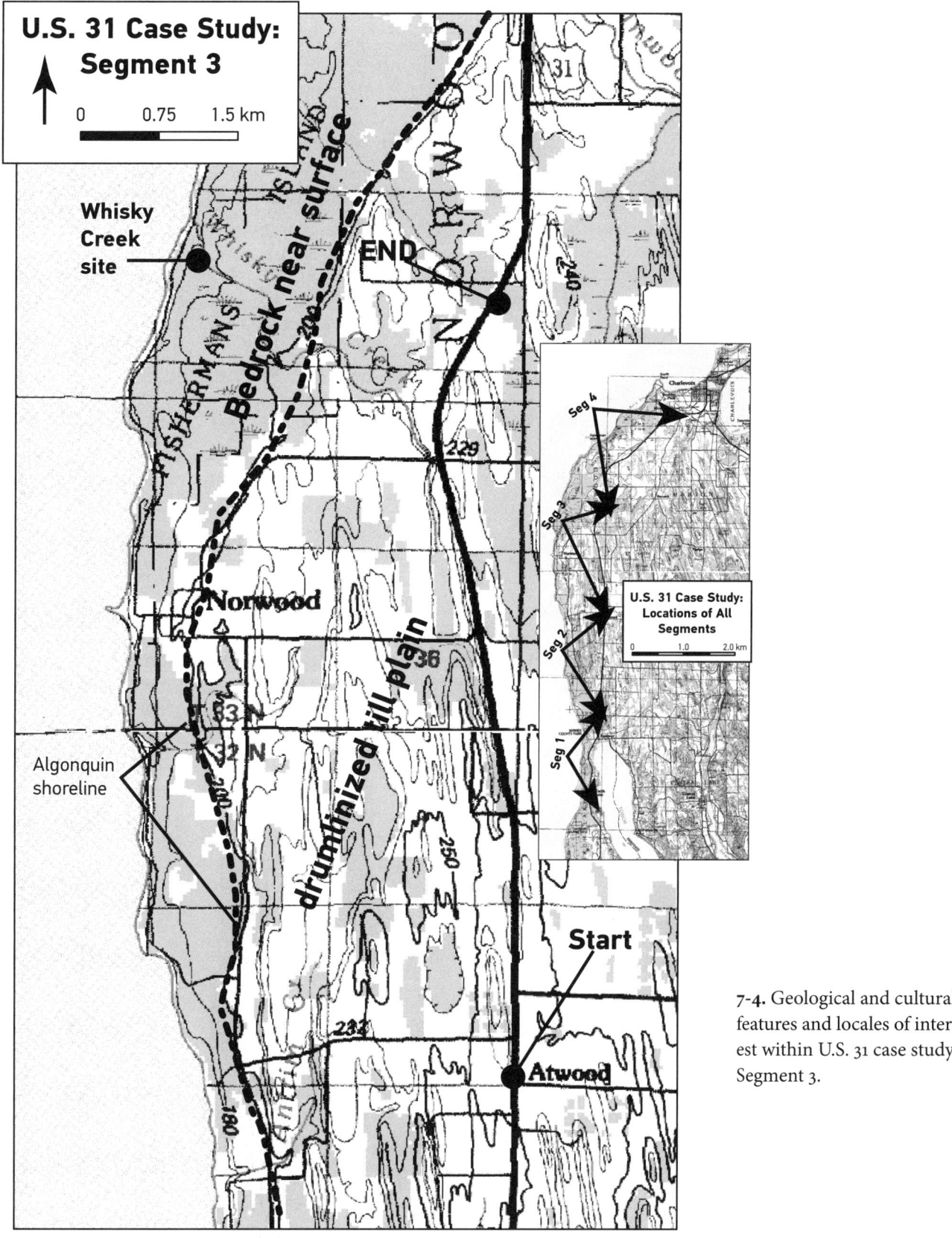

7-4. Geological and cultural features and locales of interest within U.S. 31 case study Segment 3.

circa 610 ft (185 m) asl. U.S. 31 closely follows the edge of the Algonquin terrace north of where it crosses the interior and upland portions of the Inwood Creek drainage. Several decades of concerted survey effort and research have failed to discover early sites along the Algonquin terrace.

Eolian landforms once again become evident two miles north of Whisky Creek, and continue intermittently to the northern edge of the case study transect. Unvegetated dunes fringe the coast, and are most prominent on the point of land at Fisherman's Island west of the mouth of Inwood Creek. Most of the remaining eolian landforms are vegetated to a greater or lesser extent. Of particular management import is the fact that there are two prominent dune series, one to the northeast of Inwood Creek, and one to the northeast of McGeach Creek. The intervening area has exposed gravels and bedrock, and is poorly drained. The largest dunes in each of these series greatly exceed the elevations of 611 ft (185 m) asl for Nipissing features, suggesting that they may overlay Nipissing terraces in some areas where the base of the dune lies above Nipissing elevations. OSL dating at the Fisherman's Island transect and the Solomon Seal site reveals that both of these dune series are post-Nipissing in age, had a substantial sand supply, and formed rapidly, and that the earliest age for their formation generally predates circa 3,000 years ago. Dunes closer to the lakeshore are sequentially younger, as is generally the case across all of our study transects in this northern subregion. It is these areas that also include buried and stratified sites ranging in age from Archaic to contact period. Some locations may have strandlines that can be associated with episodes of post-Nipissing stabilizations or high-water stands.

The management of dune series such as these, which also have high densities of archaeological sites interspersed between high-use park areas, poses particularly important issues. Our data reveal that site elevations are highly variable; they may occur at very low elevations as well as at the top of dune crests. There are partially stratified deposits at the Solomon Seal site at the northernmost point and highest elevation in the park, and behind the foredunes on the north bank of Inwood Creek, as well as in swales at higher elevations on the north bank of Inwood Creek but inland. Significantly, regardless of elevation, the uppermost stable buried surfaces often occur just below the modern ground surface, which means that land disturbance activities of any sort may potentially encounter cultural materials in either of these dune series. Particularly sensitive zones, however, are the swales behind foredunes; this is a sequential phenomenon, and all swales were potentially at one time behind foredunes of varying age.

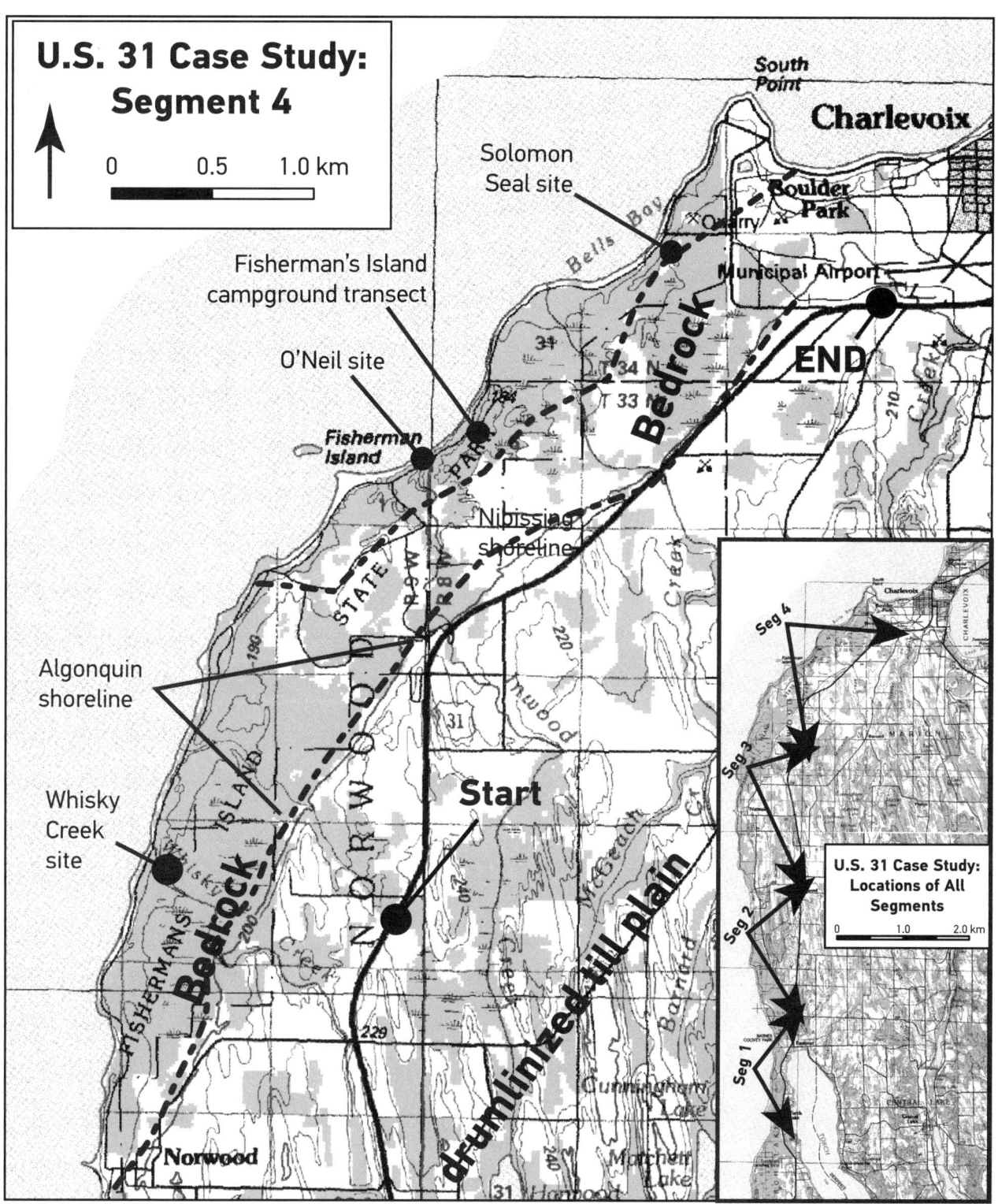

7-5. Geological and cultural features and locales of interest within U.S. 31 case study Segment 4.

APPENDIX A

Description of Methods Employed

Ceramic Residue Sampling and Other Organic Sampling Methods

Our decision to obtain absolute dates on previously undated archaeological sites resulted in the use of two related sets of procedures, one considered standard, and one of increasingly accepted utility and use. Standard or conventional ^{14}C dating is an established procedure that we employed at two locations, the Camp Miniwanca site and the Mt. McSauba site. Given that we had also sampled for OSL at both locations, this was an important comparative dating and calibration protocol for the project.

At the Camp Miniwanca site we had the good fortune to obtain an exposure of an organic horizon with cultural material buried 20 meters below the crest of a large parabolic dune. The exposure revealed a hearth feature with abundant charcoal and associated ceramics. A large sample of charcoal was removed from the hearth profile with clean metal instruments and placed in several aluminum foil envelopes. The envelopes with the samples were opened and air-dried in a fume hood at the Michigan State University Consortium for Archaeological Research. A small sample of consolidated charcoal was removed for identification of wood species by Professors Catherine Yansa and Frank Telewski of Michigan State University; it is *Picea* sp. The charcoal samples were subsequently consolidated and repackaged for shipment to Beta Analytic for a standard or conventional ^{14}C date.

The Mt. McSauba site revealed a similar parabolic dune exposure with cultural material in association with a buried organic horizon. The organic horizon was buried under multiple meters of eolian sand, and contained preserved organic material, some carbonized and some not. Similar field procedures were used for sample collection. Since we were unsure when the sample might be submitted for dating, it was kept frozen at Michigan State University until its transmittal to Beta Analytic for dating.

Carbonized food residues on the interiors of ceramic vessels have increasingly been used for a variety of purposes including absolute dating of small samples (Lovis 1990b, 1990c) and various dietary analyses (Hart et al. 2003; Morton and Schwarcz 2004; Schulenburg 2002). The issue of reservoir effects, which may affect the accuracy of Accelerator Mass Spectrometer (AMS) dating, has been raised and systematically addressed particularly for the northeastern and midwestern United States (Hart and Lovis 2007a, 2007b). We employed AMS dating of carbonized residues at two previously undated archaeological locales, the Winter site (Richner 1973) and the Scott Point site. Both sites are deep and stratified, and three residue samples were obtained from ceramics at each site. We dated identifiable rim sherds from the upper and lower parts of the sequences at both locations using procedures outlined in Lovis (1990b, 384). Only diagnostic identifiable rim sherds were used, only interior residues were collected if they were horizontally oriented accretions attributable to food preparation, and all collection was performed with clean stainless steel implements, onto laboratory glass, and then packaged for transshipment to Beta Analytic in glass vials. All sample collection was performed at the Michigan State University Consortium for Archaeological Research. We were able to augment the chronology at the Winter site with an OSL sample underlying the cultural sequence, again providing control and calibration on the different dating techniques.

All radiocarbon dates were calibrated with Calib v. 5.0.2 (Stuiver et al. 2006) employing IntCal04 (Reimer et al. 2004). All tests for contemporaneity and calculation of pooled means employing Ward's Method were undertaken with Calib 5.0.2 subroutines. Results of absolute dating procedures are presented in appendix B.

Conventions for Reporting the Ages of Samples from ^{14}C and Optical (OSL) Dating

Two different methods were used in this study and are commonly employed in the disciplines of geology, physical geography and archaeology. These dating techniques, however, are often used for different purposes and at different scales of resolution in each of the disciplines and, consequently, have different "meaning" to professionals in these fields. Although differences in the use and interpretation of dating methods across disciplines can be confusing enough, inconsistencies in reporting standards by each dating method lead to further misunderstandings. For example, the base year used for measurements of how old a sample is using ^{14}C methods is A.D. 1950, and for OSL it is typically (but not always) A.D. 2000. To minimize such potential confusion, an overview of how OSL and ^{14}C ages are typically reported is presented below. Additionally, the conventions for reporting, and the abbreviations for age notations used throughout this study, are summarized in table A-1.

Two methods were employed to determine absolute ages for episodes of sand dune formation and stability. In general intervals of activation for dunes are given by Optically Stimulated Luminescence (OSL), which generally yields the time that individual sand grains were last exposed to light (e.g., the time at which they were buried within dunes). Intervals of stabilization are given by ^{14}C age determinations on organic material derived from paleosols or cultural horizons buried within

Description of Methods Employed

TABLE A-1. Conventions for Reporting the Ages of Samples from ¹⁴C and Optical (OSL) Dating

ABBREVIATION	DATING METHOD	MEANING OF ABBREVIATION	RESULTANT AGE
ya	OSL	Years ago	Calendar years before A.D. 2000
kya (or ka)	OSL	1,000 years ago	1,000 calendar years before A.D. 2000
BP	¹⁴C	Radiocarbon years before present (uncalibrated)	Radiocarbon years before A.D. 1950 (not calendar years)
kyBP (or kBP)	¹⁴C	1,000 radiocarbon years before present (uncalibrated)	1,000 radiocarbon years before A.D. 1950 (not calendar years)
cal BP	¹⁴C	Calibrated years before present	Calendar years before A.D. 1950

the dune. Each of these methods is described below in greater detail. However, it must be recognized that each method also has different conventions for the reporting of mean ages and errors. For example, OSL age calculations yield time in calendar years without further manipulations. These ages are, by convention, reported as a mean age and 1σ error range around the mean. Typically such ages are reported as either years ago, using the abbreviation "ya" (e.g., 1500 ± 150 ya), or thousands of years ago, using the abbreviation "kya" (e.g., 1.5 ± 0.15 kya). In this book, the OSL ages are uniformly reported in years ago (ya) and errors are given as the 1σ range about the mean age.

Unfortunately no convention yet exists for the "year" that represents the "present" from which "years ago" is calculated for OSL ages. In the case of our study, two different laboratories were employed for OSL age determinations, and each laboratory uses different base ages. The University of Illinois–Chicago (i.e., UIC dates in appendix B), for example, uses 2000 as the base year. The University of Sheffield ages (i.e., Shfd in appendix B), on the other hand, are reported based on the year that the OSL date was performed (in our case, 2005 and 2006). Considering that errors in OSL age determinations are 5 percent–10 percent of the calculated ages, these differences in the reported mean age represent only a small difference and are not significant for our reporting or conclusions. These differences can, nevertheless, become confusing to other researchers. Moreover, this confusion can be particularly evident when OSL ages are compared with ¹⁴C ages, which have entirely different conventions and standards for age determination and reporting.

Carbon 14 dating results, as reported by various radiocarbon labs, determine time as "radiocarbon years before present," which are reported using the abbreviation "BP," rather than directly as calendar years. Calendar years for ¹⁴C ages are actually obtained by correcting the radiocarbon years reported by the laboratory for variations in ¹⁴C production through time. Which corrections and methods for adjusting radiocarbon ages to calendar years specifically employed for this study were discussed in the previous section. By convention, uncorrected radiocarbon ages are reported as the mean radiocarbon year BP followed by the 1σ error range about the mean age, and are usually given as radiocarbon years before present (BP; e.g., 1500 ± 40 BP). For historic reasons, the base age for calculating years BP is calendar year 1950, which is generally the year that radiometric dating was invented.

When ^{14}C ages are reported, both as uncorrected radiocarbon years BP and the calibrated calendar years before present (cal BP), they are typically both reported. However, because the actual errors for the cal BP ages are distributed uniformly about the mean calendar age, errors are reported as ranges of years and the probability that that range contains the actual year. Such calibrated ^{14}C dates are also given in years BP, but are indicated as of highest probability range and the abbreviation "cal BP" (e.g., 1520–1480 cal BP). When reporting the calibrated ages in the text of this study, we used only the age range that encompassed the highest likelihood as measured by area under the probability curve. However, for a list of all possible 2σ ages for a sample, see data in appendix B. Although it is not a standard way of reporting OSL ages, we have reported the OSL ages on illustrations as a range of the 1σ error range ya (e.g., 1470–1530 ya) so that they can be more easily compared with the range of ages reported for calibrated ^{14}C ages. However, regardless of style for listing their ages, OSL and uncalibrated ^{14}C ages have all been reported using the 1σ error range, while calibrated ^{14}C ages have been reported using the 2σ error range.

Optical (OSL) Dating (by Steven Forman, Luminescence Dating Laboratory, Department of Earth and Environmental Sciences, University of Illinois–Chicago)

Optical dating is ideal for eolian sedimentary sequences because the luminescence emission is a measure of the time between the last light exposure during grain transport and deposition and the burial period, when shielded from any further light exposure (Aitken 1998). The advent of single aliquot regeneration (SAR) protocols (e.g., Murray and Wintle 2003; Wintle and Murray 2006) for dating quartz grains provides improved accuracy and precision to yield new insights on the temporal and spatial patterns of eolian activity (e.g., Bailey et al. 2001; Ballarini et al. 2003; Forman and Pierson 2003; Clemmensen and Murray 2006). This analytical approach has been used previously to date Holocene eolian sediments on the Great Plains (e.g., Goble et al. 2004; Forman et al. 2005) and littoral sediments from the Great Lakes consistent with ^{14}C control and historic observations (Argyilan et al. 2005).

In this study only medium-to-fine eolian sand associated with primary dune bedforms was sampled for optical dating. SAR protocols (Murray and Wintle 2003) were used in this study to estimate the equivalent dose of the 250–350 or 150–250 μm quartz fraction for up to 72 separate aliquots (Murray and Wintle 2003, table 1). Each aliquot contains ~2,000 to 5,000 quartz grains and was adhered to an approximately 1 cm diameter aluminum disk. Eolian sands from coastal Michigan are mineralogically mature with SiO_2 content of > 80 percent, and are predominantly (> 80 percent) well-sorted quartz grains. The quartz fraction was isolated by density separations using the heavy liquid Na-polytungstate, and a 40-minute immersion in HF was applied to etch the outer 10+ microns of grains, which are affected by alpha radiation (Mejdahl and Christiansen 1994). Quartz grains were rinsed finally in HCl to remove any insoluble fluorides. The purity of quartz separate was evaluated by petrographic inspection and point counting of a representative aliquot. Samples that showed > 1 percent of nonquartz minerals were retreated with HF and rechecked petrographically. The purity of quartz separates was tested by exposing aliquots to infrared excitation (1.08 watts from a laser diode at 845 ± 4 nm), which preferentially excites feldspar minerals. Samples measured showed weak emissions (< 300 counts/second), at or close to background counts with infrared excitation, and ratio of emissions from blue to infrared excitation of > 20, indicating a spectrally pure quartz extract (cf. Duller et al. 2003).

An Automated Risø TL/OSL-DA-15 system (Bøtter-Jensen et al. 2000) was used for SAR analyses. Blue light excitation (470 ± 20 nm) was from an array of 30 light-emitting diodes that deliver ~15 mW/cm^2 to the sample position at 90 percent power. A Thorn EMI 9235 QA photomultiplier tube coupled with three 3-mm-thick Hoya U-340 detection filters, transmitting between 290 and 370 nm, measured photon emissions. Laboratory irradiations used a calibrated ^{90}Sr/^{90}Y beta source coupled with the Risø reader. All SAR emissions were integrated over the first 0.8 s of stimulation out of 500 s of measurement, with background based on emissions for the last 90- to 100-second interval. The luminescence emission for all quartz sands showed a dominance of a fast component (cf. Murray and Wintle 2003) with > 99 percent diminution of luminescence after 4 seconds of excitation with blue light.

A series of experiments was performed to evaluate the effect of preheating at 180°, 200°, 220°, and 240° C on thermal transfer of the regenerative signal prior to the application of SAR dating protocols (Murray and Wintle 2003). These experiments showed no preheat-based sensitivity changes, and a preheat at 220° C was used in SAR analyses. A test for dose recovery were also performed following procedures of Murray and Wintle (2003) with the initial and final regenerative dose of 2.39 grays yielding concordant luminescence response (at one sigma errors). Only aliquots with recycling ratios between 0.9 and 1.1 were retained in age calculations, indicating a minimal change in the sensitivity correction through the SAR dating protocols. Calculation of equivalent dose by the single aliquot protocols was straightforward with at least 29 aliquots measured for each equivalent dose determination (Murray and Wintle 2003, table 1). Equivalent doses for the 29+ large aliquots (1 cm plate area with 2,000 to 5,000 quartz grains) per sample showed a normal distribution with relatively low standard deviations of 4–8 percent of the mean. A determination of the environmental dose rate is needed to render an optical age. This dose rate is an estimate of sediment exposure to ionizing radiation from U and Th decay series, ^{40}K, and cosmic sources during the burial period (Murray and Wintle 2003, table 1). The U and Th content of the dose rate samples, assuming secular equilibrium in the decay series and ^{40}K, were determined by inductively coupled plasma-mass spectrometry (ICP-MS) analyzed by Activation Laboratory, Ontario, Canada. The beta and gamma doses were adjusted according to grain diameter to compensate for mass attenuation (Fain et al. 1999). A small cosmic ray component between 0.19 ± 0.02 and 0.03 ± 0.01 mGy/yr, depending on depth of sediment, was included in the estimated dose rate (Prescott and Hutton 1994). A moisture content (by weight) of 5 ± 2 percent or 10 ± 3 percent was assumed depending on current conditions. A wetter estimate was used if there was a subjacent aquitard, particularly a buried soil. The errors associated with determining equivalent dose by SAR protocols are usually < 5 percent and often yield an error in calculated age of ~10 percent or less.

Optical ages are reported in years prior to A.D. 2000 (table B-2). In turn all ¹⁴C ages for this and previous cited studies are calendar corrected (Reimer et al. 2004) to ease comparison to optical ages. There remains a small difference, 50 years, between calendar-corrected ¹⁴C ages and optical ages because of the different respective reference year, 1950 versus 2000.

Methods for Sampling of Dunes and Other Landforms for OSL Age Dating

Samples for OSL age determinations were collected to determine the age of erosion and final deposition of individual sand grains within dunes (see discussion above concerning optical dating methods). As is desired for optical samples, OSL samples were generally derived from below-soil A and B horizons (i.e., generally unweathered) and were collected in ways that avoided exposing the samples to light, which would "reset" the optical clock. Two procedures were followed to collect these samples. The first, which was employed in areas where sufficient exposure occurred to allow direct sampling of dune faces, used hand methods to clean exposure faces and collect the sand sample. The second was employed to collect deep or vertically discrete samples in places where no exposures existed and used solid-earth cores collected with a GeoProbe (model TR-54) coring device.

Where sufficient exposure existed, samples were derived by first cleaning any slump from the sand exposure to reveal a clean sand face. The faces were usually cleaned until primary bedding was apparent, which suggested that the exposure was deep enough to be below soil A or B horizons and was a fresh, intact dune profile. In some instances, shallow soil pits were excavated and the same criteria (above) were followed at these locations to ensure reliable ages for the samples. Once sample exposures were cleaned and the stratigraphy described, OSL samples were collected by first driving an optically opaque, circa 15–20 cm long, 50 cm diameter PVC pipe into the wall to fill it with sand. The exposed end was capped, the pipe extracted, and the other end capped. These samples were then sent to the OSL laboratory.

OSL samples were also collected from dunes using a GeoProbe (model TR-54) and a Dual-tube (model DT21) sampling system. With the dual-tube sampler, a core-casing is driven along with the sampler to prevent the collapse of the borehole during sampling. This procedure is ideal for colleting deep samples from below the water table or from landforms composed of very sandy sediment that lacks cohesion, such as dunes. A 122 cm long by 2.5 cm diameter, clear, PEX sample tube, which included a "core-catcher" at the end to hold the sand within the sample tube, was placed within the core-casing. This was driven the 122 cm length of the sampler into the dune. The sample tube was then extracted and the sediment described. Another section of core-casing was attached to the top of the previous casing and a new sample tube with the drive rod attached was placed into the casing. This new segment was then driven an additional 122 cm into the dune and the sample tube was extracted. In this manner, continuous cores (in 122 cm sections) were collected until the desired depth was reached.

Typically, two sets of cores were collected during the GeoProbe sampling process. The first core was used to establish the stratigraphy, determine where any buried soils occurred within the dune, and ensure that the coring did not penetrate deeper than the base of the dune. Using this set of cores, specific sample intervals were selected from which to collect OSL samples. Once the sample depths were determined, a second set of cores was collected by driving the dual-tube sampler to the appropriate depth and inserting a special PEX sample tube. This special sample tube was prepared by covering the normally clear PEX tube with "aluminum foil" duct tape to create an optically opaque sample tube, which ensured that the sample was not exposed to light. When extracted, the base of the OSL sample tube was capped, cut with a hacksaw about 50 cm from the tube base, and capped at the top. These samples were then sent to the OSL laboratory for processing. All GeoProbe bore holes were backfilled with Bentonite clay particles.

APPENDIX B

Tables of Radiocarbon and OSL Dates

TABLE B-1. ^{14}**C Ages**

SAMPLE LOCATION	LABORATORY NUMBER	ALTITUDE (M)	DEPTH (CM)	DATING METHOD	AGE* (BP)	ERROR* (YEARS)	CALIBRATED† AGE (CAL BP)	CALIBRATED RANGE†‡ (CAL BP)
Scott Point	Beta-237014	n.a.	<34	AMS	860	40	750	**689–803**; 808–831; 851–906
Scott Point	Beta-237015	n.a.	<30	AMS	870	40	780	**695–832**; 844–907
Scott Point	Beta-237016	n.a.	130	AMS	1240	40	1180	1068–1270
Winter	Beta-237017	n.a.	n.a.	AMS	1920	40	1870	1736–1763; **1769–1949**; 1963–1967
Winter	Beta-237018	n.a.	n.a.	AMS	1860	40	1820	1708–1883
Winter	Beta-237019	n.a.	n.a.	AMS	2090	40	2050	**1949–2152**; 2279–2286
Camp Miniwanca	Beta-238129	n.a.	~2000	AMS	820	40	730	**673–795**; 878–892
Cheeseman Road (South)	Beta-40288	185.6	100.0	^{14}C	2630	70	2752	2488–2644; **2665–2885**; 2908–2920
Cheeseman Road (South)	Beta-40289	185.6	140.0	^{14}C	6550	80	7461	7310–7580
Cheeseman Road (South)	Beta-40290	185.6	280.0	^{14}C	6430	70	7353	7178–7199; **7245–7474**
Crooked River (North)	Beta-40291	185.9	160.0	^{14}C	5440	70	6233	6004–6083; 6100–6160; **6170–6352**; 6366–6396
Crooked River (South)	Beta-40292	183.5	300.0	^{14}C	6450	80	7364	7178–7200; **7245–7508**; 7545–7554
Fishdam River (East)	Beta-40295	183.0	310.0	^{14}C	5330	70	6111	5944–5970; **5986–6278**
Fishdam River (East)	Beta-40296	183.0	130.0	^{14}C	400	60	441	314–523
Fishdam River (East)	Beta-40297	184.0	120.0	^{14}C	7140	80	7964	7796–7809; **7824–8161**
Manistique River (East)	Beta-40300	190.5	300.0	^{14}C	4920	80	5666	5478–5535; **5577–5798**; 5802–5893

APPENDIX B

SAMPLE LOCATION	LABORATORY NUMBER	ALTITUDE (M)	DEPTH (CM)	DATING METHOD	AGE* (BP)	ERROR* (YEARS)	CALIBRATED† AGE (CAL BP)	CALIBRATED RANGE†‡ (CAL BP)
Sturgeon River (West)	Beta-40304	187.0	260.0	^{14}C	850	60	771	**679–835**; 840–910
Torch Lake (Nipissing shore bluff)	Beta-40305	180.0	0.0	^{14}C	5380	70	6164	5993–6296
Mt. McSauba	Beta-251817	n.a.	n.a.	AMS	190	40	Modern	0–35; 70–117; **132–229**; 251–305
Portage	DIC-651	0.0	~30	^{14}C	2050	80	1995	1825–1854; **1856–2163**; 2166–2180; 2241–2303
Portage	DIC-652	0.0	~46	^{14}C	1830	120	1760	1419–1463; **1512–2043**
Portage	DIC-653	0.0	~50	^{14}C	1620	150	1535	1280–1873
Summer Island	M-1995	n.a.	n.a.	^{14}C	1700	140	1570	**1318–1898**; 1914–1918
Summer Island	M-2070	n.a.	n.a.	^{14}C	2320	140	2340	2008–2021; **2039–2735**
Summer Island	M-2071	n.a.	n.a.	^{14}C	660	100	575	508–776
Summer Island	M-2072	n.a.	n.a.	^{14}C	660	100	575	508–776
Summer Island	M-2073	n.a.	n.a.	^{14}C	1880	280	1825	**1264–2493**; 2600–2609; 2640–2680
Summer Island	M-2074	n.a.	n.a.	^{14}C	1790	130	1710	1407–1995
Summer Island	M-2014	n.a.	n.a.	^{14}C	330	100	Modern	0–32; 75–76; 83–97; 108–112; 137–224; **255–540**
Ekdahl-Goudreau	M-2311	180.3	24	AMS	870	120	780	564–590; **640–1005**; 1030–1052
Ekdahl-Goudreau	M-2312	180.3	55–58	AMS	1290	130	1195	**930–1416**; 1471–1481
O'Neil	M-2398	n.a.	~27	^{14}C	430	100	515	**287–568**; 583–649
O'Neil	M-2401	n.a.	n.a.	^{14}C	1000	140	930	**675–1182**; 1207–1233
O'Neil	M-2405	n.a.	~74	^{14}C	670	100	660	511–785
O'Neil	M-2406	n.a.	~67	^{14}C	740	100	680	**539–803**; 808–831; 852–906
O'Neil	N-1268	n.a.	n.a.	^{14}C	905	115	795	**661–1017**; 1022–1056
Wycamp Creek	M-2059	n.a.	~46	^{14}C	730	110	670	**532–832**; 846–907
Wycamp Creek	M-2060	n.a.	~84	^{14}C	240	100	Modern	0–42; 59–234; **238–483**
Wycamp Creek	M-2065	n.a.	~27	^{14}C	1320	120	1240	**964–1417**; 1467–1488; 1500–1507
Ludington Quarry #1	NSRL-3969	n.a.	600	AMS	3000	50	3195	3006–3014; 3030–3051; **3060–3346**
Ludington Quarry #1	NSRL-3970	n.a.	300	AMS	500	50	531	**475–565**; 588–643
Ludington Quarry #2	NSRL-3965	n.a.	400	AMS	150	50	Modern	0–286
Ludington Quarry #2	NSRL-3966	n.a.	500	AMS	2080	40	2048	**1947–2149**
Ludington Quarry #2	NSRL-3967	n.a.	300	AMS	2830	50	2941	**2792–3078**; 3134–3135

NOTES:

*Uncalibrated radiocarbon years. Error represents 1σ error ranges in radiocarbon years. Base year from which radiocarbon years before present (BP) is determined is A.D. 1950.

†Calibrations of ^{14}C ages after Stuiver et al. (2006) and Reimer et al. (2004) and represent 2σ age ranges. Base year from which calibrated years before present (cal BP) is determined is A.D. 1950.

‡Age ranges in bold represent intercepts with the largest area under the probability curve. Where only a single calibrated age range is present for a given date it is not bolded.

See appendix C for description of locations and exposures sampled.

TABLE B-2. OSL Ages

SAMPLE LOCATION	LABORATORY NUMBER	ALTITUDE (M)	DEPTH (CM)	DATING METHOD	AGE (YA)	ERROR (YEARS)
Wilderness State Park/Sturgeon Bay Point locale	Shfd-06138	n.a.	n.a.	OSL	920	90
Mt. McSauba	Shfd-06139	189.0	0.0	OSL	740	70
Eastport/Torch Lake (Township Hall locale)	Shfd-06140	n.a.	n.a.	OSL	5150	390
ACNA Dune 1b	Shfd-06141	n.a.	n.a.	OSL	1300	110
ACNA Dune 3	Shfd-06142	n.a.	n.a.	OSL	1950	180
ACNA Dune 2	Shfd-07001	n.a.	n.a.	OSL	2150	170
Winter site	UIC-2133	n.a.	~168	OSL	2180	215
Moran Dunefield (Sandpit locale)	UIC-2134	202.7	200	OSL	1950	205
Ekdahl-Goudreau	UIC-2135	180.3	147	OSL	575	70
Round Lake Dune locale	UIC-2136	198.1	0.0	OSL	930	90
Wilderness State Park (Roadcut locale)	UIC-2137	201.2	0.0	OSL	1930	225
Fisherman's Island State Park Solomon Seal site (lower)	UIC-2138	196.0	~300	OSL	3280	265
Fisherman's Island State Park Solomon Seal site (upper)	UIC-2139	196.0	~900	OSL	3380	300
Manistique Quarry Beach locale	UIC-2140	190.5	0.0	OSL	4280	390
Manistique Quarry Cliff Swallow locale	UIC-2141	193.6	0.0	OSL	2080	185
Manistique Quarry Beach locale	UIC-2142	190.5	0.0	OSL	1610	170
Fisherman's Island State Park South Loop Core 1 Campsite 66 Site 3	UIC-2143	191.1	~130–250	OSL	3240	260
Fisherman's Island State Park South Loop Core 1 Campsite 66 Site 3	UIC-2144	191.1	~500–700	OSL	2315	220
Fisherman's Island State Park South Loop Core 1 Site 4	UIC-2145	187.1	0.0	OSL	2420	240
Fisherman's Island State Park (high dune most inland)	UIC-2146	204.2	0.0	OSL	3260	305
TBNP Core 3 most inland dune below Nipissing	UIC-2147	185.0	400–500	OSL	3160	280
TBNP Core 2 2nd most inland dune	UIC-2148	192.0	~930	OSL	2490	230
TBNP Core 2 2nd most inland dune	UIC-2149	192.0	~240–360	OSL	2765	225
TBNP Core 1 lakeward dune	UIC-2150	179.5	~70–130	OSL	540	65
TBNP Core 4 Midslope West Aspect	UIC-2151	185.9	0.0	OSL	910	90
Camp Miniwanca Unit 1	UIC-2178	n.a.	n.a.	OSL	920	80
Camp Miniwanca Unit 2	UIC-2179	n.a.	n.a.	OSL	870	80
Ludington State Park Unit 1	UIC-2207	n.a.	~1950	OSL	2360	260
Ludington State Park Unit 2	UIC-2208	n.a.	~1800	OSL	2820	210
Ludington State Park Unit 3	UIC-2209	n.a.	~1600	OSL	2500	220
Ludington State Park Unit 4	UIC-2227	n.a.	~1500	OSL	2480	200
Ludington State Park Unit 5	UIC-2228	n.a.	~700	OSL	1510	130

NOTE: See chapter 5 and appendix C for description of locations and exposures sampled.

APPENDIX C

Descriptions of Sample Locales

This compendium of sample locale descriptions and summaries contains many of the primary observations on both the archaeological and nonarchaeological locales sampled, reviewed, reassessed, or reanalyzed from 2006 to 2008 as part of the ISTEA buried site taphonomy/dune activation and cycling project. The presentation is organized geographically around the Michigan locations in the Lake Michigan basin. The first site summarized, Summer Island, is the most western site along the northern coastline of Lake Michigan. Site descriptions then proceed eastward to the Straits of Mackinac, and then southward to our most southern sample point in Mason County. Six other sites have already been presented in chapter 5. As will be evident, the specific content of each site or locale summary is variable, although each contains a synopsis of past and current research, a detailed statement of the depositional sequence and formation processes, and any absolute ages (AMS, ^{14}C, OSL) associated with either the geological deposits (paleosols, sands) or the cultural residues (features, artifacts). Dependent on the availability of reports or published materials, other aspects of the summary may vary. For example, the Wycamp Creek site, among others, has never been published and only has field notes associated with it. This and other sites are thus less well described than others. At the opposite extreme the Summer Island site has numerous formal publications and has been summarized extensively in other venues; we chose not to be overly redundant with the available literature.

Summer Island Site (20DE4), Delta County
(with contributions by Marieka Brouwer)

Site Location and Description

The Summer Island site is located in Section 27, T37N, R19W, Fairbanks Township, Delta County, Michigan (figure C-1). This site is the southernmost of all we

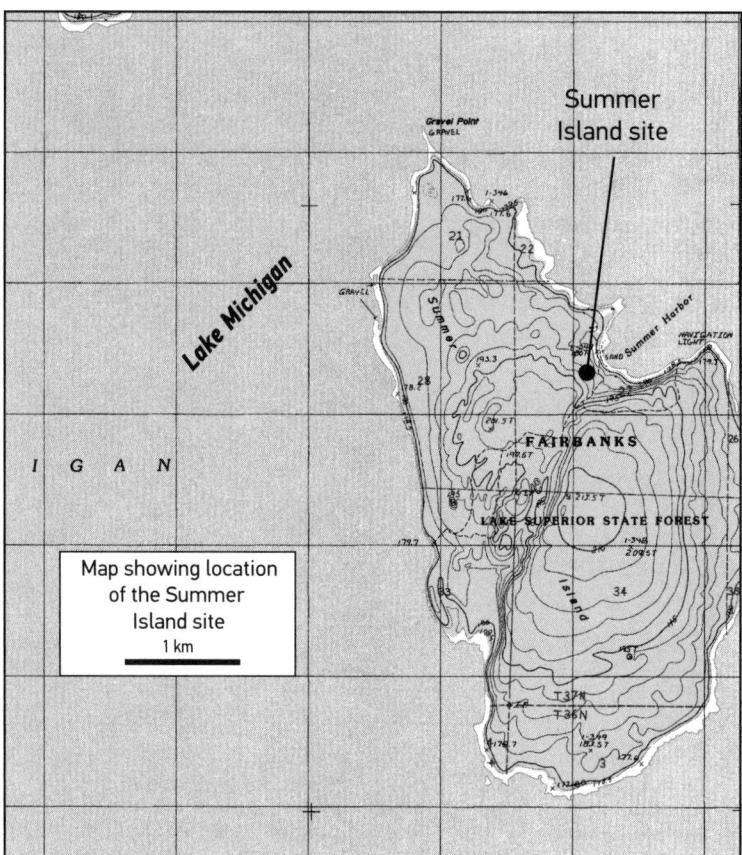

C-1. The Summer Island site location.

investigated along the northwestern part of the Lake Michigan basin. It is also just north of the "hinge line" or zero Isobase in the Lake Michigan basin. The site is located on an embayment on the east coast of Summer Island, the northernmost of a chain of eight outlying islands that span the mouth of Green Bay. Topographically, the Summer Island site is a fairly level sandy meadow, dipping downslope from the edge of the forest to a rather steep sand bank that drops about 4 m to a wide sand beach running for 50 m along the harbor. Although elevation varies a great deal across the site, the middle of the site designated Area "B" by the investigators lies at an elevation of about 181.2 m (594 ft) asl.

Not only has the Summer Island site been known since at least the 1850s (Schoolcraft 1851), it has been the subject of substantial research during the past half century, and numerous publications both primary (Binford and Quimby 1963; Brose 1964, 1970a, 1970b, 1970c) and secondary (Brose and Hambacher 1999; Fitting 1975; Martin 1985) are available detailing work performed at the site. There is also a substantial radiocarbon record available from the site (Crane and Griffin 1958, 1959, 1960, 1961, 1963, 1964, 1965), which is presented in appendix B in this book. Given this abundance of detailed information, the site will be given only a limited summary here.

George I. Quimby of the UMMA visited the Summer Island site in 1959, made surface collections, and excavated several test pits with a view toward understanding the stratigraphic sequence (Binford and Quimby 1963, 277–307). Quimby's visit was followed in 1963 by two survey parties from the UMMA, both part of the Upper Peninsula Archaeological Survey. Barry Kent and David Griffin excavated three test pits at Summer Island, while G. Richard Peske and Dan Higgins excavated a 5- by 10-foot profile trench and two test pits (Brose 1970a, 18). The site was clearly stratified and contained a major Upper Mississippian component reminiscent of occupations on the Door Peninsula, Wisconsin. In 1965 a field crew under James Fitting and John Speth surface-collected the deflated dune areas at Summer Island.

Given the evident regional significance of the Summer Island site, major excavations were planned by the UMMA. In spring 1967 David Brose undertook preliminary site survey, followed by formal excavations between 2 July and 25 August, 1967 under Brose's direction.

Stratigraphy and Depositional History

The Holocene soil stratigraphy of the Summer Island site, which overlies a dolomite foundation, is deep, complex, and also well summarized in Brose (1970a). The first 10 stratigraphic units defined by Brose as Units A through J, precede the human occupation of the site and present a complex sequence initiating with the Valders retreat from the Michigan basin. This sequence is significant only inasmuch as it documents the deposition of sand units at the site during Algonquin and Nipissing high-water stands, sand that ultimately underwent eolian activation and periodic stabilization. Of particular interest to this study is the presence of buried A1b (Unit G) and A2b (Unit J) horizons in close vertical proximity. The first evidence of human occupation and use of the Summer Island site occurs in Unit K (figure c-2).

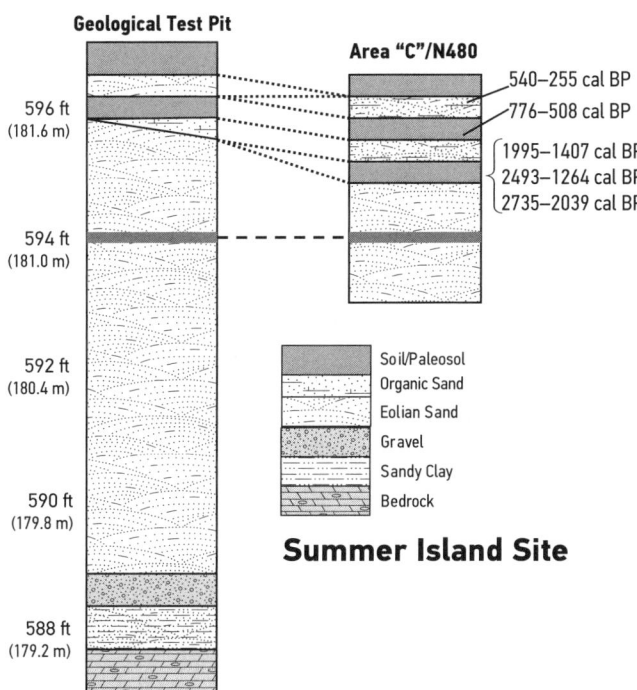

c-2. Generalized, composite stratigraphic column of the Summer Island site. Ages of various units shown. See appendix A for discussion of how ages are reported. *Note:* unit depths and boundaries are approximate.

- *Unit K*—This unit consisted of three discontinuous but homogeneous levels of medium sands. The sands of Unit K were stained black by organic materials and, according to Brose, the unit displayed substantial humic acid content, while the soil matrix further revealed variably sized fragments of bone, shell, charcoal, other charred organics, small fragments of ceramics, fire-cracked igneous rock, and small flakes of crypto- and microcrystalline raw materials. High frequencies of grit-tempered ceramics, animal bone, chipped-stone tools and debitage, fire-cracked rock, and charcoal were found in this matrix, both in spatial concentrations and distributed more uniformly throughout the unit. Pits or postholes emanating from this depositional unit frequently intruded into lower stratigraphic units (see Brose 1970a, figures 7 and 8). Spatially, Unit K was confined to the area designated Area "C" of the site, with a maximum thickness of 12 cm toward the centers of uniformly level lenses that thinned toward the margins of the deposit (Brose 1970a, figure 4). Three conventional radiocarbon assays reveal occupation circa 2,700–1,300 years ago.

  As chapters 2 and 5 reveal, Unit K is representative of human occupation potentially inducing stabilization of eolian deposits through the formation of Anthrosols. Unit K does not appear to be a natural depositional unit, but rather appears to be a sheet midden indicative of intensive human occupation introducing a variety of organic and inorganic, burned and unburned materials into the eolian surface. According to Brose, eolian sand comprised about half of the Unit K deposit, the remainder being culturally introduced materials present in the form of refuse on the living floors (Brose 1970a:34–37).
- *Unit L*—Unit L represents an intervening eolian deposit, generally culturally sterile, and probably a backslope dune formed when the occupation of Unit K ended (Brose 1970a, 37).
- *Unit M*—Unit M consisted of multiple lenses of dark, grayish-brown, medium sands lying unconformably upon the surface of Unit L. Unit M contained evidence of human occupation in the form of shell-tempered Oneota ceramics, fire-cracked rock, and chipped stone tools and debitage; faunal remains were infrequent. In some instances the cultural pits and postholes originating in Unit M intruded as deep as Unit I. Brose's microscopic and distributional analyses of Unit M sand particles revealed eolian redeposition of what were originally lacustrine deposits on a deflated Unit L surface (Brose 1970a, 37).
- *Unit N*—Unit N occurs across both Area "B" and Area "C" at Summer Island as a deposit of yellowish-brown to very dark brown medium-fine sands occurring in discontinuous cross-bedded lenses. Again, as discussed elsewhere, such discontinuous surfaces appear to be common in Lake Michigan coastal dune occupations. The depositional environment as reconstructed by Brose prompted analogies with the sandy meadow present during the fieldwork, that is, deflated areas on the backslopes of low eolian dunes. Small, dark soil lenses within Unit N may indicate local vegetation-stabilized areas. As with other depositional units with evidence of human occupation, cultural materials were incorporated into the Unit N deposits. These include grit-tempered pottery, chipped flint projectile points and debitage, and historic European trade goods. Immediately overlying Unit N, and across the entire site south of Area "A" at depths of 4 to 13 cm below surface there was a clearly defined plow zone (Ap).

This plow zone had the effect of leveling the site surface, at times truncating higher parts of Unit N (Brose 1970a, 37–38).
- *Unit O*—This unit consisted of the present humus (A horizon), disturbed by plowing with subsequent deflation, eolian redeposition, and disturbance from historic logging operations. Unit O contained cultural materials related to all of the site occupations (Brose 1970a, 38).

Assemblage Composition and Site Significance

The Middle/Initial Woodland Laurel-like occupation appears to have been a series of annual spring through early autumn occupations, estimated by Brose to be composed of between 25 and 40 individuals organized into six families of a small, exogamous, patrilocal band. Three conventional ^{14}C dates reveal that this occupation most likely occurred circa 2,700–1,300 years ago. Primary site economic activity in spring relates to its use as a fishing station where spawning sturgeon were both netted and speared. During late summer, bear and beaver were hunted or trapped, and individualized line fishing was practiced. Brose opines that the seasonal abandonment of Summer Island coincides with the onset of wild rice availability.

Ceramic styles clearly relate to the Northern Tier Middle/Initial Woodland tradition; consistent with the ^{14}C assays. The use of bipolar, block, and pebble core techniques most likely relate to extensive utilization of small packages of locally available chert. The remainder of the Middle/Initial Woodland assemblage is typical of this time period, consisting of stemmed and side-notched projectile points; small unifacial "thumbnail" scraper varieties; ceremonial, ornamental, and utilitarian annealed copper artifacts; bone needles, awls, and net-shuttles; antler harpoon-heads and drifts; and ground stone artifacts including net-sinkers, hammers, anvils, and mortars.

According to Brose the Late Woodland occupation is indicative of a series of short-term midsummer occupations by small family groups utilizing Summer Harbor as a temporary base camp sustained by mollusks and fish extraction along with deer and small mammals. Two conventional ^{14}C dates reveal that these occupations occurred circa 780–500 years ago. The ceramic assemblage is reminiscent of Oneota-related ceramics in the circum–Green Bay area that combine the resident early Late Woodland tradition with an intrusive Upper Mississippian tradition. Interestingly, the chipped stone industry does not change markedly from its Middle/Initial Woodland antecessors, consisting of bipolar, block, and pebble cores reduced from local chert nodules, triangular projectile points classified as Late Woodland Madison points along with a side-notched Cahokia point, high proportions of unifacial "thumbnail" scrapers, and limited frequencies of bone tools.

The final occupation of Summer Island, chronologically associated with a radiocarbon date of circa 575 years ago, took place from the summer through the fall, probably by about 20 to 30 individuals. The site may have had multiple functions both within the local seasonal economic cycle, for example as a residential hunting camp, and within the larger arena of the postcontact economy as an aggregation locale for those engaged in a European-stimulated fur trade.

Perhaps as a consequence of this larger economic arena the ceramics display stylistic relationships with several late precontact groups in the areas around the Lake Michigan basin. Additionally, the presence of other styles more common to southern Ontario may indicate broader-ranging trade relations, increased regional mobility, and exchange system expansion, or even postcontact population displacements.

In contrast with the ceramic assemblage, however, the chipped stone industry displays continuity, emphasizing local raw materials, the production of triangular projectile points, a dominance of small "thumbnail" scrapers, along with limited copper-working including ceremonial, ornamental, and utilitarian items. Finally, groundstone celts, inferred to be woodworking tools, are present. By far the most distinctive aspect is the assemblage of European-manufactured materials, including glass beads, tinkling cones, bells, and a brass ring, along with iron awls, needles, knives, brass kettles, and a brass thimble. Gunflints, lead shot, and other gun parts round off the assemblage. Brose laments that "the lack of precise historical documentation and the population displacements caused by the fur trade have prohibited" directly linking the occupants with specific ethnolinguistic groups (1970a, 218).

Manistique Sand Quarry, Schoolcraft County

The Manistique sand quarry is located in the NE ¼, NE ¼, Section 7, T41N, R15W, Manistique Township, Schoolcraft County. It occurs about 1 km northeast of the city of Manistique on the north shore of Lake Michigan (figure C-3). The quarry is on the south side of Tannery Road and is on the northern margin of a dune field that extends from the north side of U.S. 2 to a point about 0.5 km inland (figure C-3). Dunes in this field consist of several poorly defined ridges, transverse dunes, and parabolic dunes that range in height from 1 m to 10 m. Quarry operations at the site have exposed about 11 m of sediment (figures C-4 and C-5). The lower portion of the exposure consists of probable lacustrine sands that are capped by an about 15 cm thick gravel lens (figure C-5). Overlying this gravel is a sand unit that is about 30 cm thick. This sand unit is, in turn, capped by yet another gravel bed that is about 25 cm thick. This uppermost gravel is buried by about 7 m of eolian sand that contains no buried soils. Formed in the uppermost part of the dune is a truncated Spodosol with E/Bhs/Bs/C horizonation. Although quarry operations appear to have stripped the A horizon and perhaps the upper ½ of the E horizon, the remaining soil is nevertheless about 1 m thick.

In an effort to reconstruct the depositional chronology at the Manistique sand quarry, three OSL ages were obtained from sediments in the exposure (figure C-5). The lowermost sample was derived from the sand unit that lies between the gravel beds in the lower part of the exposure. This sample provided an age of 4280 ± 390 ya (UIC-2140). In addition to this age assessment, a pair of samples was acquired from the eolian sands that cap the uppermost gravel bed. The lowermost of these samples was obtained from the base of the dune and yielded an age of 1610 ± 170 ya (UIC-2142). The uppermost age sample was extracted from a depth of 2 m and provided an age of 2080 ± 185 ya (UIC-2141).

Ages obtained from the Manistique sand quarry indicate that the following depositional chronology likely occurred. The basal sands and gravels were apparently

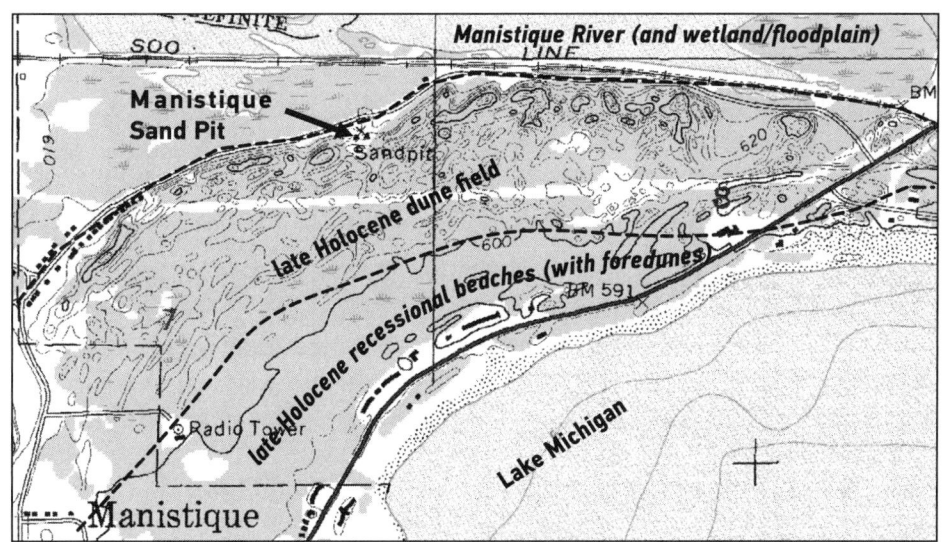

c-3. The Manistique Sand Pit location.

c-4. The two sample locales at the Manistique Sand Pit locale.

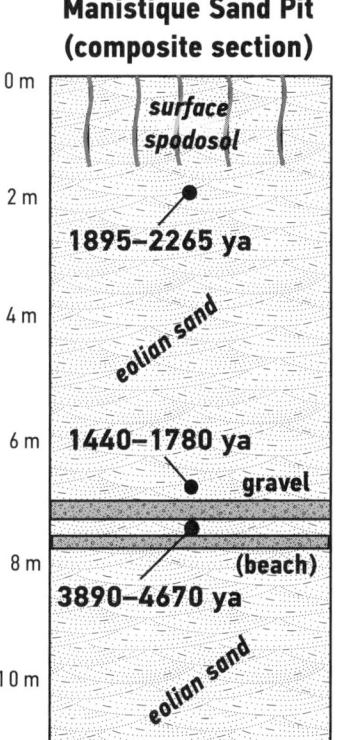

c-5. Generalized, composite stratigraphic column of the Manistique Sand Pit. Ages of various units shown. See appendix A for discussion of how ages are reported. Unit depths and boundaries are approximate.

deposited during the Nipissing phase of ancestral Lake Michigan, as the age of circa 4,300 years ago from the sand bed that lies between the two gravel layers suggests. These gravel beds, in turn, imply that the active beach must have been at this locality at least twice during this high stand.

After the Nipissing stage ended, lake level dropped to an elevation below the uppermost gravel bed. This regression marks the last time that lake level reached this elevation. The OSL ages from the overlying eolian sands indicate that the dune grew quickly circa 2,100–1,600 years ago. The ages are unfortunately stratigraphically inverted, with the youngest of the pair being in the deeper position in the exposure. Both of the ages lie within the same range of statistical probability, however, which supports the overall age estimate for the timing of sand deposition. The dune has been stable since that time, as evidenced by the strong development of the surface soil.

The chronostratigraphy at the Manistique sand quarry raises an interesting question regarding the circa 1,500 years between the deposition of the uppermost gravel bed circa 4,300 years ago and the onset of dune formation circa 1,800 years ago. Given that the dune sands directly overlie the gravel, it is likely that some kind of erosional episode occurred during this intervening time, one that removed perhaps a thin unit of eolian sand that had previously buried the gravel shortly after the Nipissing high stand. Otherwise some kind of deposit, or a soil, should be preserved from this interval of time. Given its apparent absence, the most logical scenario is that some kind of erosional event occurred before the dune formed. It is possible that this missing unit was remobilized during the early stages of dune growth circa 1,800 years ago.

Moran Dune Field, Mackinac County (Moran and Round Lake Dunes Sample Locales)

The Moran dune complex is located in the southeastern part of the Upper Peninsula along the northern shore of Lake Michigan (figures C-6 and C-7). The center of this dune cluster is about 10 km northeast of St. Ignace in the southern part of Moran Township. The dune field is between 8 and 10 km wide, with the western margin located about 2 km north of Pointe aux Chenes. Dunes in the field consist of a variety of widely spaced ridges and individual dunes. The dune ridges range from sublinear landforms that show no evidence of reactivation to ridges with numerous parabolic dunes embedded within them. Individual dunes usually consist of parabolic dunes with arms that are generally oriented to the southwest.

In an effort to ascertain the general age of the dunes in the Moran dune field, a pair of OSL ages was obtained from representative landforms on the eastern and western sides of the cluster. The easternmost of these ages was acquired from a sand quarry or pit in the NW ¼, Section 27, T41N, R4W, Moran Township, which is about 0.25 southwest of Hay Lake (figure C-7). This quarry is in the southeastern edge of a prominent dune ridge that extends about 5 km to the northwest. This ridge contains numerous parabolic dunes that are embedded within the landform. These embedded dunes have limbs that are oriented to the southwest.

Exposed in the western wall of this quarry is about 4.5 m of eolian sand (figures C-8 and C-9). Other than the surface soil, which has A/E/Bhs/Bs/C horizonation, no soils were observed in the exposure. This stratigraphy indicates that the sand

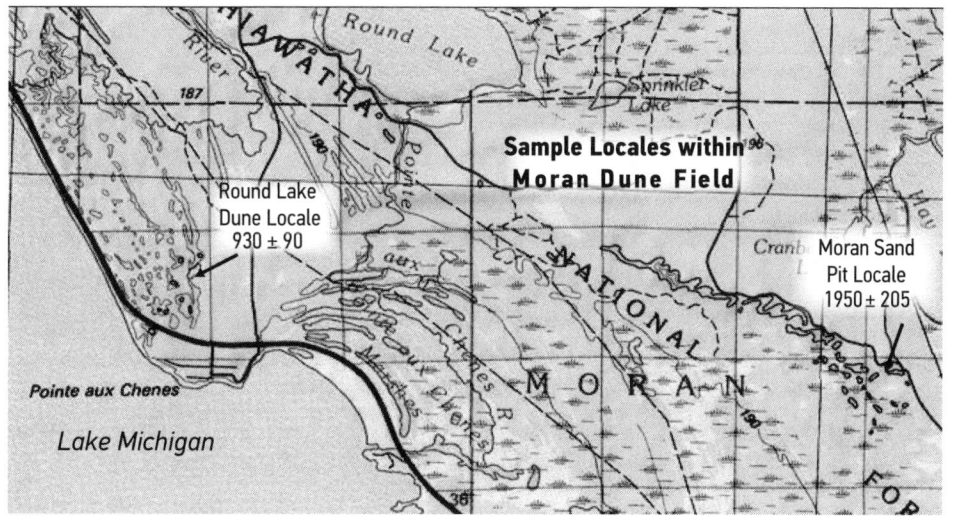

c-6. The Moran and Round Lake dunefield sample locales.

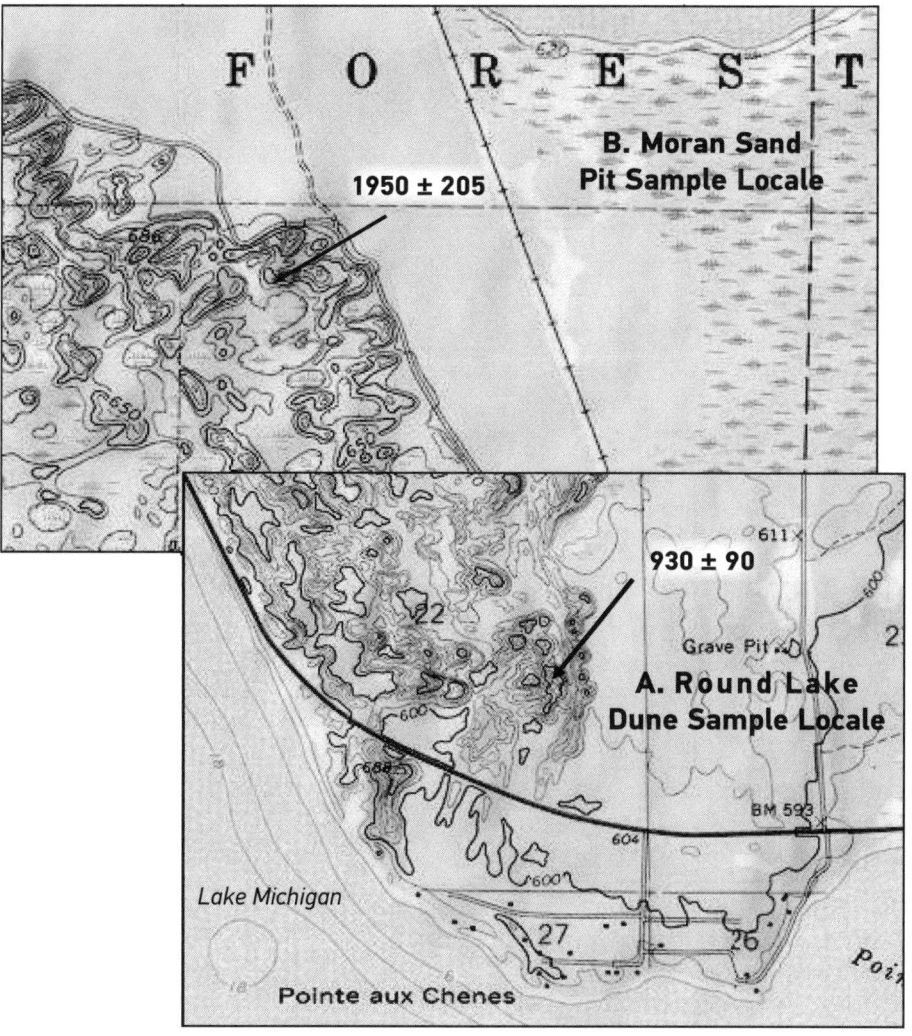

c-7. Detailed maps of the Moran Sand Pit and Round Lake dune sample locales.

C-8. The Moran dunefield sites showing the Moran Sand Pit and Round Lake sample locales.

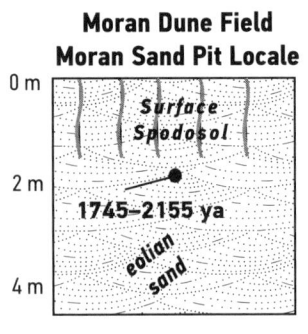

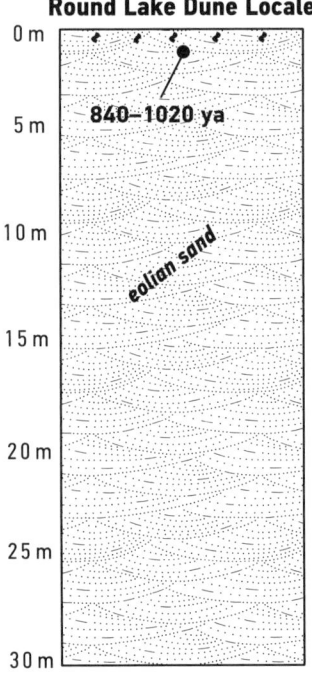

C-9 (LEFT). Generalized, composite stratigraphic column of the Moran Sand Pit and Round Lake dune samples locales. Ages of various units shown. See appendix A for discussion of how ages are reported. Unit depths and boundaries are approximate.

accumulated in one depositional event. In an effort to estimate when this event occurred, a sample for OSL dating was obtained from a depth of 1.85 m and yielded an age of 1950 ± 205 ya (UIC-2134).

The second OSL age estimate was derived from the western edge of the Moran dune field (i.e., Round Lake Dune, figures C-8 and C-9). This age was obtained from a soil pit excavated in the crest of a large parabolic dune in the NE ¼, Section 22, T41N, R5W, Moran Township. This dune is embedded within a well-defined dune ridge that is about 3 km long (figures C-7 and C-9). The sample was collected from a depth of 1.5 m within the pit and provided an age of 930 ± 90 ya (UIC-2136).

The topographic relationships of the Moran dune field, coupled with the associated OSL ages derived from dunes on either side of the area, suggest the following chronology of dune landform evolution in this part of Upper Michigan. It appears that much of the area was submerged during the Nipissing high stand of ancestral Lake Michigan. During this time, it seems that no dunes were constructed in the area. Lake levels subsequently fell, resulting in a complex wetland landscape. The first documented episode of dune formation in the Moran area occurred about 2,000 years ago when the easternmost ridge in the complex developed. Given the distinct parabolic forms embedded within this ridge, however, it is highly possible that the ridge was initially constructed sometime before and that the OSL age derived from it represents the most recent reworking of the landform.

Following the episode of eolian activity circa 2,000 years ago, the development of dunes apparently slowed or stopped entirely for at least 1,000 years. This hiatus may be related to high lake levels circa 1,700 years ago that may have raised local water tables sufficiently to retard dune formation. After this hiatus, water levels dropped and the western side of the dune complex became the focal point of eolian sand mobilization and dune growth. An OSL age of circa 1,000 years ago was derived from the easternmost ridge in this part of the Moran dune field (i.e., Round Lake dune locale; figures C-6 and C-7). Given the parabolic dunes embedded within the dune ridge, this date may reflect secondary reworking of a ridge that actually formed shortly before. Dunes to the west of this ridge have likely formed since 1,000 years ago.

Wilderness State Park Dune Field, Emmet County

The Wilderness State Park dune field is a large complex of dunes in the northwest part of Lower Michigan (figure C-10). This dune field is approximately 15 km² in size and generally lies between Cross Village, on the southern margin of the dune complex, to Cecil Bay at the northern end of the field. At its widest point, just north of Wycamp Lake, the dune field extends from the coast to a point about 3.2

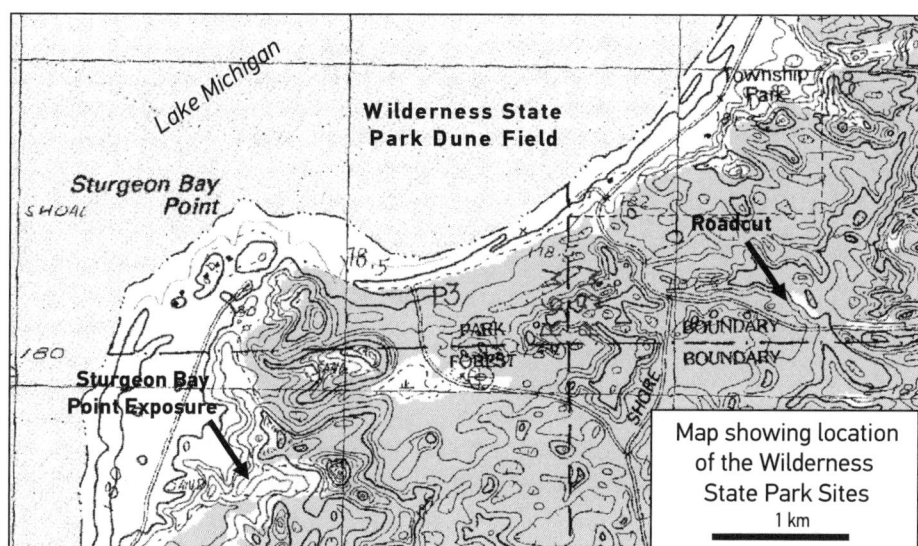

C-10. Wilderness State Park sample locations.

C-11. Wilderness State Park sample exposures.

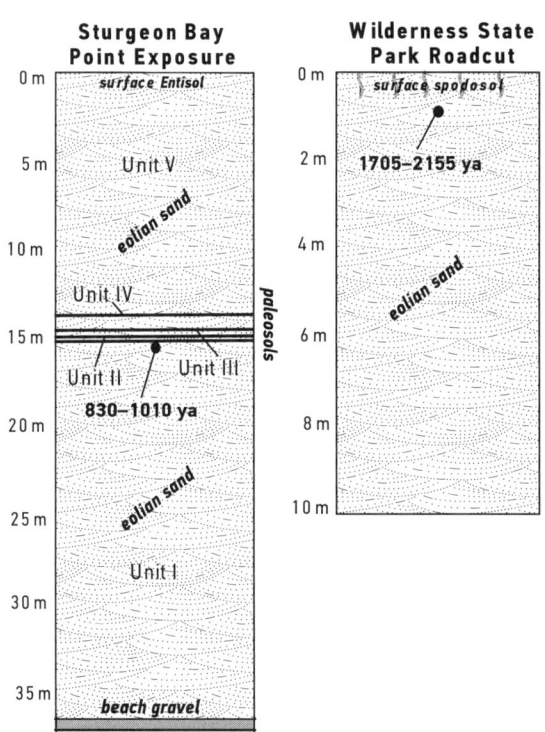

C-12. Generalized, composite stratigraphic columns of the Wilderness State Park exposures. Ages of various units shown. See appendix A for discussion of how ages are reported. *Note:* unit depths and boundaries are approximate.

km inland. A variety of dune forms occur in the complex, including dune ridges, parabolic dunes, transverse dunes, and shadow dunes. Although most of the dunes are stabilized with forest, many are active with large areas of unvegetated sand.

In an effort to determine the general age of the dunes, two OSL ages were acquired from within the dune field (figures c-11 and c-12). The most northerly of these sample sites was a road-cut exposure in the SE ¼, SE ¼, SW ¼, Section 18, T38N, R5W, Bliss Township, along the north side of Sturgeon Bay Trail. This site, which is about 20 m high, is located in the approximate middle of the dune field along an east/west transect, about 800 m southeast of the coast. Although much of the exposure was disturbed due to off-road traffic, it appears that the dune contains a single unit of eolian sand because the surface soil is a moderately developed Spodosol with A/E/Bs1/Bs2/C horizonation. The OSL sample was obtained from the C horizon below this soil at a depth of 0.9 m and yielded an age of 1930 ± 225 ya (UIC-2137).

In addition to the Sturgeon Bay Trail site, another section was described and sampled in the Wilderness State Park dune field. This site, which is referred to as the Sturgeon Bay Point exposure (figures c-10, c-11, and c-12), is located about 600 m east of the coast in the SW ¼, NE ¼, NW ¼, Section 24, T38N, R6W, Cross Village Township. This exposure faces west and lies in the core of a large parabolic dune. The dune overlies beach gravels that were probably deposited about 3,200 years ago. Overlying this deposit is about 37 m of eolian sand that contains five depositional units separated by buried soils (figure c-11). Unit I directly overlies the beach gravels and is about 21 m thick. Formed in the top of this unit is a weakly developed Entisol with A/C horizonation. Immediately above this unit are two thin deposits (Units II, III) of eolian sand that are each about 25 cm thick. Each of these units is capped by an Entisol with A/C horizonation. Unit IV covers these deposits and is about 1.4 m thick. This unit is also capped by an Entisol with A/C horizonation. The uppermost unit of eolian sand is Unit V, which is about 14.7 m thick and extends to the top of the dune. This unit contains a very weakly developed surface soil with A/C horizonation. To test the basic age of the dune an OSL sample was collected from the uppermost part of Unit I at a depth of about 16 m. This sample lies just below the 5Ab horizon and provided an age of 920 ± 90 ya (Shfd-06138).

A combination of topographic assessment and two OSL ages provides a general chronological framework for the dunes in the Wilderness State Park dune field. Given the location of the Sturgeon Bay Trail road cut in the approximate midpoint of the dune field, it appears that the dunes east of this site are older than circa 2,000 years ago. Given the landscape position of the easternmost dunes, they are estimated to be circa 3,500 years old. Dunes in the central part of the field are apparently circa 2,000 years old. Based on observations elsewhere in this study, we believe that these dunes accumulated fairly quickly in distinct depositional events.

The age and stratigraphy at the Sturgeon Bay Point exposure indicate a slightly different depositional history for this part of the dune field. The onset of dune formation apparently occurred about 1,000 years ago, resulting in the deposition of 21 m of eolian sand. Shortly after this unit was deposited the dune stabilized for a brief period of time, resulting in the formation of the Entisol in the uppermost part of the deposit. Following this period of stability deposition of eolian sand occurred episodically, resulting in the accumulation of an additional pair of sand units that contain Entisols. Given the weakly developed nature of these soils, it is believed that these units accumulated over 200 to 300 years. The uppermost unit of sand was most likely deposited within the past 500 years. Since this time, the focus of dune formation has been in the nearshore environment.

The Wycamp Creek Site (20EM4), Emmet County

Location and Background

In July 1927, an archaeological survey of the Lake Michigan shoreline was conducted by Emerson F. Greenman, Museum of Anthropology, University of Michigan. The survey extended from the mouth of Wycamp's (or Wycamp) Creek, just north of Cross Village, to Seven Mile Point in Emmet County, Michigan, along the northeast shoreline of Lake Michigan. Among the several results of this investigation was the discovery of an extensive prehistoric archaeological site encompassing both the north and south banks of the mouth of Wycamp's Creek where it empties into Lake Michigan (figure C-13). A series of undisturbed so-called pedestals in an area of deflated coastal sand dunes yielded both precontact ceramics and chipped stone, and historic period artifacts, some dated to the eighteenth century. Subsequent to the survey, in 1929, a description of the site and a summary of the surface collections was prepared (Greenman 1927). The materials he collected were accessioned into the UMMA collections under catalogue numbers 1860–1871.

Following Greenman's brief report on what became known as the Wycamp Creek site this area received little attention from archaeologists for close to half a century. The Wycamp Creek site was not directly identified by Hinsdale (1931) in his *Archaeological Atlas*, although he recorded other as yet unrelocated villages in the immediate vicinity. It is possible that Hinsdale's attempt to mask specific site locations is responsible for this confusion. A renewed interest in the collections from this site arose when Alan McPherron (1967, 272–273) employed them as comparative materials for his analysis of the assemblage from the substantial and deeply stratified Late Woodland Juntunen site on Bois Blanc Island in the Straits of Mackinac.

During the 10-week period from 15 June through the end of August 1967, MSUM field parties were engaged in a settlement pattern study of the northwest Lower Peninsula of Michigan, including the shoreline of Emmet County. This work was under the direction of Charles E. Cleland, Curator of Anthropology at the MSUM. One of the results of this investigation was the relocation of a remnant of the Wycamp Creek site on the north bank of Wycamp's Creek. Surface collection distributions clearly indicated that the prehistoric occupation component at the site was quite extensive, and that few undisturbed areas remained on the south bank of the creek. Additionally, the hewn log foundation incorporates the race of a nineteenth-century mill attributed to Father Wycamp.

The Wycamp Creek site proper is located on the north bank of Wycamp Creek, 1⅛ miles north of Cross Village, Michigan, in the NE ¼, NE ¼, NE ¼, Section 34, T38N, R6W, Emmet County, Michigan (figure C-13). Due to the preliminary nature of the 1967 field research conducted by the MSUM only 1,295 ft² (ca. 120 m²) of test excavation were undertaken, partially financed by a National Science Foundation Research Grant to Cleland. At the time, the property was owned by the Forester

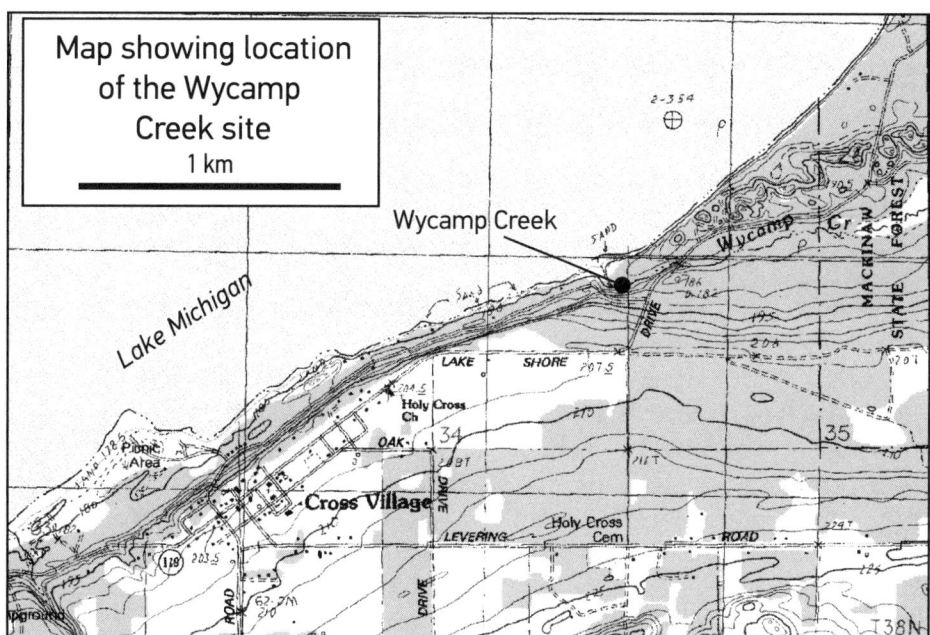

c-13. The Wycamp Creek site location.

family of Ecorse, Michigan, but ownership has changed several times in the past 40+ years. The Wycamp Creek site was eventually nominated and listed on the National Register of Historic Places. Despite this listing, this important site has recently been heavily disturbed by vandalism and construction, which may have destroyed much of the site's research value. Several attempts by MSU to return to the site and perform more extensive controlled excavation on undisturbed portions of the occupation, and salvage remnants of the disturbed areas, have met with no success. Most recently, a house has been constructed on the property although the extent of its impact on the archaeological site has not been assessed.

Although the site collections and field notes have been referenced in numerous professional publications, no comprehensive site report has yet been published. Much of the published material has been extracted from a draft report prepared by Lovis (n.d.). The present summary is also extracted from that draft material, and will provide the most comprehensive published statement on the site stratigraphy and depositional history.

The excavations at the Wycamp Creek site were designed to answer questions of a descriptive nature. As the subsequent sections on stratigraphy will further detail, the Wycamp Creek site is situated on several landforms interpreted as former terraces of Lake Michigan. The 1,295 ft^2 (120.4 m^2) block excavated in 1967 sampled evident terraces at elevations of 586 ft (178.6 m) and 592 ft (180.4 m) amsl. These are clearly post-Nipissing stage elevations; a well defined Nipissing terrace is identifiable inland of the site proper. The upper, or 592 ft (180.4 m) terrace was sampled with 740 ft^2 (68.8 m^2) of excavation, and the lower or 586 ft (178.6 m) terrace was subject to 550 ft^2 (51 m^2) of excavation. An estimate of the spatial extent of the occupation of Wycamp Creek site, based on combined data from surface collection and excavation, would conservatively include 173,800 ft^2 (16,163.4 m^2) of area. This estimate does not take into account the area collected by Emerson Greenman on the south bank. Thus, the excavations conducted by the MSUM in 1967 sampled approximately 0.0075 percent of the part of the site that remains on the north bank of the creek. While such a limited sample can only provide a narrow picture of the spatial organization of the various occupations, the distribution of the excavations across the site provide a rather clear perspective on the nature of site formation and occupation sequencing.

Stratigraphy and Deposition

To properly understand the geomorphic formation processes at the Wycamp Creek site it is essential to understand the relationship of the larger macroscale regional processes as they relate to the formation of the current landscape. The more general processes are presented elsewhere in this book (chapter 4). In the Wycamp Creek area, isostatic rebound has raised the fossil beaches of Lake Nipissing to elevations of circa 620 ft (189 m) amsl. Leverett and Taylor (1913) claim that this postglacial formation runs inland of the present Lake Michigan shoreline near Wycamp Lake, the water body from which Wycamp Creek derives (figure c-13). If this is indeed the case, then the steep lakeshore bluff inland of Wycamp Creek is not the main Nipissing terrace but rather may relate to either a recessional (perhaps Nipissing II?) stage, or a remnant wave terrace of Lake Nipissing that would have formed a deep inland embayment including present-day Wycamp Lake. A strong 620 ft (189 m) terrace can be traced on the margins of Wycamp Lake and marsh area, and either the Nipissing II or wave terrace alternatives seem likely interpretations for the current lakeshore bluff. If this scenario is correct, then drainage of the area would have proceeded from the channel between the stranded Algonquin stage Readmond and Levering islands (Larks Lake vicinity), and thence into the Wycamp embayment.

Subsequent reductions in the water levels of the Michigan-Huron basin to post-Nipissing and ultimately modern planes had major consequences in the Wycamp Creek area. Drainage of the Wycamp embayment and the late- or post-Nipissing feature exposed sandy bottom sediments. North of Wycamp Creek and inland from Sturgeon Bay Point these sediments became subject to wind action and dune formation, and eventual vegetation stabilization. With the formation of dunes along the late Nipissing stage feature, waters draining from the Larks Lake area became trapped in the former embayment, forming the present Wycamp Lake and adjacent marshes; a process similar to that described by Dorr and Eschman for the Herring Lakes embayment further south (1970, 213–217, figure IX-39; 223 citing Scott and Dow 1937). Drainage of the Larks Lake / Wycamp Lake area took place via Wycamp Creek, which has cut a deep channel through the Nipissing features en route to Lake Michigan.

Recessional, post-Nipissing elevations at 595 ft (181.4 m) were probably lapping the base of the Nipissing wave terrace/Nipissing II terrace, with little if any beach between the base of the bluff and the lake. Any prehistoric occupation at this time would probably have occurred at higher elevations—none is evident on the Wycamp Creek site. Modern lake levels of 580 ft were likely attained by circa 2,200 years ago (Monaghan and Lovis 2005), exposing a broad strip of sandy beach suitable for habitation. Current topography at the Wycamp Creek site, however, includes a remnant beach terrace at elevations ranging between 584 ft (178 m) and 594 ft (181

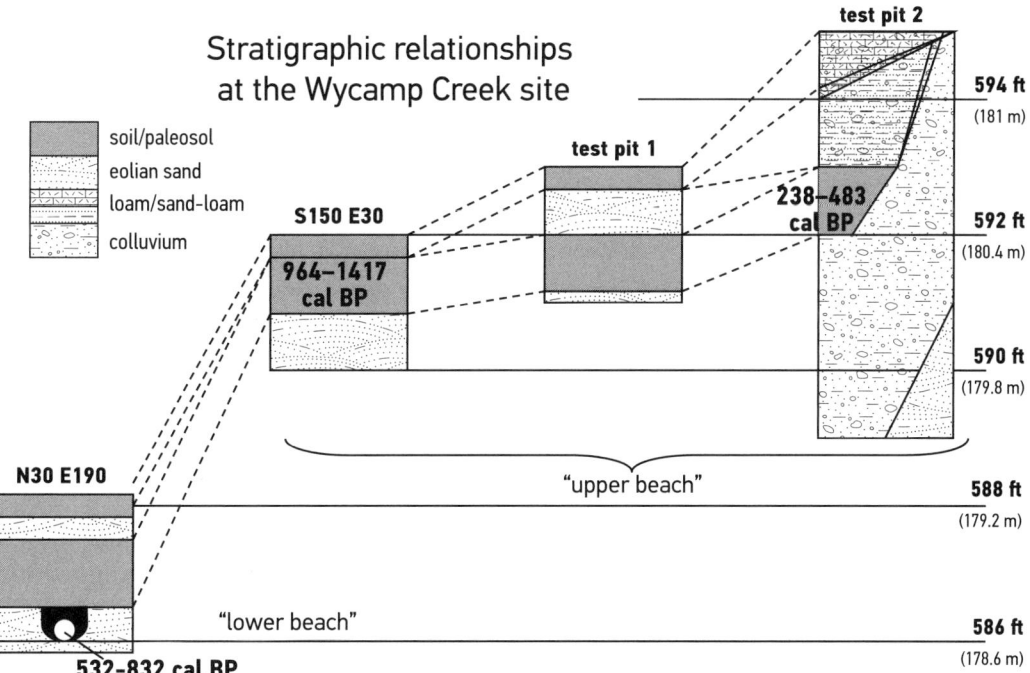

C-14. Generalized, composite stratigraphic column of the Wycamp Creek site. Ages of various units shown. See appendix A for discussion of how ages are reported. Unit depths and boundaries are approximate.

m). This may, in fact, be a storm beach relating to seiche activity, similar to that described by Mason (1965) at the Mero site on the Door Peninsula of Wisconsin, or a late resurgent high-lake stage of basin-wide proportions. Further evidence pertaining to the possible age of this feature is presented in succeeding discussion.

The majority of the stratigraphic variability (figure C-14) present at the Wycamp Creek site was observed in excavations on the "upper," 592 ft (180.4 m) beach terrace. "Lower" beach units consistently displayed the following stratigraphy, essentially composed of two depositional units with soils developing within each unit.

- *2Bsb horizon*—This is the basal depositional unit (Unit I) excavated at the Wycamp Creek site. The weak 2Bsb deposits contained within this horizon are well-compacted eolian sands with evidence for a sesquioxide enriched Spodosol development. Once the 2Bsb was encountered during excavation, evidence of cultural activity rapidly diminished. The overlying 2Ab horizon developed within this unit.
- *2Ab horizon*—This is a well-developed soil within Unit I. It is a black/dark brown organic horizon with little indication of any B horizon soil development. Substantial evidence of human occupation of this surface is present. Cultural features were discernable at the base of this horizon, having apparently originated from within it. A conventional radiocarbon date (M-2059) derived from a feature at the base of this zone produced an age of 670 cal BP (appendix B).

- *B/E horizon*—This is a stratum of loosely compacted gray sands, probably eolian, with low organic content.
- *A horizon*—This is a thin layer of humus stabilizing the present fragile site surface.

Although fundamentally similar to the lower terrace sequence, the upper beach excavations encountered three types of stratigraphic variation (figure c-14), one of which is identical to that described above. Further variability is due to the presence or absence of intervening inorganic eolian sand layers between the humic A horizon and the well-developed A2 occupation surface. This variation is well illustrated by the stratigraphy in excavation units s150 w30 and Test Pit 2. For example, unit s150 w30 revealed a sequence consisting of overlying humus or A horizon in direct contact with a well-developed and highly organic zone containing occupation materials, which in turn overlay the basal eolian sands. Test Pit 1, in turn, duplicated the lower-terrace sequence described above.

Test Pit 2, however, south and east of these units, had an overlying thick layer of gray sandy loam that dipped markedly lakeward. Underlying the loam was a light brown sand layer containing lenses of compacted black loams, and occasionally overlain by tan inorganic eolian sand. This brown layer presented a sharp contact with the underlying basal C horizon of yellow/tan Spodosols, similar to the 2Bsb described above. Both the upper loam layer and the underlying light brown sand rise and converge to the north, that is, in the direction of the foot of the Nipissing stage feature. This sequence is illustrated in figure c-14.

Correlation of lower- and upper-terrace stratigraphic sequences is rather straightforward for all excavation units except Test Pit 2 (figure c-14). In viewing this relationship it should be noted that as one approaches the shoreline of Lake Michigan, a gray sandy loam layer intervenes between the humic A horizon and the occupation zone (variably characterized as a 2Ab horizon, or possibly a developed E horizon). Test Pit 2, on the other hand, possesses a thick loam stratum dipping in elevation toward the north. Thus, explanation for the differential sand loam deposition is in order. Lack of bedding, sorting, and no erosion of artifacts would eliminate water activation. With the exception of Test Pit 2, most of the remaining unit stratigraphy can be explained by modern phenomena. Currently, approximately the first 60 feet (20 m) of lakeshore is devoid of vegetation, consisting of gravels and sand. Due to the location of the site, subject to onshore wind, this sand supply is moved onshore and collects in hollows as well as being trapped in areas with light dune vegetation. It is most likely, therefore, that for the most part these are eolian sands.

Test Pit 2, however, is situated near the base of a steep bluff, and presents a different problem. This unit did not contain a depositional unit that could be immediately recognized as an occupied stable surface. The layer corresponding with the occupation stratum dips markedly in elevation toward the beach, and thickens as the dip of the stratum diminishes. In section, the brown sand stratum illustrates accretion at the point where its dip flattens, whereas it thins as the basal slope approaches 45 degrees. This suggests colluvial processes. Furthermore, this layer is overlain by eolian sand and loamy sand. Greenman (n.d.) noted a similar phenomenon at a site near Goodhart, Michigan, located at the foot of a similar bluff formation. Whether the Wycamp Creek situation is due to the proximity of

Test Pit 2 to the base of the bluff, or whether this area is a deep depression that underwent infilling, is conjectural. In either case, both the sand and loamy sand overlying this stratum have eolian characteristics.

A series of three ^{14}C dates from the Wycamp Creek site, in conjunction with a historic feature, allows assignment of chronology to the depositional sequence of the area. The earliest date on the occupied 2Ab horizon was obtained on a series of charred seeds from a depth of 27 cm in unit S150 E30, located on the upper terrace. This age of 1320 ± 120 BP (M-2065, 2σ 1417–964 cal BP, intercept 1240 cal BP), indicates stabilization of the upper beach by that time. A second date was obtained on unidentifiable charcoal from a hearth emanating from the base of the occupation zone A2 horizon in unit N30 E190, on the lower terrace. The date obtained, 730 ± 110 BP (M-2059; 2σ 832–532 cal BP, intercept 670 cal BP) may well illustrate the fact that this stabilization was a long-lived phenomenon if both terraces stabilized at similar points in time.

During excavation, the wall foundation trenches of a late 1800s structure were defined. This trench was excavated by the builders through the eolian sand layer of the upper terrace, and penetrated the occupation zone into the basal C horizon. A humus layer then developed above the eolian sands and the filled trench. In at least one area the eolian sands are located above the remains of this structure, and were not disturbed by construction. Thus, the process of eolian deposition, which began in the late nineteenth and early twentieth centuries, was occurring both prior to and after construction of this building.

The final ^{14}C date from the Wycamp Creek site was obtained from a depth of 82 cm in Test Pit 2, on charcoal from the upper part of the colluvially deposited brown sand zone. An age of 240 ± 100 BP (M-2060, modern, 2σ 483–238 cal BP) would lead to the inference that the colluvial infilling of this part of the site had terminated at this time or slightly later. Certainly, the fact that no trace of this light brown sand wash was observed in the superposed loam layer would indicate that slope stabilization had occurred, or that erosion had moved the bluff face further inland, so that deposition in the vicinity of Test Pit 2 did not continue.

Discussion and Archaeological Implications

The Wycamp Creek site assemblage has never been fully described, and the discussion here is too brief to undertake this task. However, based on an unpublished report by Lovis (n.d.), the following summary will give an indication of occupation sequencing, function, and spatial organization at the site. Rim and decorated sherds can be grouped into several major categories conforming to a number of existing classifications (Brose 1970a; Janzen 1968; Wright 1967; McPherron 1967; Mason 1966; Hambacher 1992). These include various Laurel Middle/Initial Woodland (29 sherds representing 21 minimal vessels), Mackinac Ware var. punctate (31 sherds representing seven minimal vessels), Bowerman Ware var. plain (47 sherds representing 15 minimal vessels), Bois Blanc Ware var. beaded (six sherds representing two minimal vessels), Traverse Wares (18 sherds representing nine minimal vessels), Oneota-like (28 sherds representing two vessels), and afinis Lawson Incised Ontario Iroquois (16 sherds representing eight vessels). Almost 90 percent of the 121 items in the formal lithic tool assemblage is composed of projectiles, end and side scrapers, and preforms/knives, with 40 percent of the assemblage ($n = 48$) being projectile points. There is a limited copper assemblage. The overall assemblage appears to be a hunting/retooling/processing-related activity set, but in the context of residential activity by a mixed gender and age group.

Broadly speaking, the occupations can be spatially separated; Late Woodland projectiles are found on the lower terrace with more recent radiocarbon dates, and Laurel Middle Woodland forms are found on the upper beach in association with ceramics of similar age, and the earlier radiocarbon date. The earlier occupation appears to be situated on a beach related to the 1700 BP high-lake-level fluctuation, but given the absolute date may have been occupied later, after the subsidence of lake elevations to modern levels. In terms of eolian cycling and burial, it is of importance that both the older Laurel occupation surface on the upper beach, and the Late Woodland occupied surface on the lower beach, were sealed by eolian sands.

Portage Site (20EM22), Emmet County

Location and Background

The Portage site was discovered during the course of the MSUM survey of the Inland Waterway Project, a National Science Foundation funded research project initiated during 1974 and 1975 (Lovis 1978a, 1978b). Survey during 1974 revealed an extensive and relatively undisturbed ceramic producing site in the dune complex between Round Lake and Lake Michigan at altitudes between 597 and 600 ft asl (figure C-15). Lake Michigan lies 60 m to the west of the main site area. Due to its location between Lake Michigan and the westernmost lake of an interior lake chain it was named the Portage site, reflecting its potential function. The site is in private ownership within the L'Arbre Croche Subdivision, proximal to Petoskey State Park, in the W ½, SW ¼, Section 22, T35N, R5W, Little Traverse Township, Emmet County (figure C-15).

Due to the potential for stratified deposits the Portage site was test excavated in 1974. Initial suspicions about the site's structure were vindicated when stratified eolian deposits were encountered in association with substantial Middle Woodland as well as Late Woodland occupation. Subsequently, in 1975, the site was more fully excavated. A total of 775 ft^2 of the site was excavated, with approximately 150 ft^2 presenting stratified contexts. To date, the only published analysis of the site has focused on the ceramic sequence (Lovis et al. 1998), although S. Martin refers to the faunal assemblage in her 1985 dissertation. The faunal assemblage from the Portage site was analyzed and reported by Beverley Smith (Smith 1983). While there was a substantial amount of both boney fish (Osteichthyes sp.) as well as mammal species in the assemblage, what stands out is the restricted range of identifiable remains. These include beaver, turtle, and lake sturgeon, as well as a woodchuck and probable smallmouth bass. Martin (1985) opines that this assemblage reflects a warm season occupation.

Stratigraphy

The stratified area of the Portage site produced evidence for prehistoric occupation of two highly organic paleosols in eolian deposits (figure C-16). Field designations

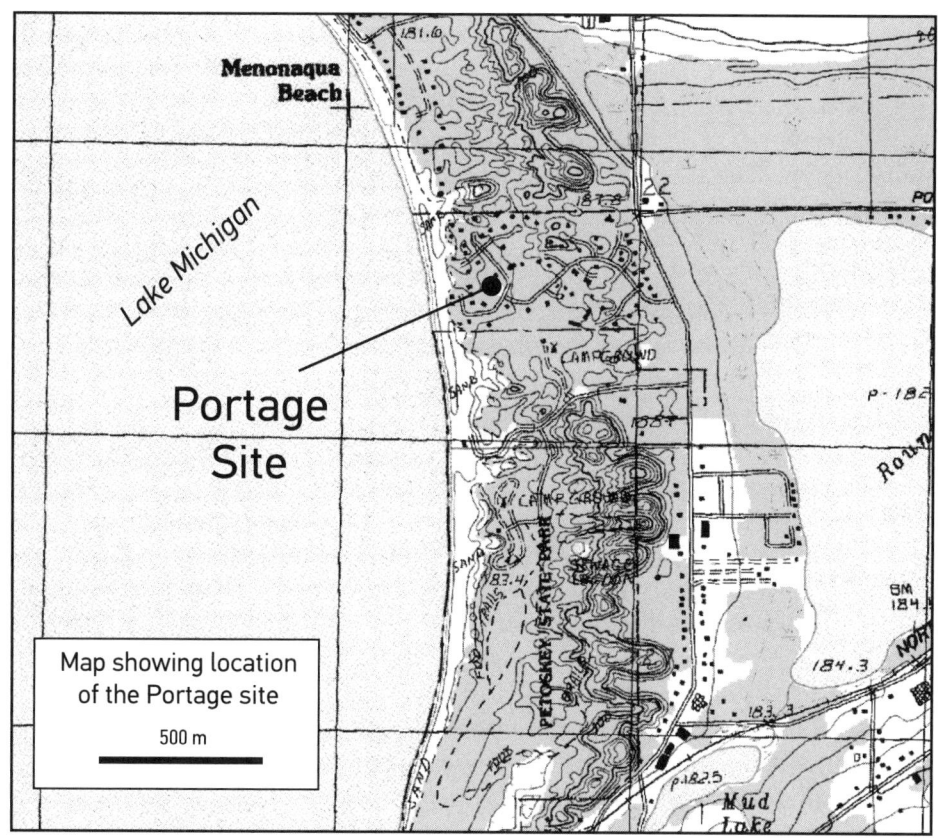

c-15. The Portage site location.

c-16. Generalized, composite stratigraphic column of the Portage site. Ages of various units shown. See appendix A for discussion of how ages are reported. Unit depths and boundaries are approximate.

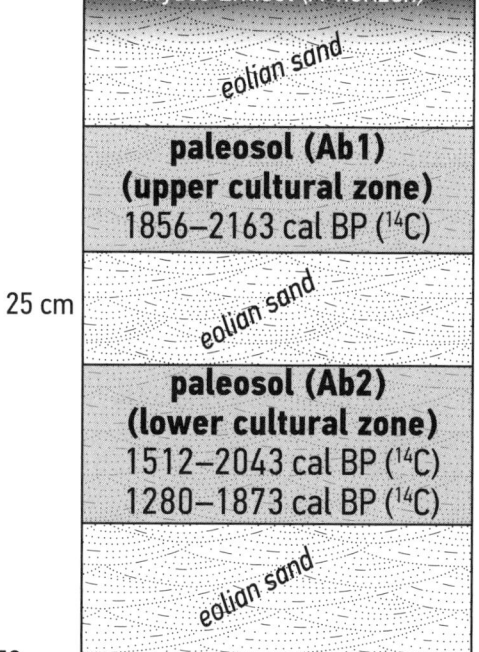

for the occupied paleosols were Upper Occupation Zone and Lower Occupation Zone. The Upper Occupation Zone and Lower Occupation Zone were separated by culturally sterile eolian sand. The Upper Occupation Zone lay immediately below a lens of eolian sand that was the C horizon of the surface soil. The following more detailed discussion of the Portage site stratigraphy is drawn largely from MSU field notes (1974, 1975).

- *Unit I—basal eolian sand.* The paleosol designated Lower Occupation Zone sits directly on eolian sand deposits of substantial thickness. These deposits are at elevations above that of the Algoma stage, and below that of the Nipissing stage maximum. They are most likely Nipissing lakebed deposits that, when exposed, became subject to eolian activation.
- *Ab2—Paleosol (Lower Occupation Zone) formed in Unit I.* The basal paleosol is a highly organic and even compact deposit of variable thickness, but is no thicker than about 10 cm. This zone has a "greasy" texture, and incorporates Laurel ceramics and other cultural material including hearths and postmolds. On its margins the Lower Occupation Zone comes into contact with, and ultimately merges with, the Upper Occupation Zone. Two radiocarbon dates were obtained from this zone. Charcoal from a small pit or postmold at the base of this zone was dated to 1830 ± 120 BP (DIC-652; 2σ 2043–1512 cal BP, intercept 1760 cal BP), and charred wood or pitch from a hearth feature was dated to 1620 ± 150 BP (DIC-653; 2σ 1873–1280 cal BP, intercept 1535 cal BP). This time span most likely dates the period of stabilization of the dune. The dates are also substantially younger than those from the Upper Occupation Zone, calling into question the age of DIC-651.
- *Unit II—Eolian sand.* An inorganic eolian sand deposit of variable thickness intervenes between the Upper and Lower Occupation Zone paleosols. This deposit was formed in a swale or basin-shaped depression. There are no absolute dates from this deposit. The 2Ab is formed within Unit II.
- *Ab1—Paleosol (Upper Occupation Zone).* The paleosol designated the Upper Occupation Zone variably lies either directly beneath and in contact with the A Horizon, or is separated from it by Unit III eolian sand. It was formed within Unit II. The 2Ab has higher organic content and lower sand content than the A horizon, and incorporates both Initial and Late Woodland cultural materials. The Upper Occupation Zone was radiocarbon-dated with charcoal derived from a suspected hearth feature within the paleosol that contained Laurel ceramics. The radiocarbon age of 2050 ± 80 BP (DIC-651; 2σ 2163–1856 cal BP, intercept 1995 BP) suggests an early stage for stabilization of the dune deposit. However, two more recent ^{14}C age determinations from the Lower Occupation Zone, in combination with comparative ceramic cross-dating, suggest that this date is too old. The presence of Pine River Ware (Holman 1978, 1984) and Mackinac Ware (McPherron 1967) in this zone would make an age of circa 1,400 to 1,300 years ago more likely.
- *Unit III—Lens of eolian sand.* The A horizon is formed in the uppermost part of Unit III.
- *A Horizon*—The A horizon is a variably thick loamy sand formed in Unit III and overlying the Upper Occupation Zone. There is no evidence for B horizon development.

Reconstructing the depositional history and the timing of site formation at the Portage site is, therefore, rather straightforward, although it would be useful to have additional comparative dates from the Upper Occupation Zone. The original sand supply at the Portage site apparently derives from lake-bottom sands of the Nipissing stage circa 5,000 years ago. As lake elevations in the Michigan basin underwent recession to lower Algoma and modern levels by circa 2,500 years ago, these lakebed sands became subject to activation and dune formation. By circa 1,800 to 1,500 years ago the more inland parts of the dune complex in the vicinity of the Portage site stabilized, forming a soil horizon that was occupied by Laurel-affiliated populations who camped on this stable surface. Subsequently, the lakeward dunes became sufficiently active to mantle hollows on this stable surface with eolian sands. Apparently these sandy areas were not suitable for human occupation. Based on ceramic cross-dating, these infilled and active areas once again stabilized by circa 1,400 to 1,300 years ago, remaining relatively stable until the present. Soil formation on this surface continued, albeit with a slightly higher sand content than during prior stabilization episodes.

Discussion and Archaeological Significance

The significance of the Portage site lies in the fact that it is both partially stratified and also spans a period of upper Great Lakes occupation history that is not well represented in the archaeological record: the transition from the Initial Woodland to the Late Woodland periods. While the lithic assemblage does not stand out as distinctive, the ceramic assemblage from the site is of singular importance. The major published work on the site takes advantage of both of these characteristics to address issues of ceramic change, cultural affiliation, and increasing regional identity in the northwestern Lower Peninsula of Michigan (Lovis et al. 1998).

Among the more significant conclusions from the numerical taxonomic analysis performed on the ceramic assemblage was to partition ceramic styles temporally, and then assess this chronological seriation through evaluation of the stratigraphic positions of the different ceramic groups. The early Middle or Initial Woodland in this scenario begins as a highly generalized Laurel complex ceramic assemblage, dominated by banked stamp ceramics typical of many other sites across the northern Lake Michigan and Lake Superior drainage basins. These ceramics are associated with the Lower Occupation Zone (3Ab), and datable on that basis to circa 1,800 to 1,500 years ago. This represents the earliest dated presence of ceramic-producing Laurel occupation in the region.

The transition to more localized regional ceramic styles and regional identity occurs during the late Middle or Initial Woodland. While Laurel banked stamp ceramics continue, they are joined by distinctive linear corded types as well as styles reflecting more southern, Havana, influence. Lovis et al. (1998) provide the appellation "Portage Complex" to this heterogeneous mix and equate it to other regional traditions such as Nokomis and North Bay to the west in Wisconsin. The Portage Complex forms the foundation of subsequent ceramic developments represented both at the Portage site and elsewhere in northwestern Lower Michigan during the transition to the Late Woodland.

The transition between the Initial (or Middle) Woodland and the Late Woodland in the region has been termed the Pine River Phase (Holman 1984; Hambacher

1992). The distinctive ceramics of this phase, Pine River Ware, share features of Laurel paste, and the use of superimposed punctates, with that of vessel shapes common to early Late Woodland Mackinac Wares, such as more flared rims, and coupled with sparse decoration. Dated to circa 1,400 to 1,300 years ago, both Pine River Wares and Mackinac Wares are present in the Upper Occupation Zone at the Portage site. Thus, analysis of the Portage site ceramic data leads to two significant conclusions. First, there is an unbroken chain of ceramic style development in the region from early Middle Woodland circa 1,800 years ago through the Mackinac Phase Late Woodland as late as 1,000 years ago. Second, it is during the latter part of the Middle Woodland period that local ceramic expressions begin to symbolize more discrete regional identities in the northern lower peninsula of Michigan.

Fisherman's Island State Park, Charlevoix County

Fisherman's Island State Park is located in the northwestern part of Lower Michigan in Charlevoix County (figure C-17). The northern end of the park is approximately 3.5 km southwest of Charlevoix. From there the park extends along the shore for about 15 km to a point just north of Norwood, Michigan. Two locales were investigated in Fisherman's Island State Park in this study, including (1) the dune and associated archaeological site designated the Solomon Seal site, at the northern end of the park near South Point, and (2) the Fisherman's Island transect (figures C-18 and C-19) at the southern campsite loop north of Inwood Creek and the O'Neil site.

Fisherman's Island Transect: South Campground Loop

The Fisherman's Island transect consists of a series of OSL ages obtained from a suite of dunes that lie about 2.5 km southwest of the Solomon Seal site. This study transect is located in the NW ¼, Section 6, T33N, R8W, Norwood Township (figure C-17), and extends about 500 m from a large (ca. 30 m high) dune ridge on the east to much smaller (~5 m high) dunes near the coast. The transect lies within a small campground and is bisected by a pair of access roads. In an effort to determine the history of eolian sand deposition in this portion of the park, four OSL ages were obtained from select points across the transect (figures C-17 and C-18). The easternmost sample was acquired from a soil pit excavated in the crest of the easternmost dune ridge. This ridge is located east of an access road and contains a variety of embedded parabolic and subparabolic dunes. The sample was collected from a depth of 1.5 m in the highest point of the largest parabolic dune in the ridge and yielded an age of 3260 ± 305 ya (UIC-2146). An additional pair of samples was collected with a GeoProbe from a small dune about 25 m west of the easternmost access road. One of the samples was obtained from a depth of 5.5 m and provided an age of 3240 ± 260 ya (UIC-2143). Another age estimate of 2315 ± 220 ya (UIC-2144) was obtained on sands at that site from a depth of 2.0 m. The last of the four samples was collected within a small dune about 10 m west of the westernmost access road. This sample was obtained from a depth of 2.0 m and yielded an age of 2420 ± 240 ya (UIC-2145).

The combined ages obtained from Fisherman's Island State Park suggest a simple history of eolian sand deposition and dune formation along this portion of Lake

Descriptions of Sample Locales

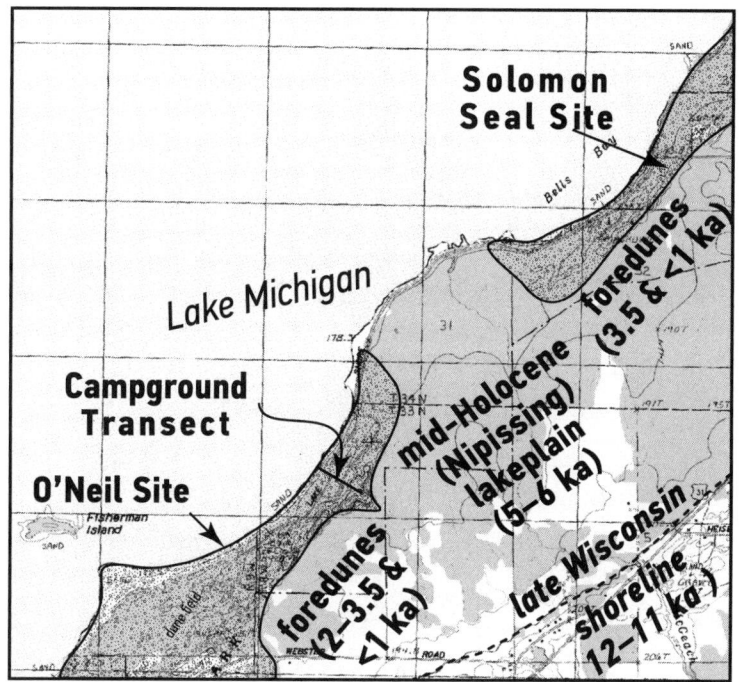

C-17. Fisherman's Island State Park archaeological sites and dune sample locations.

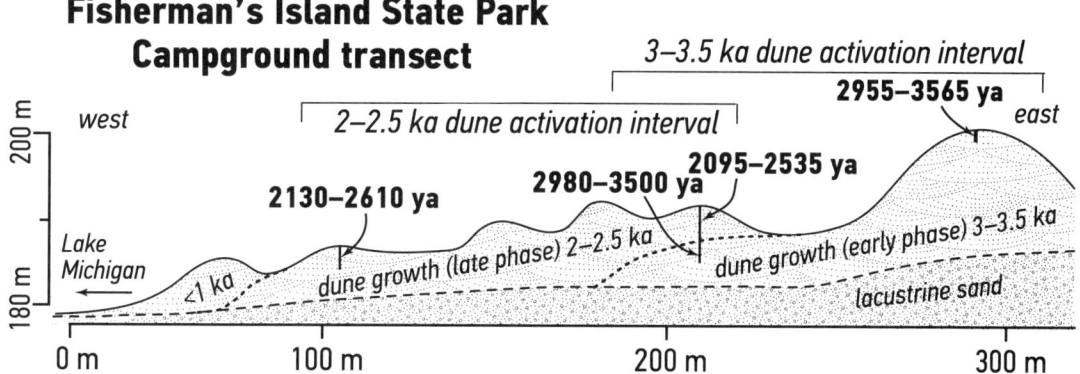

C-18. Fisherman's Island transect across increasingly younger dune ridges. Locations shown on figure C-17.

Michigan's northeastern shore. OSL ages from the dune ridges at the Solomon Seal site and the Fisherman's Island transect suggest that sand was supplied to the most easterly dunes in the park at circa 3,200 years ago. Given that these dunes are by far the largest dunes in the park, it is reasonable to conclude that the supply of eolian sand in the area was higher during this period of deposition than at any other time in the late Holocene. Ages from the Fisherman's Island transect suggest that the deposition of eolian sand continued until circa 2,400 years ago, resulting in the formation of progressively younger dunes in the western part of the transect. Aside from the age of the dunes, another question to consider is

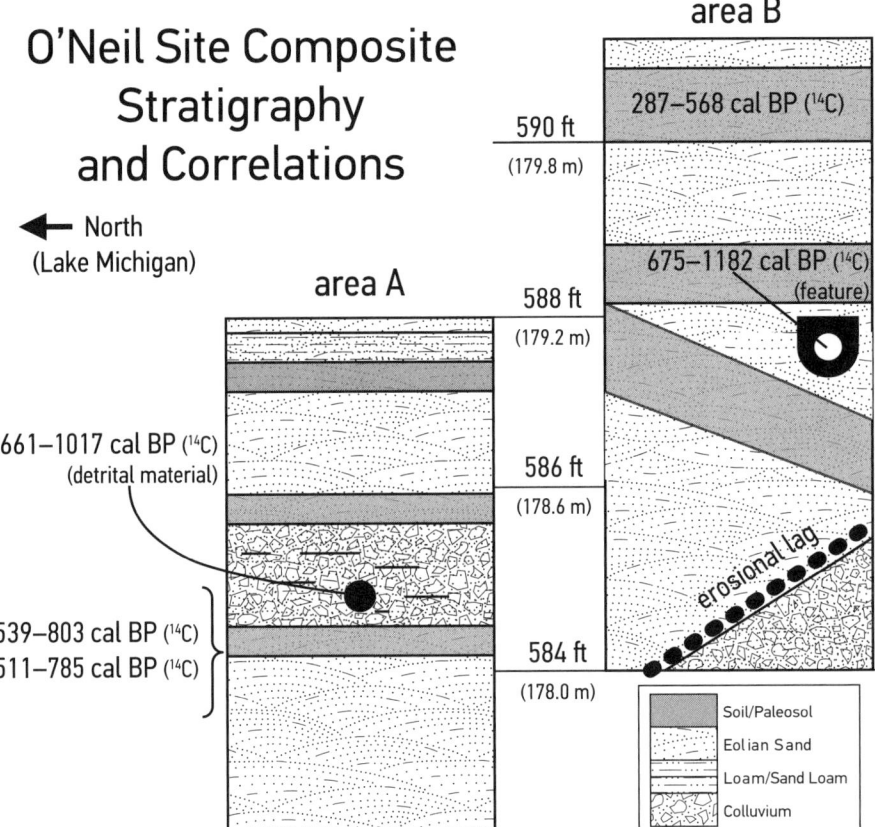

C-19. Generalized, composite stratigraphic column of the O'Neil site, Fisherman's Island State Park showing probable correlation of archaeological and sedimentary units. Ages of various units shown. See appendix A for discussion of how ages are reported. Unit depths and boundaries are approximate.

why dunes are located in these specific places. Visual assessment of topographic relationships indicates that the dunes associated with both the Solomon Seal site and Fisherman's Island transect are centered within small embayments bounded by headlands. These embayments may have served as local depositional centers of littoral sediment that was ultimately reworked into dunes.

O'Neil Site (20CX18), Charlevoix County

Location and Background

The O'Neil site is located on the north bank of Inwood Creek in the NE ¼, SE ¼, Section 1, T33N, R9W, Norwood Township, Charlevoix County, Michigan, in proximity to a broad shallow bay of Lake Michigan approximately three miles south of Charlevoix (figure C-17). The topography of the site consists of stabilized dunes at elevations between 589 ft and 592 ft asl (180–180.5 masl). The O'Neil site was discovered in 1969 by an MSUM field crew who noticed both lithics and ceramics in disturbed off-road vehicle tracks at the site. Discovery of the site prompted test excavation in 1969, initially on what was presumed to be a Late Archaic component situated on a raised Nipissing stage terrace at 610–612 ft asl (~186 masl), but subsequently determined to be a Late Woodland component located near the Lake Michigan shoreline.

During the 1969 season 1,750 ft² of the site was excavated. Due to the presence of stratified deposits, excellent faunal preservation, as well as development threats the MSUM returned to the site in 1971, when additional major excavations totaling 2,100 ft² were conducted. At the close of excavation 3,850 ft² of the site had been fully excavated, in some cases to depths exceeding five feet. Research at the site revealed a well-stratified series of Late Woodland period occupations incorporated into both water-laid and eolian deposits. Results of field research at the O'Neil site formed the basis for both a dissertation (Lovis 1973), and a subsequent paper on the site (Lovis 1990b), which collectively form the core of the following discussion. The O'Neil site is listed on the National Register of Historic Places and is currently in State of Michigan ownership as part of Fisherman's Island State Park.

This site is situated close to a broad range of potentially exploitable habitats (Holman 1978). Analysis of the flora and fauna reveal that the majority of Late Woodland occupation can be attributed to the spring and summer. The O'Neil site is located only a few kilometers north of the Pi-wan-go-ning or Norwood Quarry site (Cleland 1973). It is not surprising, therefore, to find that throughout its prehistoric occupation lithic reduction was a primary activity of the residents. Importantly, the O'Neil site is one of the few sites in northwestern Michigan that has evidence for cyclic occupation and use by both Straits of Mackinac sequence (McPherron 1967) and Grand Traverse Bay sequence (Hambacher 1992) populations (also see Brashler et al. 2000; Monaghan and Lovis 2005).

Stratigraphy

The O'Neil site was excavated in several partially contiguous blocks coupled with more expansive test excavation. This work revealed both sheet midden deposits associated with A horizon development, as well as stratified deposits both with and without associated cultural material and datable organics. Stratified datable sequences of singular importance were recorded from both the Area A and Area B block excavations (figure C-19). Area A yielded three primary cultural horizons labeled I/Ib, II, and III, sequentially from the surface. Area B produced a shallow and more poorly defined sequence. Much of the following stratigraphic discussion is summarized from Lovis (1973, 1990b).

The Area A stratigraphy at O'Neil consisted of the following:

- *Lakebed sands.* This is the basal deposit upon which the upper part of the sequence was deposited, and most likely dates to the Nipissing transgression circa 5,000 years ago.
- *Bedded gravels.* Overlying the basal sand deposits was a thin layer of cross-bedded gravels exposed at altitudes between 178 m and 179 m, dipping in elevation lakeward. It is likely that these gravels are associated with lakeshore activity during the post-Nipissing stage regression, and its recession to modern elevations circa 3,200 years ago, but alternatively could date as late as circa 900 to 500 years ago during a period of late period of lake transgression (Monaghan and Lovis 2005).
- *Eolian sand.* Directly overlying the bedded gravel deposits was an eolian sand layer of highly variable thickness. The source of this sand deposit was probably

lacustrine sediments that were subaerially exposed. The sands form the base of dune hollow oriented parallel to the lakeshore.
- *Paleosol (Occupation Zone III).* This is an organic deposit ranging from 0.5 ft to 0.3 ft in thickness, with the thicker deposits positioned at the base of the dune hollow. Cultural material, including hearth features, was abundant on this stable surface. Two ^{14}C dates were obtained from a hearth feature in this zone: 670 ± 100 (M-2405; 2σ 785–511 cal BP, intercept 660 cal BP) and 740 ± 100 (M-2406; 2σ 803–539 cal BP, intercept 680 cal BP).
- *Ponded/lacustrine sands.* Occupation Zone III was inundated subsequent to occupation, most likely having to do with local-level coastal processes associated with the outlet of Inwood Creek. Bedding planes incorporating organic laminae into sand grains of small and uniform size were present in this at times substantial sand deposit. Lack of displacement of cultural materials on the surface of Occupation Zone III suggests a relatively calm or shallow energy environment. Carbonized organics incorporated into these sands were dated to 905 ± 115 BP (N-1268; 2σ 1017–661 cal BP, intercept 795 BP), inconsistent with dates from the underlying stratum.
- *Paleosol (Occupation Zone II).* This is a relatively level organic horizon with incorporated cultural materials directly overlying the lacustrine sands. While not directly dated, ceramic refits with vessels from dated contexts in Area B suggest an age of circa 600 to 500 years ago for this zone. This is an occupied stable paleosol.
- *Eolian sand.* Occupation Zone II is sealed by a layer of culturally sterile eolian sand. This deposit is not dated.
- *Paleosol (Modern A horizon, Occupation Zone I/Ib).* This is the modern A horizon. Numerous historic period cultural artifacts dated to the late seventeenth or early eighteenth century A.D. (ca. 300 years ago) were recovered from Occupation Zone I/Ib.

From a geomorphic perspective, the archaeological stratigraphy can be restated as follows:

- *Unit 1*—Lacustrine sands
- *Unit 2*—Bedded gravels
- *Unit 3*—Eolian sand, with a 3Ab horizon formed within the topmost part of the unit, conforming to Occupation Zone III
- *Unit 4*—Mixed lacustrine and eolian sand, with a 2Ab horizon formed within the topmost part of the unit, conforming to Occupation Zone II
- *Unit 5*—Eolian sand, with an A/Ab horizon formed within the topmost part of the unit, conforming to Occupation zone I/Ib

The Area B stratigraphy at the O'Neil site varied somewhat from that observed in Area A, as follows:

- *Unit 1*—The deepest deposits encountered in Area B were colluvial materials derived from glacial deposits. Unit 1 dipped lakeward, becoming deeper toward the west.

- *Unit 2*—Directly overlying Unit I was a variable but generally thin layer of bedded and sorted gravels. These are most likely shallow shoreline deposits.
- *Unit 3*—Unit III deposits consist of eolian sands. These eolian sands are probably reworked lacustrine deposits and are of variable thickness. They mantle the gravel beds in Area B and tend to be thin inland due to the rise in elevation of Units 1 and 2.
- *3Ab*—A thin paleosol was formed at the top of and within Unit 3, and contained low densities of early Late Woodland cultural materials.
- *Unit 4*—This eolian sand unit infilled a swale behind the second foredune at the site, with the base of the swale underlain by the 3Ab. Thus, Unit 4 is of variable thickness, tending to be thicker and deeper inland, toward the east. A hearth, Feature 3, isolated in the upper part of the sand deposits was dated to 1000 ± 140 BP (M-2401; 2σ 1182–675 cal BP, intercept 930 cal BP), suggesting that eolian activity was still progressing during the eleventh century A.D.
- *2Ab*—This paleosol is a thin zone of organic-enriched sand formed at the top of Unit 4 and containing cultural materials. It indicates a brief period of stabilization and is generally level where encountered across the site.
- *Unit 5*—This eolian sand unit is culturally sterile, generally level, and of relatively uniform thickness, and separates the 2Ab and Ab paleosols.
- *Ab*—The Ab unit is formed within Unit 5, and is the uppermost paleosol, containing abundant cultural material. Feature 8, a hearth associated with this horizon, was dated to 430 ± 100 BP (M-2398, 2σ 568–287 cal BP, intercept 515 cal BP). As noted above, ceramic sherds from this dated context refit with sherds from Area A, Occupation Zone II, suggesting contemporaneity of deposition.
- *Unit 6*—Unit 6 is a thin mantle of eolian sand that appears to be of relatively recent origin that clearly postdates circa 515 cal BP.

The depositional sequence at the O'Neil site provides significant information regarding cyclic patterns of inundation, recession, eolian activation, and dune stabilization. Nipissing terraces are present above the O'Neil site at altitudes of 186 m, whereas the Algoma terrace has largely been obscured through cyclic burial by eolian sand. It is likely that the Nipissing transgression is responsible for the basal lacustrine sands at the site.

Bedded gravel deposits at 178 m, however, would suggest raised levels of the Michigan basin and associated nearshore activity. It is possible that these deposits are associated with either the Algoma stage recession from the Nipissing stage circa 3,200 to 2,500 years ago, or with subsequent lake transgression circa 2,900 to 2,500 years ago. The latter event has been documented in both the Michigan basin (Mason 1965) and the Huron basin (Lovis et al. 2005; Monaghan and Lovis 2005; Speth 1972). The dating and elevation of the transgressive event is supported by the presence of beach gravels underlying a Middle Woodland occupation at the Wycamp Creek site, at an altitude of 179.8 m (C. Larsen personal communication; Branstner and Cleland 1994). The overlying occupation has been dated to circa 1,320 years ago. While this earlier age is preferable, there is also a slight possibility that the gravel zone may be a product of a transgression after circa 1,200 years ago documented by McPherron (1967) at the Juntunen site.

Subsequent recession of the basin to below 178 m apparently resulted in eolian activation, and the onset of human occupation at the site by circa 900 years ago. By about 800 to 700 years ago a deep and partially stable hollow had formed inland of the foredune and was occupied. These occupations, and the dune hollow, were inundated and sealed by low-energy deposits to elevations of 179 m. A similar event is documented at the Beyer site in St. Ignace, where a sealed hearth feature yielded a ^{14}C age of 680 ± 90 BP (N-1726; Fitting and Clarke 1974). At the O'Neil site it is likely that the context for this sedimentation at the O'Neil site was the back-flooding of Inwood Creek caused by higher basin elevations. The inconsistent ^{14}C date from this stratum is probably a consequence of partial erosion of earlier occupations and the incorporation of organics into the back-flood deposits.

Regression of the Lake Michigan basin allowed reoccupation of the O'Neil site by circa 500 to 400 years ago on a partially stabilized surface. This surface was then capped by eolian sand that sealed the occupation surface. Ultimately, the entire surface of the site stabilized by about 250 years ago, based on historic artifacts recovered from the A horizon. The depositional history of the O'Neil site has been reinterpreted in accord with our standardized nomenclature. This reinterpretation is reflected in the stratigraphic section (figure C-19), which associates eolian sand units with paleosol formation on an episodic basis.

Discussion and Archaeological Significance

The O'Neil site displays cyclic seasonal occupation by Straits of Mackinac populations of the Mackinac and Juntunen phases, as well as by populations from the Grand Traverse Bay region during the Skegemog and Traverse phases. The site was used for both smaller activity-specific logistical purposes and large multiple--activity residential purposes. The Skegemog component at O'Neil is clearly logistic in nature, whereas the Mackinac, Juntunen, and Traverse occupations are all residential based on the presence of structures and a broad range of activity sets. Grand Traverse Bay–centered groups changed their use of the O'Neil site over time from logistic to residential, whereas the site was used residentially by more northern groups throughout the Late Woodland period. This pattern of mutual use argues for low and permeable spatial boundaries among Late Woodland social groups in the region.

The Mackinac Phase occupation and other early Late Woodland occupations are the least well represented components at the O'Neil site. The Mackinac phase component displays three activity areas all of which have evidence of domestic activity in the form of ceramic containers. Whether they were occupied sequentially or are contemporary is moot. Spring spawning fish are present in this and other parts of Area A. The reduction of Norwood chert cores appears to have occurred largely within the Area A dune hollow, with more generalized activities inland of this location. Multiple activity sets suggest a residential use of the site.

The Skegemog Phase occupation of the O'Neil site, on the other hand, is more spatially restricted and functionally discrete, and therefore contrasts with that of the Mackinac Phase. Seasonality of this occupation cannot be confidently assessed, although it appears to have been a transient logistic locale.

There are two major episodes of Juntunen Phase occupation at the O'Neil site variously spanning both Area A and Area B. The earlier of the two occupations is largely arranged around a series of hearths within the dune hollow of Area A.

Dated to circa 800 to 700 years ago, the fauna from this occupation is dominated by spring-spawning fish. The presence of large ceramic vessels, hearths, faunal remains, and chert reduction speaks to a generalized series of domestic and food-processing activities. This occupation can be interpreted as a spring / early summer residential occupation associated with local-level extraction of food, as well as chert procurement from the nearby Pi-wan-go-ning Quarry.

Between circa 600 to 500 years ago the O'Neil site was once again visited and occupied by Juntunen Phase groups. This occupation is spatially more extensive, incorporating both the dune hollow of Area A and the more stable terrace inland of the backdune. Much like the earlier Juntunen occupation there is a series of hearths in the dune hollow, with an array of fauna, lithics, and ceramic vessels arrayed around the hearths. Chert reduction activities and either animal-processing or refuse disposal are spatially well defined during this occupation. Of particular interest is the presence of an oval structure on the stable surface in Area B, which despite the lack of clear association with Juntunen Phase diagnostics, is most likely associated with this occupation. Structurally this occupation is quite similar to the earlier Juntunen occupation and likely is a residential site supporting local-level extractive forays for food and chert procurement.

While the Traverse Phase occupation at the O'Neil site occupies a fairly large area, no seasonal data are available to assess the timing of use by this more southern group. However, the two areas of Traverse occupation reveal domestic activity, and highly generalized tool inventories. These occupations are most likely intermittent residential occupations of the O'Neil site.

The turn of the eighteenth-century Native American occupation is highly constrained spatially, and comparatively dated on the basis of distinctive artifact styles including beads, knife blades, and other materials. This occupation appears to be a transient residential campsite with hunting-related activities. The fact that cultural materials from this occupation are found in the modern A horizon reveals that by at least circa 300 years ago the backdune surface at the O'Neil site had stabilized.

In summary, spatial analysis of the O'Neil site reveals regular patterning and organization despite the presence of overlapping activity sets (Lovis 1990b): "When features could be associated with discrete occupations and/or activities, these distributions apparently center on a hearth or series of hearths. Where stable occupation surfaces are present, the orientation of such distributions may vary. Where unstable surfaces are present, and/or where topographic features such as dune hollows or lake terraces are linearly oriented, the distribution of activity areas is oriented with respect to these features" (Lovis 1990b, 209).

Antrim Creek Natural Area, Antrim County

The Antrim Creek Natural Area (ACNA) is a small township park in the northwestern part of Antrim County that is located in the NW ¼, Section 14, T32N, R9W, Banks Township (figure c-20). The park extends along the Lake Michigan coast for about 1 km and is bordered circa 200 m to the east by a prominent (ca. 12 m high) bluff that is apparently associated with the Nipissing high stand. Along this bluff, a narrow terrace is cut into the exposure about 8 m below the top of the surface. Between this bluff and the beach is a sequence of three dune ridges (figures c-21

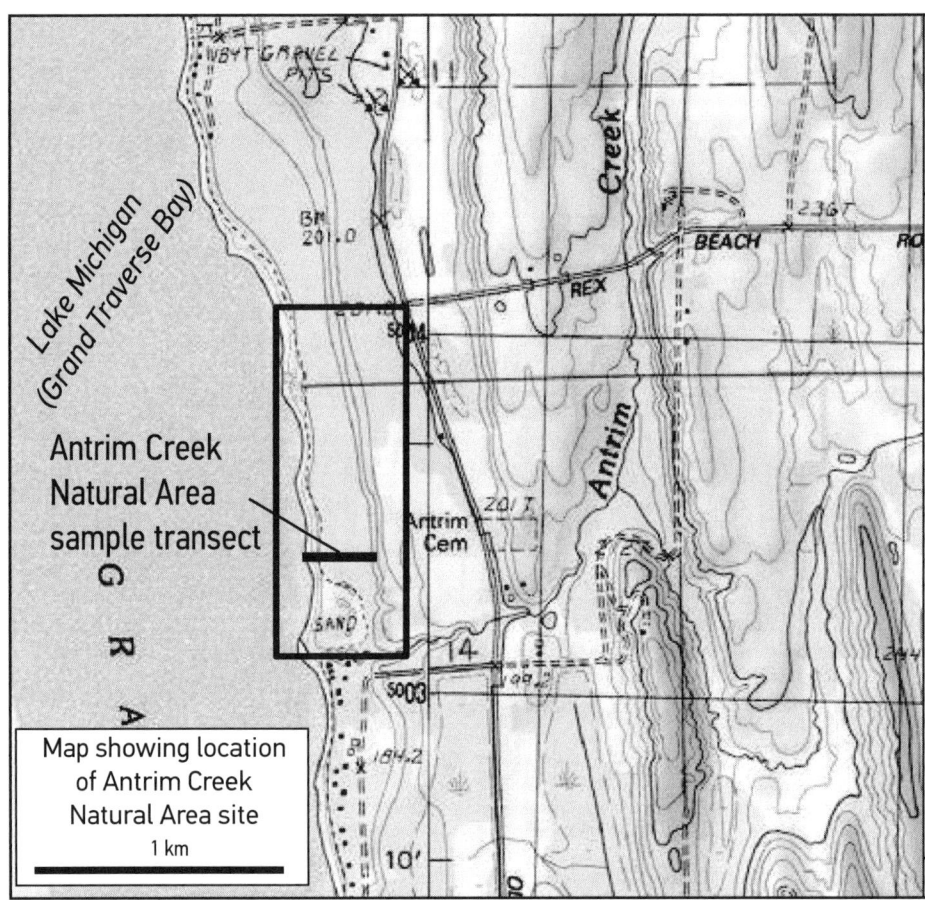

C-20. The Antrim Creek Natural Area sample location.

C-21. A dune swale that includes an ephemeral buried cultural horizon at the Antrim Creek Natural Area.

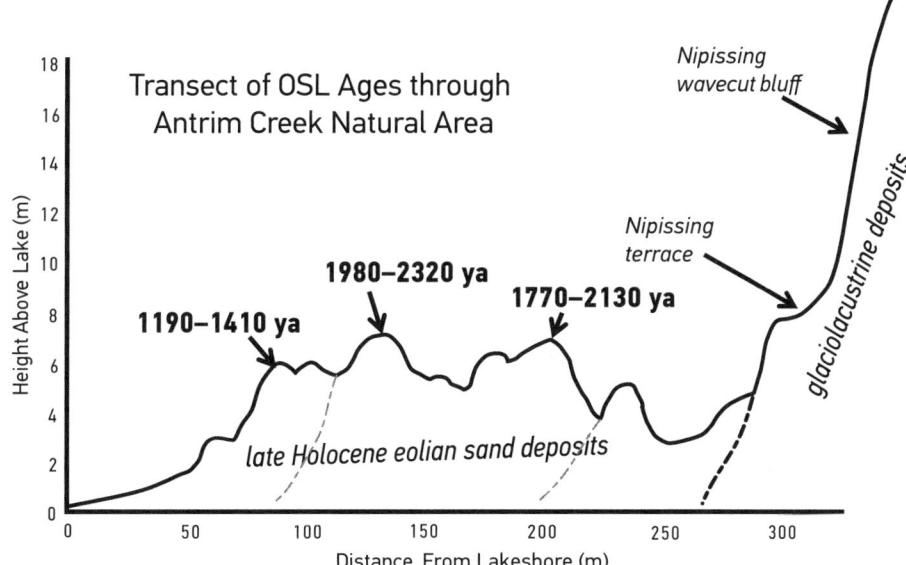

C-22. Sample transect through dune ridges between the Nipissing wave terrace and the Lake Michigan beach.

and C-22) that overlie beach gravels. According to Cleland (2002), archaeological remains from two cultural periods occur in the park. He interprets the oldest site as the Nipissing 2 Terrace site, which is a Late Archaic site confined to the remnants of a Nipissing 2 stage terrace forming a bench at an elevation several meters below that of the Nipissing 1 terrace. A ^{14}C date of circa 600 years ago (Raviele 2006, 28) suggests, however, that there are either occupations of several ages on the terrace, or that it was the location of an aceramic Late Woodland lithic reduction station. In addition to the Nipissing 2 Terrace site, an extensive, long, and linear Late Woodland site occurs within the dunes and is characterized by relatively near surface deposition and the presence of ceramics and diagnostic lithics. This site has not been dated.

In an effort to determine the age of the dunes, OSL ages were acquired from a depth of about 1.5 m in soil pits excavated in each of the dune ridges in a transect that extends west-east across the park. The most easterly dune ridge returned a date of 1950 ± 180 ya (Shfd-06142), whereas the sample from the middle ridge provided an age of 2150 ± 170 ya (Shfd-07001). The most westerly of the three ridges returned a date of 1300 ± 110 ya (Shfd-06141).

Geomorphic research conducted at ACNA suggests that the evolution of the landscape is directly tied to the history of late Holocene lake-level fluctuations in Lake Michigan. The landscape began to evolve during the Nipissing phase between circa 5,500 and 4,000 years ago when a prominent wave-cut bluff was formed. After the Nipissing 1 transgression, lake levels fell to about 178 m before rising to 180 m during the Nipissing 2 transgression circa 4,800–4,300 years ago (Monaghan and Lovis 2005). The basal beach gravels were probably deposited at this time.

The formation of dunes at ACNA is apparently related to lake-level fluctuations after circa 2,500 years ago. The topography of the dune field indicates that the dunes are most likely ancient foredunes that probably formed during relatively low lake stages. OSL ages indicate that eolian sand probably first accumulated

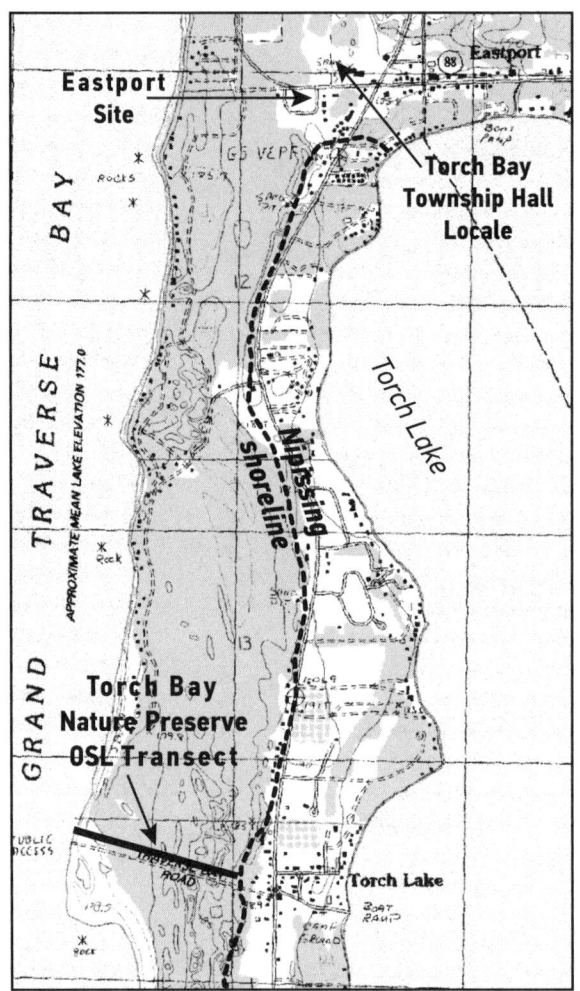

C-23. Sample locations near Torch Bay and Eastport.

C-24. GeoProbe sampling a dune in the Torch Bay Nature Preserve.

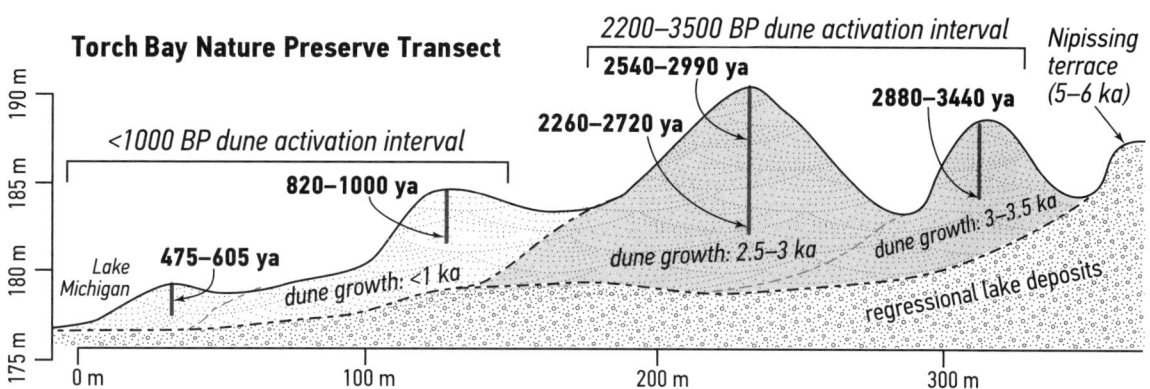

C-25. Sample transect through the Torch Bay Nature Preserve between the Nipissing shoreline and Lake Michigan.

somewhere between circa 2,500 and 2,000 years ago, forming the easternmost and middle dune ridges. The westernmost ridge apparently formed circa 1,300 years ago as lake levels fell after a relatively high lake stage circa 1,700 years ago.

Torch Bay Nature Preserve Transect and the Eastport Site (20AN24), Antrim County

The Torch Bay Nature Preserve transect is a composite sequence of dunes that were dated by OSL near the villages of Eastport and Torch Lake in Antrim County (figure C-23). The Eastport site is located immediately west of the Torch Bay Township Hall in the SE ¼, Section 1, T31N, R9W, Torch Lake Township (figure C-23) and lies on the Algonquin lake plain. The remaining dunes on the Torch Bay transect are located west of Torch Lake in the NW ¼, Section 24, T31N, R9W (figure C-23) and consist of a series of compound dune ridges that are spatially associated with a small embayment along the east coast of Traverse Bay (figures C-24 and C-25). These dunes extend about 400 m between the Nipissing bluff on the east and the modern beach. They were accessed along Traverse Bay Road.

The Eastport site has already been mentioned in the history of coastal zone archaeology presented earlier. It is located almost due west of the Torch Bay Township Hall in the Arrowhead Subdivision. The site area had been leveled of any dunes that were present. Eastport is a well known and presumably among the earliest of the Archaic sites in the region based on test excavation by a group including E. Gillis, G. Davis, and a UMMA field crew consisting of J. B. Griffin and M. L. Papworth. A description and analysis of the lithic assemblage was subsequently published by M. Papworth with L. R. Binford. This report also reconstructed both the age of the site based on geology and geomorphology, as well as the reduction sequence for triangular cache blades common during the Archaic period (Binford and Papworth 1963). The Eastport site has figured prominently in various interpretations of the Michigan Archaic since its publication.

As revealed by the site report (Binford and Papworth 1963), three test pits were excavated at the Eastport site. The site geomorphology included a well-developed spodic sequence with a substantial artifact-bearing zone embedded within it. This artifact zone largely lacked organic materials, and sloped markedly suggesting it was not on a level surface when deposited. The lack of organics prohibited obtaining a ^{14}C date from the site. The assemblage was exclusively banded varieties of Norwood chert from the nearby Pi-wan-go-ning Quarry primary source, and is primarily comprised of chipping debris from the early stages of chipped stone tool manufacture.

The artifact-bearing zone, presumed to be associated with the occupation surface, varied in both depth/elevation as well as density. Contemporary interpretations of the information presented in the 1963 report suggest that the artifact pavement may be related to preservation of dune backslopes or slip faces, or swales behind foredunes, by eolian activity and burial. The combined information related to local elevation of the Nipissing stage, the timing of its maximum altitude and subsequent regression as interpreted through the work of Hough (1958), and reconstruction of the relative altitude of the occupation zone suggested that the Eastport site had to date after circa 4,200 years ago. Comparisons with artifact assemblages from the Saginaw basin of Michigan compelled Binford and Papworth (1963) to conclude that Eastport dated between circa 3,300 and 2,500 years ago. We believed our research in the Eastport site vicinity could further clarify the age of the site and its depositional context.

Stratigraphy and Depositional History

When viewed holistically, OSL ages from the dunes in the Torch Bay transect provide a record of eolian sand mobilization in this part of Antrim County. An organic sample within beach deposits formed during the Nipissing transgression preserved in shore bluff deposits south of the transect (Beta-40305; appendix B) provides a limiting age for the Nipissing shoreline, and dune formation in the area, of 5993–6296 cal BP. Such a delimiting age for dune formation is also supported by the OSL age from a dune at the Torch Bay Township locale (figure C-23). Here, an OSL age of 5150 ± 390 ya (Shfd-06140) was obtained from a depth of 2.25 m deep within a circa 3.0 m high dune (figure C-23). The dune is near and part of the landform that comprises the Eastport site, which was discussed in chapter 5. An additional five samples were acquired with a GeoProbe from the dune ridges west of the village of Torch Lake (figure C-23 and C-25). The easternmost of these ridges yielded an age of 3160 ± 280 ya (UIC-2147) from a depth of 4.3 m. Approximately 30 m west of this dune ridge lies another, more prominent ridge that is about 15 m tall. Two samples were obtained from this ridge for age determination. The deepest sample was collected from a depth of 8.5 m and provided an age of 2490 ± 230 ya (UIC-2148). Another sample was collected from a depth of 3.0 m and yielded an age of 2765 ± 225 ya (UIC-2149). The remaining OSL ages in the Torch Lake transect were obtained from a pair of smaller ridges that lie to the west of the tallest dune ridge. The more inland of this latter pair of ridges is about 5 m tall and yielded an age of 910 ± 90 ya (UIC-2151) from a depth of 3.0 m. In contrast, the most westerly of the dune ridges is about 2.0 m high and provided an age of 540 ± 65 ya (UIC-2150) at a depth of 1.0 m.

Assessment of topographic relationships and OSL ages indicates that deposition of eolian sand began near Eastport about 5,100 years ago. This period of time nicely correlates to the Nipissing high stand, when the bluff along Traverse Bay Road was cut. The dunes are small and scattered on the Algonquin lake plain, which indicates that sand supply was low during this period of mobilization. Lake level subsequently fell, exposing the Nipissing lakebed west of the village of Torch Lake. Deposition of eolian sand apparently began circa 3,200 years ago, resulting in the most easterly dune ridge. The supply of sand apparently continued to be high until about 2,300 years ago. Sometime during this period of time, deposition of eolian sand shifted slightly westward and the largest dune ridge formed. Embedded within this ridge are some subtle parabolic forms, suggesting that the ridge was slightly reworked after it initially formed.

After the largest ridge formed, sand supply at the site clearly diminished for about 1,000 years. Following this hiatus eolian sand began to be supplied again about 900 years ago, resulting in the development of a third dune immediately west of the largest ridge. The supply of sand continued at least until 500 years ago, when a fourth ridge formed, one that is now closest to the lake. This last ridge is the smallest of the four, suggesting that the supply of eolian sand tapered off in the past few hundred years.

Discussion and Archaeological Implications

Of particular importance to the current project's archaeological direction is the information illuminating the age of the Eastport site. Based on the OSL ages from both the Eastport vicinity in close proximity to the archaeological site, as well as from the Torch Bay Nature Preserve transect, it would appear that the Eastport site is best placed chronologically between circa 5,300 and 3,200 years ago, and we believe that it probably dates to the earlier part of that interval. This would make Eastport the earliest of the intact Archaic sites in the region with the exception of the Late Paleoindian/Early Archaic Samel's Field site. It would appear that the organic-poor artifact-bearing zone was in a hollow or on the back side of a former dune that has been leveled by modern activity. Based on the data available it cannot be ascertained whether the occupation is relatively intact, or whether it is a lag pavement that has been collapsed, reworked, and secondarily redeposited.

Porter Creek South Site (20MN100), Mason County

Location and Description

The Porter Creek South site (20MN100) is located in the E ½ Section 23, T20N R17W, Grant Township, Mason County, Michigan, within the Manistee National Forest Lake Michigan Recreation Area immediately north of the Nordhouse Dunes Management Area (figure C-26). The site is situated on the south bank of Porter Creek, a small stream that flows through the substantial coastal dunes in the area. The site is approximately 35 m from the Lake Michigan shoreline atop a prominent, so-called "bluff." Since its discovery in 1978, substantial histories of the site and its investigations have been presented in multiple locations, most notably in Holman and Hambacher's initial report on their fieldwork at the site (1989), and subsequently in a more accessible form in *The Michigan Archaeologist* (Hambacher and Holman 1995), which also included published information on

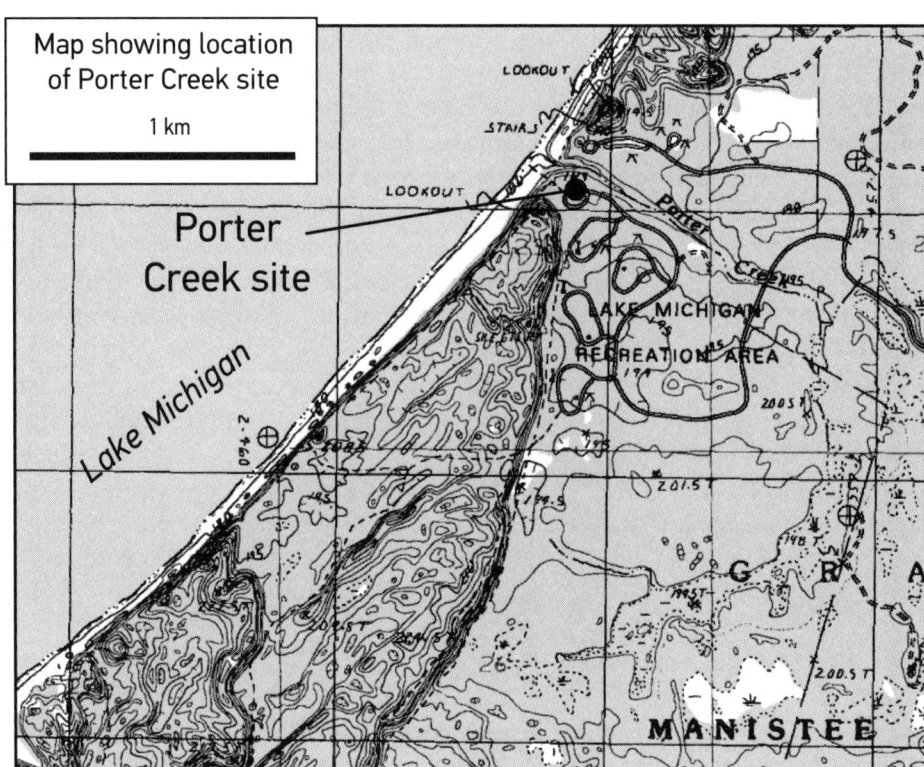

c-26. The Porter Creek site location.

the flora and fauna recovered from the investigations (Egan 1995; Smith 1995). The 1988 fieldwork at 20MN100 and its companion site 20MN31 encompassed an area of 12 acres, specifically designed to evaluate the sites for inclusion on the National Register of Historic Places. At the time, it was known that the site had produced early Late Woodland ceramics ranging in age from circa 1,400 to 1,000 years ago, and had a buried paleosol, but site depth, integrity, presence of features, or the ability to preserve subsistence materials were unknown.

Excavations at the site incorporated shovel testing, as well as the excavation of 25 formal test units of varying size. The excavation units were grouped into one larger block of 15 units, four blocks of two units each, and two isolated or individual units. Of significance was the presence of buried and stratified paleosols at multiple locations across the site, with stratigraphy as described in more detail below. Twelve features, some buried by eolian sand and others derived from occupation horizons, were recorded along with four concentrations of ceramics. Collectively, the latter speaks to excellent site integrity, and perhaps rapid burial. From our larger project perspective of the chronology of dune activation and cycling and our expectations regarding the preservation of sites of varying age, the ceramic and lithic assemblages from Porter Creek South site are highly instructive.

Based on the information provided in Hambacher and Holman (1995, 64–79), chronologically diagnostic artifacts include Levanna, Juntunen Triangular, and affinis Raccoon Notched projectile points, and typologically distinct pottery including Spring Creek, Skegemog, and Traverse Wares. The authors interpret the site as having its primary occupation during the early Late Woodland period circa 1,400 to 1,000 years ago, but with a second less intensive or extensive occupation

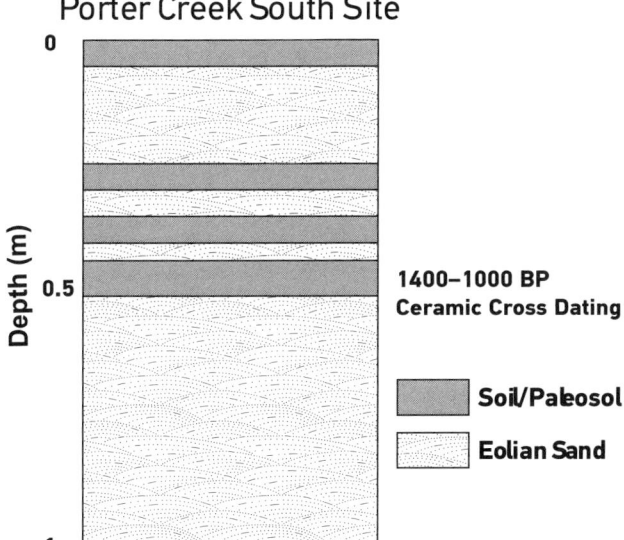

C-27. Generalized, composite stratigraphic column of the Porter Creek site.

circa 900 to 650 years ago (1995, 88). Given this solid chronology, we did not feel it was necessary to obtain absolute ages from this site. Egan's analysis of site flora reveals that occupation probably occurred between July and early September (1995), an assessment largely corroborated by the presence of beaver, turtle, and small fish in the faunal assemblage (Smith 1995).

Site Stratigraphy and Deposition

Stratigraphic sections of Units 1, 2, and 3, and Units 16 and 17 at the Porter Creek South site are presented in Hambacher and Holman (1995, 57, figure 3). These sections are arranged in common archaeological fashion as levels, which here translate to color and texture distinctions that have depositional and/or soil development implications. The most complex sequence presented in figures and text is the one for Unit 17, which displays four distinct organic paleosols, the basal one containing the primary early Late Woodland occupation (figure C-27). In our recording nomenclature this sequence translates to four depositional units as follows, from basal to current A horizon. The following stratigraphic discussion is reinterpreted from information provided in the published report (Hambacher and Holman 1995).

- *Unit 4*—This consists of eolian sand displaying the effects of vertical leaching of organics, with color sequences ranging from brown (7.5YR 5/6) in the basal part of the unit to light yellowish brown (10YR 6/4 to 6/3) in the upper part of the unit. These two color distinctions conform to Level 3 in the original description. A 4Ab soil horizon was formed in the top of Unit 4. The latter conforms to Level 2 in the original site description, and is also labeled as "Prehistoric A (A2)." From the base of the excavation to the top of the 4Ab Unit 4 is as much as 50+ cm thick.
- *4Ab*—This is a thick, approximately 15 cm thick soil formed within Unit 4,

ranging in color from 10YR 4/4 to 4/5. It contains early Late Woodland cultural material, and could potentially date as early as circa 1,500 years ago. However, correlation of stratigraphy with Pottery Concentration A in Unit 16 suggests a more likely age for burial circa 1,400 to 1,000 years ago (Hambacher and Holman 1995, 58).
- *Unit 3*—This is a sand unit overlying Unit 4. Its basal component is a thin zone lying directly atop the 4Ab. The 3Ab is formed within the top of the unit, ranging in color from 10YR4/1 to 4/2. The entire unit including the 3Ab averages 15 cm in thickness.
- *3Ab*—This is a thin, approximately 10 cm thick, soil formed within the uppermost part of Unit 3.
- *Unit 2*—Unit 2 is a zone of eolian sand slightly thicker than that found at the base of Unit 3. It, as do the other eolian sand units, displays a 10YR 6/3 color. The 2Ab is formed within the upper part of Unit 2. Including the 2Ab, Unit 2 is about 20 cm in thickness.
- *2Ab*—The 2Ab is labeled in the original report as an additional "A Horizon" in Level 1, with a color range from 10YR 4/1 to 4/2. It is approximately 10 cm in thickness, and formed within the upper part of Unit 2.
- *Unit 1*—Unit 1 at Porter Creek South is the uppermost depositional unit in what was originally designated as Level 1. It consists of a 20+ cm thick lens of eolian sand with a consistent 10YR 6/3 color. The modern A horizon is formed at the surface of Unit 1.
- *A horizon*—The modern A horizon is formed within Unit 1, and is approximately 5 cm thick.

In summary, the section at Unit 17 of the Porter Creek South site, which displays the most complex cycling of stabilization and activation, reveals four stabilization episodes. Each episode of stabilization was followed by greater or lesser amounts of eolian deposition, although the Unit 1 deposit is more substantial than the two that precede it. The 4Ab, the basal paleosol at the site, contains the early Late Woodland occupation, and while it could date as early as circa 600 years ago, it more likely dates to circa 1,000 years ago. This chronology is entirely consistent with episodes of Medieval Warm stabilization and activation cycles dated at other locales in our sample.

Discussion

The Porter Creek South site is one of the most southern sites in our distribution of coastal locales to display sequential burial and stratification of cultural deposits. In the context of this study, of particular interest is the mix of paleosols both with and without cultural material. The warm season occupation, as assessed from the fauna and flora recovered, is consistent with those of other coastal sites, as are the contents of the assemblages. We find parts of the concluding descriptions of the site particularly consistent, as well, with observations from other locales reported in our research.

"While a relatively large area south of the creek and east of the main barrier dunes contains evidence of occupation, it appears that the low spots across the site were the favored locales for occupation. Not only do these areas contain the highest density of material, but they also contain remnants of intact living surfaces. ... Although a long series of occupations is represented, the pattern of intact living surfaces with high densities of material in low spots separated from one another by areas without intact prehistoric surfaces and with low densities of prehistoric materials suggests the likelihood that discrete occupation episodes may be spatially isolated from one another" (Hambacher and Holman 1995, 94). These low areas are most likely areas inland of dune crests fronting the Lake Michigan shoreline, and are variably preserved due to local conditions of activation, deflation, and redeposition.

Ludington State Park, Mason County

Ludington State Park is a very large dune field that encompasses virtually all of Big Sable Point in northwestern Mason County (figure c-28). The dune field extends about 16 km from Nordhouse Dunes north of the park to a point just northwest of the city of Ludington. The dune field is about 2.5 km wide at it broadest point, ranging from the shore of Lake Michigan to west side of Hamlin Lake. Although a variety of dune forms occur within the park, the most prominent eolian landforms are clearly the large (ca. 30 m high) transverse dune ridges that dominate the landscape. Ridges in the eastern parts of the park tend to be forested or partially forested, whereas those in the center of the park, and close to the lakeshore, tend to be relatively free of vegetation and thus highly active.

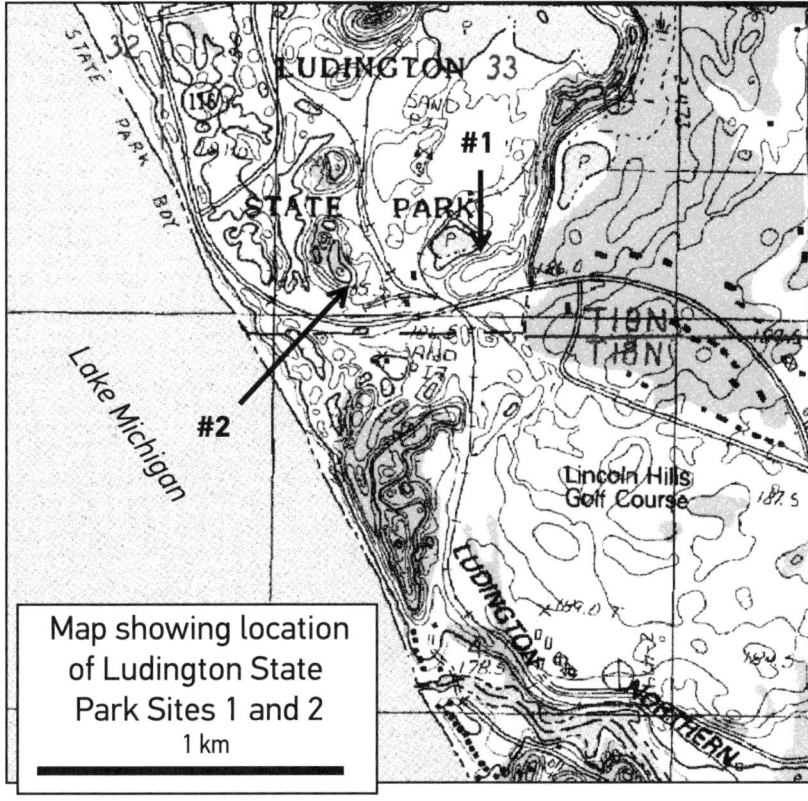

c-28. Ludington State Park sample locations, Ludington Sand Quarry Site 1 and Ludington Sand Quarry Site 2.

C-29. Ludington State Park exposures in Ludington Sand Quarry Site 1 and Ludington Sand Quarry Site 2.

Given the size of the park and the active nature of the dunes, it would be virtually impossible to sample the dunes for OSL dating with a GeoProbe. In addition no natural exposures occur anywhere within the majority of the park, making stratigraphic comparisons difficult as well. Fortunately, two sand quarries are located in the south end of the park (s ½, Section 33, T19N, R18W; figures C-28 and C-29) that expose dune sediments sufficiently for descriptive purposes and age assessments. These quarries were investigated previously in 1997 and again in association with this study.

Ludington Sand Quarry Site 1

Ludington Sand Quarry Site 1 is a sand pit located east of the old rail line in the south end of Ludington State Park (figures C-28 and C-29). It was revisited by this project in October 2007 by Arbogast and Lovis. Exposed on the south side of this quarry are about 11 m of eolian sand that overlies an organic lens at the base of the pit (figure C-30). Contained within the dune sand are three pedostratigraphic units. Unit I is about 6.5 m thick and overlies the basal organics, which provided an uncalibrated AMS age of 3000 ± 50 BP (NSRL-3969; 2σ 3346–3060 cal BP; intercept 3195 cal BP). Formed in this unit is a weakly developed Spodosol with A/E/Bs/C

Descriptions of Sample Locales

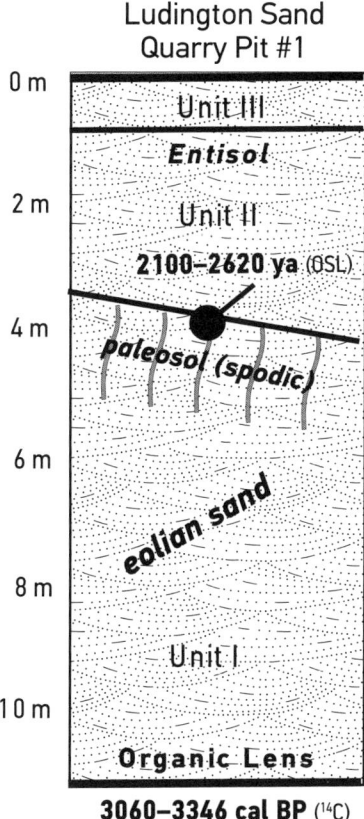

C-30. Generalized, composite stratigraphic column of Ludington Sand Quarry Site 1. Ages of various units shown. See appendix A for discussion of how ages are reported. Unit depths and boundaries are approximate.

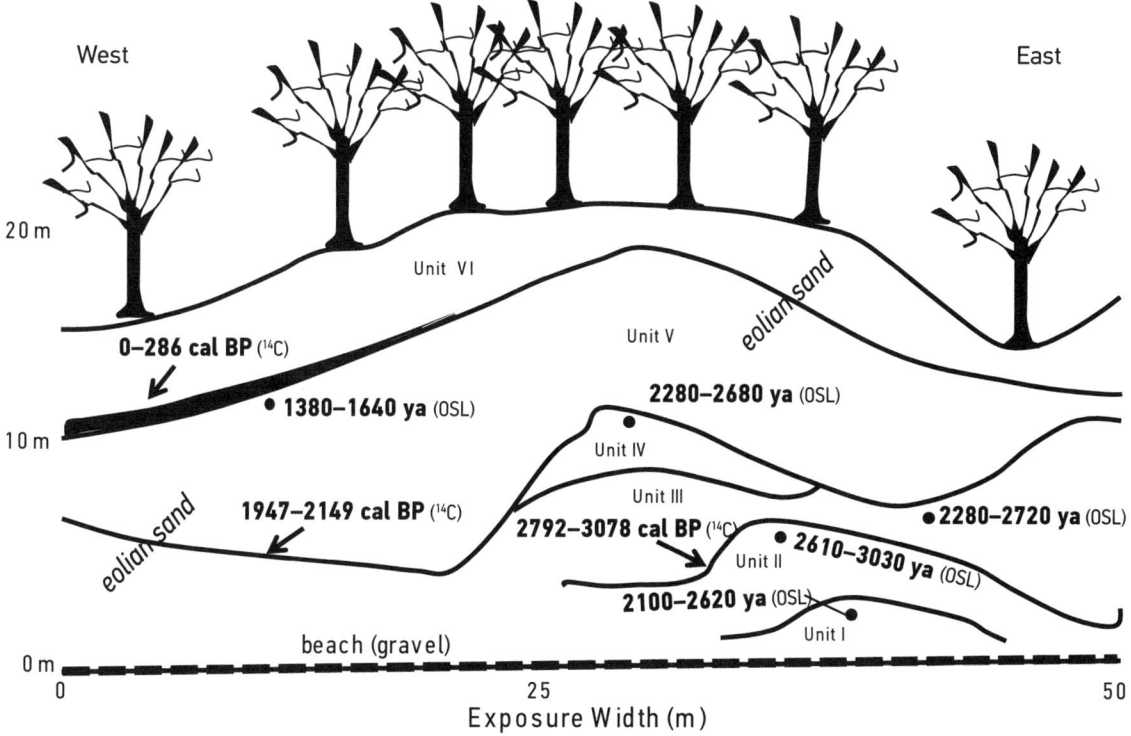

C-31. Stratigraphy of Ludington Sand Quarry Site 2.

horizonation that yielded an uncalibrated AMS age of 500 ± 50 BP (NSRL-3970; 2σ 565–475 cal BP; intercept 531 cal BP) on charcoal. This soil is in turn overlain by Unit II, which is about 3 m thick at its maximum thickness. This unit is capped by a very weakly developed Entisol (A/C horizonation) that is visible intermittently and contained no charcoal. As a result, it was not dated. This soil is overlain by about 1 m of eolian sand.

Ludington Sand Quarry Site 2

Ludington Sand Quarry Site 2 is located about 300 m west of Ludington Sand Quarry Site 1 (figures C-28 and C-29) and is visible from the road that enters the park. The south-facing wall of the quarry is an excellent exposure that is about 20 m high and about 60 m wide (figure C-31). Contained within this exposure are six pedostratigraphic units in eolian sand that overlie basal beach gravels at an elevation of 183 m. The depositional units within the dune are bounded by paleosols that are traceable across much of the exposure. As a result, two-dimensional topographic changes can be seen over time as the dune evolved. For simplicity, the description largely focuses on the central part of the exposure, where the units are generally thickest. In an effort to reconstruct the history of eolian-sand deposition at the site, three AMS ages and five OSL ages were obtained from the exposure.

The lower four depositional units appear to reflect a growing dune crest at a central location. The basal eolian unit in this sequence is Unit I, which overlies the beach gravels. This unit is about 1 m thick at its thickest point and represents the first preserved deposit of eolian sand at this site. In an effort to determine when this unit was deposited, a sample was collected from within the unit that provided an age of 2360 ± 260 ya (UIC-2207). Formed in the top of this unit is a very weakly developed Entisol with A/C horizonation. Overlying this soil is Unit II, which is a deposit of eolian sand that is about 4 m thick at its thickest point. Two ages were obtained from this unit, with one being an OSL age of 2820 ± 210 ya (UIC-2208) directly from the sands. An AMS age was also obtained from the weakly developed Entisol formed in the top of the unit. This soil contained charcoal that provided a ^{14}C age of 2830 ± 50 BP (NSRL-3967; 2σ 3078–2792 cal BP; intercept 2941 cal BP). Unit II is covered immediately by Unit III, which is about 3 m thick at its thickest point. An OSL age of 2500 ± 220 ya (UIC-2209) was obtained from the upper part of this unit, just below the weakly developed Entisol (4Ab horizon) formed in the top of the deposit. The 4Ab is, in turn, covered directly by Unit IV, which is a deposit that occurs only in the central part of the exposure and thus represents localized accumulation of about 2.5 m of eolian sand as the dune grew. An OSL age of 2480 ± 200 ya (UIC-2227) was obtained from the uppermost part of this unit. As with the depositional units below, the upper part of Unit IV is also bounded by a weakly developed Entisol (3Ab horizon) formed within it.

In contrast to the lower four depositional units, the upper pair of sand deposits at Ludington Sand Quarry Site 2 clearly extends across the entire exposure. The lowermost of these two units is Unit V, which is about 5 m thick at its thickest point. This unit overlay the 3Ab horizon formed in the uppermost part of Unit IV in the central part of the exposure and in Unit III in the eastern and western parts of the face. In an effort to estimate when this soil was buried, a ^{14}C age of 2080 ± 40 BP (NSRL-3966; 2σ 2149–1947 cal BP; intercept 2048 cal BP) was obtained from charcoal within the 3Ab horizon in the western part of the exposure. Approximately 6 m of eolian sand subsequently accumulated as Unit V across much of the dune following this burial. To determine when this pulse of eolian sedimentation ceased, an OSL sample was obtained from the upper part of Unit V. This sample provided an age of 1510 ± 130 ya (UIC-2228).

Formed in the uppermost part of Unit V is a prominent buried soil that is visible across the entire exposure. This paleosol is best developed in the western part of the face, where it consists of an overthickened (ca. 50 cm) Ab horizon formed in a slight depression. The character of this soil, coupled with its paleotopographic expression, implies that accumulation of Unit V slowed sufficiently at the end of this depositional interval for organic enrichment of the sand to begin and continue until sand supply ceased. This period of combined slow deposition and soil formation apparently continued for several hundred years and may have even ceased entirely at the end for a period of time. Stability ended when sand supply increased and deposition of Unit VI began. In an effort to determine when this most recent depositional interval began, a piece of wood from the uppermost part of the 2Ab horizon was submitted for AMS dating and provided an uncalibrated age of 150 ± 50 ya (NSRL-3965; 2σ 286–0 cal BP). Approximately 4 m of eolian sand has accumulated since that time, resulting in Unit VI.

In summary, two sand quarries in the southern end of Ludington State Park were investigated to reconstruct the history of dune formation in this part of the park. A combination of AMS and OSL ages suggests that dune formation began in the area circa 3,000 years ago when eolian sand spread inland. At Ludington Sand Quarry Site 1 this episode of dune construction resulted in the deposition of one unit of eolian sand. Farther west, at Ludington Sand Quarry Site 2, deposition of eolian sand was clearly episodic between about 3,000 and 2,500 years ago with periods of stability and soil formation that resulted in the development of Entisol with A/C horizonation. These stable episodes were clearly brief, as the age of the deposits in which the lower four paleosols occur all fall with 1 standard deviation.

While sand continued to accumulate at the site of Ludington Sand Quarry Site 2, the dune at Ludington Sand Quarry Site 1 apparently stabilized for a relatively long period of time. This period of stability is indicated by the presence of a poorly developed Spodosol that is buried in the exposure. This soil is similar to the Holland Paleosol discussed by Arbogast et al. (2004) in dunes to the south. This soil apparently continued to develop during a depositional event that occurred about 1,500 years ago at Ludington Sand Quarry Site 2 to the west, which resulted in the further growth of that dune. At Ludington Quarry Site 1, the Spodosol was apparently buried circa 500 years ago, which is consistent with burial dates of the Holland Paleosol (Arbogast et al. 2004). At the same time, deposition of eolian sand at Ludington Quarry Site 2 was sufficiently slow to allow a cumlic A horizon to form in Unit V. Rapid sand accumulation then began at this latter site about 150 years ago.

References

Aitken, M. J. 1998. *An Introduction to Optical Dating: The Dating of Quaternary Sediments by the Use of Photon-Stimulated Luminescence.* Oxford University Press, New York.

Alley, R. B. 2000. The Younger Dryas cold interval as viewed from central Greenland. *Quaternary Science Reviews* 19:213–226.

Alley, R. B., D. A. Meese, C. A. Shuman, A. J. Gow, K. C. Taylor, P. M. Grootes, J. W. C. White, M. Ram, E. D. Waddington, P. A. Mayewski, and G. A. Zielinski. 1993. Abrupt increase in Greenland snow accumulation at the end of the Younger Dryas event. *Nature* 362:527–529.

Anderton, J. B., and W. L. Loope. 1995. Buried soils in a perched dunefield as indicators of late Holocene lake-level change in the Lake Superior basin. *Quaternary Research* 44:190–199.

Andrews, J. T. 1970. Differential crustal recovery and glacial chronology (6,700 to 0 BP), West Baffin Island, N.W.T., Canada. *Arctic and Alpine Research* 2:115–134.

Arbogast, A. F. 2000. Estimating the time since final stabilization of a perched dune field along Lake Superior. *Professional Geographer* 52:594–606.

Arbogast, A. F. 2009. Sand dunes in Michigan. In *Michigan: Geography and Geology*, edited by R. J. Schaetzl, J. T. Darden, and D. Brandt. Pearson Custom Publishers, Upper Saddle River, NJ.

Arbogast, A. F., E. C. Hansen, and M. D. Van Oort. 2002. Reconstructing the geomorphic evolution of large coastal dunes along the southeastern shore of Lake Michigan. *Geomorphology* 46:241–255.

Arbogast, A. F., and W. L. Loope. 1999. Maximum-limiting ages of Lake Michigan coastal dunes: Their correlation with Holocene lake level history. *Journal of Great Lakes Research* 25:372–382.

Arbogast, A. F., R. J. Schaetzl, J. P. Hupy, and E. C. Hansen. 2004. The Holland Paleosol: An informal pedostratigraphic unit in the coastal dunes of southeastern Lake Michigan. *Canadian Journal of Earth Sciences* 41:1385–1400.

Argyilan, E. P., S. L. Forman, J. W. Johnston, and D. A. Wilcox. 2005. Optically stimulated luminescence dating of late Holocene raised strandplain sequences adjacent to Lakes Michigan and Superior, Upper Peninsula, Michigan, USA. *Quaternary Research* 63:122–135.

Baedke, S., and T. A. Thompson. 2000. A 4,700 year record of lake level and isostasy for Lake Michigan. *Journal of Great Lakes Research* 26:416–426.

Bagnold, R. A. 1941. *The Physics of Blown Sand and Desert Dunes*. Chapman and Hall, London.

Bailey, S. D., A. G. Wintle, G. A. T. Duller, and C. S. Bristow. 2001. Sand deposition during the last millennium at Aberffraw, Anglesey, North Wales as determined by OSL dating of quartz. *Quaternary Science Reviews* 20:701–704.

Ballarini, M., J. Wallinga, A. Murray, S. van Heteren, A. P. Oost, A. J. J. Bos, and C. W. E. van Eijk. 2003. Optical dating of young coastal dunes on a decadal time scale. *Quaternary Science Reviews* 22:1011–1017.

Bianchi, T. H. 1974. *Description and Analysis of the Prehistoric Ceramic Materials Recovered on the Winter Site*. Unpublished report, Western Michigan University, Kalamazoo. Copy on file at the Division of Anthropology, MSU Museum, Michigan State University, East Lansing.

Binford, L. R., and M. L. Papworth. 1963. The Eastport site. In *Miscellaneous Studies in Typology and Classification*, by A. M. White, L. R. Binford, and M. L. Papworth. Anthropological Papers No. 19. Museum of Anthropology, University of Michigan, Ann Arbor.

Binford, L. R., and G. I. Quimby. 1963. Indian sites and chipped stone materials in the northern Lake Michigan area. *Fieldiana: Anthropology* 36(12):277–307.

Blumer, B. E. 2008. *Application of OSL Dating to a High-Perched Dune Field in Northwestern Lower Michigan*. MA thesis, Department of Geography, Michigan State University, East Lansing.

Booth, R. K., and S. T. Jackson. 2003. A high-resolution record of late-Holocene moisture variability from a Michigan raised bog, USA. *Holocene* 13:863–876.

Bøtter-Jensen, L., E. Bulur, G. A. T. Duller, and A. S. Murray. 2000. Advances in luminescence instrument systems. *Radiation Measurements* 32:523–528.

Branstner, C. N., and C. E. Cleland. 1994. *A Laurel High?* Paper presented at the Consortium for Archaeological Research Symposium, Michigan State University, East Lansing. On file at the Division of Anthropology, MSU Museum, Michigan State University, East Lansing.

Brashler, J. G. 1972. *Summer 1971 Archaeological Survey of Emmet, Antrim, and Charlevoix Counties, Michigan*. Unpublished report on file at the Division of Anthropology, MSU Museum, Michigan State University, East Lansing.

Brashler, J. G., E. B. Garland, M. B. Holman, W. A. Lovis, and S. R. Martin. 2000. Adaptive strategies and socioeconomic systems in northern Great Lakes riverine environments: The Late Woodland of Michigan. In *Late Woodland Societies: Tradition and Transformation Across the Midcontinent*, edited by T. Emerson, D. McElrath, and A. Fortier, 543–582. University of Nebraska Press, Lincoln.

Bretz, J. H. 1951a. The stages of Lake Chicago—their causes and correlations. *American Journal of Science* 259:401–429.

Bretz, J. H. 1951b. Causes of the glacial lake stages in Saginaw Basin. *Journal of Geology* 59:244–258.

Bretz, J. H. 1953. Glacial Grand River, Michigan. *Michigan Academician* 38:359–382.

Bretz, J. H. 1955. *Geology of the Chicago Region, Part II—The Pleistocene*. Bulletin 65. Illinois

State Geological Survey, Urbana.

Bretz, J. H. 1963. The prehistoric Great Lakes of North America. *American Scientist* 51:84–109.

Bretz, J. H. 1964. Correlation of glacial lake stages in the Huron-Erie and Michigan basins. *Journal of Geology* 72:618–627.

Broecker, W. S. 1966. Glacial rebound and the deformation of the shorelines of pro-glacial lakes. *Journal of Geophysical Research* 71:7089–7105.

Brose, D. S. 1964. Infra-red photography: An aid to stratigraphic interpretation. *Michigan Archaeologist* 10(4):69–74.

Brose, D. S. 1970a. *The Archaeology of Summer Island: Changing Settlement Systems in Northern Lake Michigan*. Anthropological Papers No. 41. Museum of Anthropology, University of Michigan, Ann Arbor.

Brose, D. S. 1970b. *The Summer Island Site: A Study of Prehistoric Cultural Ecology and Social Organization in the Northern Lake Michigan Area*. Studies in Anthropology No. 1. Case Western Reserve University, Cleveland, Ohio.

Brose, D. S. 1970c. Summer Island III: An early historic site in the upper Great Lakes. *Historical Archaeology* 4:3–33.

Brose, D. S. 1975. The Fisher Lake site (20 Lu 21): Multicomponent occupations in northwest Michigan. *Michigan Archaeologist* 21:71–90.

Brose, D. S., and M. J. Hambacher. 1999. The Middle Woodland in northern Michigan. In *Retrieving Michigan's Buried Past, The Archaeology of the Great Lakes State*, edited by J. R. Halsey, 173–192. Bulletin 64. Cranbrook Institute of Science, Bloomfield Hills, MI.

Buckler, W. R. 1979. *Dune Type Inventory and Barrier Dune Classification, A Study of Michigan's Lake Michigan Shore*. Report of Investigation 23. Michigan Geological Survey, Michigan Department of Natural Resources Geological Survey Division, Lansing.

Buckler, W. R., and H. A. Winters. 1983. Lake Michigan bluff recession. *Annals of the Association of American Geographers* 73:89–110.

Buckley, S. B. 1974. *Study of Post-Pleistocene Ostracod Distribution in the Soft Sediment of Southern Lake Michigan*. PhD dissertation, Department of Geology, University of Illinois, Urbana-Champaign.

Calkin, P. E., and B. H. Feenstra. 1985. Evolution of the Erie-Basin Great Lakes. In *Quaternary Evolution of the Great Lakes*, edited by P. F. Karrow and P. E. Calkin, 149–170. Special Paper 30. Geological Society of Canada, St. John's, Newfoundland.

Champion, E. 1926. *Berrien's Beginnings*. Berrien County, MI.

Clark, J. A., H. S. Pranger, J. K. Walsh, and J. A. Primas. 1990. A numerical model of glacial isostasy in the Lake Michigan basin. In *Late Quaternary History of the Lake Michigan Basin*, edited by A. F. Schneider and G. S. Fraser, 111–123. Special Paper 251. Geological Society of America, Boulder, CO.

Clark, R. H., and N. P. Persoage. 1970. Some implications of crustal movement in engineering planning. *Canadian Journal of Earth Sciences* 7:628–633.

Cleland, C. E. 1967. *A Preliminary Report on the Prehistoric Resources of North Manitou Island*. Report submitted to the William R. Angell Foundation, on file at the Division of Anthropology, MSU Museum, Michigan State University, East Lansing.

Cleland, C. E. 1970. *Summary of Lake Charlevoix Survey, October 1–4, 1970*. On file at the Division of Anthropology, MSU Museum, Michigan State University, East Lansing.

Cleland, C. E. 1973. The Pi-wan-go-ning prehistoric district at Norwood, Michigan. In *Geology and the Environment: Man, Earth, and Nature in Northwestern Lower Michigan*, 85–87. Michigan Basin Geological Society, Ann Arbor.

Cleland, C. E. 2002. *Archaeological Survey of the Antrim Creek Natural Area, Antrim*

County, Michigan. Report prepared for the Antrim Natural Area Commission, Bellaire, Michigan. Copy on file at the Division of Anthropology, MSU Museum, Michigan State University, East Lansing.

Cleland, C. E., and G. R. Peske. 1968. The Spider Cave site. In *The Prehistory of the Burnt Bluff Area*, assembled by J. E. Fitting, 20–60. Anthropological Papers No. 34. Museum of Anthropology, University of Michigan, Ann Arbor.

Clemmensen, L. B., and A. Murray. 2006. The termination of the last major phase of aeolian sand movement, coastal dunefields, Denmark. *Earth Surface Processes and Landforms* 31:795–808.

Coolidge, Judge O. W. 1906. *A Twentieth Century History of Berrien County, Michigan*. Lewis Publishing Company, Chicago.

Cordoba-Lepczyk, X., and A. F. Arbogast. 2005. Geomorphic history of dunes at Petoskey State Park, Petoskey, Michigan. *Journal of Coastal Research* 21:231–241.

Cowles, E. B. 1871. *Berrien County Directory and History: Containing Historical and Descriptive Sketches of the Villages and Townships within the County, and the Names and Occupations of Persons Residing Therein*. Edward B. Cowles, Berrien, MI.

Cowles, H. C. 1899. The ecological relations of the vegetation on the sand dunes of Lake Michigan, Part I. Geographical relations of the dune floras. *Botanical Gazette* 27:95–117, 167–202, 281–308, 361–391.

Crane, H. R., and J. B. Griffin. 1958. University of Michigan radiocarbon dates III. *Science* 128:1117–1123.

Crane, H. R., and J. B. Griffin. 1959. University of Michigan radiocarbon dates IV. *American Journal of Science*, Radiocarbon Supplement 1:173–198.

Crane, H. R., and J. B. Griffin. 1960. University of Michigan radiocarbon dates V. *American Journal of Science*, Radiocarbon Supplement 2:31–48.

Crane, H. R., and J. B. Griffin. 1961. University of Michigan radiocarbon dates VI. *Radiocarbon* 3:105–125.

Crane, H.R. and J. B. Griffin. 1962. University of Michigan radiocarbon dates VII. *Radiocarbon* 4:183–203.

Crane, H. R., and J. B. Griffin. 1963. University of Michigan radiocarbon dates VIII. *Radiocarbon* 5:228–253.

Crane, H. R., and J. B. Griffin. 1964. University of Michigan radiocarbon dates IX. *Radiocarbon* 6:1–24.

Crane, H. R., and J. B. Griffin. 1965. University of Michigan radiocarbon dates XI. *Radiocarbon* 8:256–285.

Crane, H. R., and J. B. Griffin. 1972. University of Michigan radiocarbon dates XIV. *Radiocarbon* 14:155–194.

Davis, G. W., and E. V. Gillis. 1959. Grand Traverse Bay Archaic. *Coffinberry News Bulletin* 6(6). Wright L. Coffinberry Chapter, Michigan Archaeological Society, Grand Rapids.

Dorr, J. A., and D. F. Eschman. 1970. *Geology of Michigan*. University of Michigan Press, Ann Arbor.

Dow, K. W. 1937. The origin of perched dunes on the Manistee Moraine, Michigan. *Michigan Academy of Science Arts and Letters* 23:427–440.

Dreimanis, A. 1958. Beginning of the Nipissing phase of Lake Huron. *Journal of Geology* 66:591–594.

Dreimanis, A. 1969. Late-Pleistocene lakes in the Ontario and the Erie basins. *Proceedings of 12th Conference on Great Lakes Research*, 170–180. Ann Arbor, MI.

Dreimanis, A., and R. P. Goldthwait. 1973. Wisconsin glaciations in the Huron, Erie, and

Ontario lobes. In *The Wisconsinan Stage,* edited by R. F. Black, R. P. Goldthwait, and H. B. Willman, 71–106. Memoir 136. Geological Society of America, Washington, DC.

Drexler, C. W., W. R. Farrand, and J. D. Hughes. 1983. Correlation of glacial lakes in the Superior basin with eastward discharge events from Lake Agassiz. In *Quaternary Evolution of the Great Lakes,* edited by P. F. Karrow and P. E. Calkin. Special Paper 26: 303–323. Geological Association of Canada, St. John's, Newfoundland.

Duller, G. A. T., L. Botter-Jensen, and A. S. Murray. 2003. Combining infrared and green-laser stimulation sources in single-grain luminescence measurements of feldspar and quartz. *Radiation Measurements* 37:543–550.

Egan, K. C. 1995. Archaeobotanical remains from the Porter Creek South site (20MN100): A Late Woodland occupation in Mason County, Michigan. *Michigan Archaeologist* 41(2–3):95–100.

Eschman, D. F., and P. F. Karrow. 1985. Huron basin glacial lakes: A review. In *Quaternary Evolution of the Great Lakes,* edited by P. F. Karrow and P. E. Calkin, 79–93. Special Paper 30. Geological Society of Canada, St. John's, Newfoundland.

Fain, J., S. Soumana, M. Montret, D. Miallier, T. Pilleyre, and S. Sanzelle. 1999. Luminescence and ESR dating Beta-dose attenuation for various grain shapes calculated by a Monte-Carlo method. *Quaternary Science Reviews* 18:231–234.

Farmer, J. 1926. *Map of the Surveyed Parts of the Territory of Michigan.* V. Balch and S. Stiles, Engravers, Utica, MI.

Farrand, W. R. 1962. Postglacial uplift in North America. *American Journal of Science* 260:181–199.

Farrand, W. R., and C. W. Drexler. 1985. Late Wisconsinan and Holocene history of the Lake Superior basin. In *Quaternary Evolution of the Great Lakes,* edited by P. F. Karrow and P. E. Calkin, 17–32. Special Paper 30. Geological Society of Canada, St. John's, Newfoundland.

Farrand, W. R., R. Zahner, and W. S. Benninghoff. 1969. *Cary-Port Huron Interstade—Evidence from a Buried Bryophyte Bed, Cheboygan County, Michigan.* Special Paper 123. Geological Society of America, Washington, DC.

Farrell, J. P., and J. D. Hughes. 1985. *Long-term Implications, from a Geomorphological Standpoint of Maintaining H-58 in Its Present Location at Grand Sable Lake.* Contract Report PX60000-4-0839. National Park Service, Munising, MI.

Fillon, F. H. 1972. Possible causes of the variability of postglacial uplift in North America. *Quaternary Research* 1:522–531.

Finamore, P. F. 1985. Glacial Lake Algonquin and the Fenelon Falls outlet. In *Quaternary Evolution of the Great Lakes,* edited by P. F. Karrow and P. E. Calkin, 127–132. Special Paper 30. Geological Society of Canada, St. John's, Newfoundland.

Fisher, T. G., and W. L. Loope. 2005. Aeolian sand preserved in Silver Lake: A new signal of Holocene high stands of Lake Michigan. *Holocene* 15:1072–1078.

Fisher, T. G., W. L. Loope, W. Pierce, and H. M. Jol. 2007. Big lake's record preserved in a little lake's sediment: an example from Silver Lake, Michigan, USA. *Journal of Paleolimnology* 37: 365–382.

Fitting, J. E. 1967. The camp of the careful Indian: An Upper Great Lakes chipping station. *Papers of the Michigan Academy of Science, Arts, and Letters* 52 (Part II: Social Science): 237–242.

Fitting, J. E. 1970. *The Archaeology of Michigan.* Natural History Press, New York.

Fitting, J. E. 1975. *The Archaeology of Michigan.* 2nd ed. Cranbrook Institute of Science, Bloomfield Hills, MI.

Fitting, J. E., and W. S. Clarke. 1974. The Beyer site (SIS-20). In *Contributions to the Archaeology of the St. Ignace Area,* edited by J. E. Fitting. *Michigan Archaeologist* 20:227–277.

Forman, S. L., L. Marín, J. Pierson, J. Gómez, G. H. Miller, and R. S. Webb. 2005. Eolian sand depositional records from western Nebraska: Landscape response to droughts in the past 1500 years. *Holocene* 15:973–981.

Forman, S. L., and J. Pierson. 2003. Formation of linear and parabolic dunes on the eastern Snake River Plain, Idaho in the nineteenth century. *Geomorphology* 56:189–200.

Fraser, G. S., C. E. Larsen, and N. C. Hester. 1975. Climatically controlled high lake levels in the Lake Michigan and Lake Huron basins. *Anais. Academia Brasileira Ciencias* 47:51–66 (supplement).

Fraser, G. S., C. E. Larsen, and N. C. Hester. 1990. Climatic control of lake levels in the Lake Michigan and Lake Huron basins. In *Late Quaternary History of the Lake Michigan Basin,* edited by A. F. Schneider and G. S. Fraser, 75–89. Special Paper 251. Geological Society of America, Washington, DC.

Fullerton, D. S. 1980. *Preliminary Correlation of Post-Erie Interstadial Events (16,000–10,000 Radiocarbon Years Before Present), Central and Eastern Great Lakes Region, and Hudson, Champlain, and St. Lawrence Lowlands, United and Canada.* Professional Paper 1089. United States Geological Survey, Washington, DC.

Gerety, K. M., and R. Slingerland. 1983. Nature of the saltating population in wind tunnel experiments with heterogeneous size density sands. In *Eolian Sediments and Landforms,* edited by M. E. Brookfield and T. S. Ahlbrandt. Developments in Sedimentology 38. Elsevier, Amsterdam.

Gilbertson, D. D., J.-L. Schwenninger, R. A. Kemp, and E. J. Rhodes. 1999. Sand-drift and soil formation along an exposed north Atlantic coastline: 14,000 years of diverse geomorphological, climatic and human impacts. *Journal of Archaeological Science* 26:439–469.

Goble, R. J., J. A. Mason, D. B. Loope, and J. B. Swinehart. 2004. Optical and radiocarbon ages of stacked paleosols and dune sands in the Nebraska Sand Hills, USA. *Quaternary Science Reviews* 23:1173–1182.

Goldthwait, J. W. 1891. High level shores in the region of the Great Lakes and their deformation. *American Journal of Science* 41:201–221.

Goldthwait, J. W. 1908. A reconstruction of water planes of extinct glacial lakes in the Lake Michigan Basin. *Journal of Geology* 16:459–476.

Greeley, R., S. H. Williams, and J. R. Marshall. 1983. Velocities of windblown particles in saltation: Preliminary laboratory and field measurements. In *Eolian Sediments and Landforms,* edited by M. E. Brookfield and T. S. Ahlbrandt. Developments in Sedimentology 38. Elsevier, Amsterdam.

Greenman, E. F. 1926. *Fort Hill at St. Ignace, Michigan.* On file at the Museum of Anthropology, University of Michigan, Ann Arbor.

Greenman, E. F. 1927. *Museum Report: Archaeological Investigations in Emmet County, Michigan, July 22–30, 1927.* On file at the Museum of Anthropology, University of Michigan, Ann Arbor.

Greenman, E. F. 1941. *Field notes on the Eastport Site, Antrim County.* On file at the Museum of Anthropology, University of Michigan, Ann Arbor.

Halsey, J. R. 1999. *Retrieving Michigan's Buried Past: The Archaeology of the Great Lakes State.* Bulletin 64. Cranbrook Institute of Science, Bloomfield Hills, MI.

Hambacher, M. J. 1992. *The Skegemog Point Site: Continuing Studies in the Cultural Dynamics of the Carolinian-Canadian Transition Zone.* PhD dissertation, Department of Anthropology, Michigan State University, East Lansing. University Microfilms International, Ann Arbor.

Hambacher, M. J., and M. B. Holman. 1995. Camp, cache, and carry: The Porter Creek South site (20MN100) and cache pits at 20MN31 in the Manistee National Forest. *Michigan Archaeologist* 41(2–3):47–94.

Hansel, A. K., and D. M. Mickelson. 1988. A reevaluation of the timing and causes of high lake phases in the Lake Michigan Basin. Quaternary Research 29:113–128.

Hansel, A. K., D. M. Mickelson, A. F. Schneider, and C. E. Larsen. 1985. Late Wisconsinan and Holocene history of the Lake Michigan basin. In *Quaternary Evolution of the Great Lakes,* edited by P. F. Karrow and P. E. Calkin, 39–53. Special Paper 30. Geological Society of Canada, St. John's, Newfoundland.

Hansen, E. C., A. F. Arbogast, S. C. Packman, and B. Hansen. 2003. Post-Nipissing origin of a backdune complex along the southeastern shore of Lake Michigan. *Physical Geography* 23:233–244.

Hansen, E. C., A. F. Arbogast, and B. Yurk. 2004. The history of dune growth and migration along the southeastern shore of Lake Michigan: A perspective from Green Mountain Beach. *Michigan Academician* 35:455–478.

Hansen, E.C., T. G. Fisher, A. F. Arbogast, and M. Bateman. 2010. Geomorphic history of low perched, transgressive dune complexes along the southeastern shore of Lake Michigan. *Aeolian Research* 1: 111–127.

Hart, J. P., and W. A. Lovis. 2007a. The freshwater reservoir and radiocarbon dates on cooking residues: Old apparent ages or a single outlier? Comments on Fischer and Heinemeier (2003). *Radiocarbon* 49:1403–1410.

Hart, J. P., and W. A. Lovis. 2007b. A multi-regional analysis of AMS and radiometric dates from carbonized food residues. *Midcontinental Journal of Archaeology* 32:201–260.

Hart, J. P., R. G. Thompson, and H. J. Brumbach. 2003. Phytolith evidence for early Maize (Zea Mays) in the northern Finger Lakes region of New York. *American Antiquity* 68:619–640.

Hesp, P. A. 1999. The beach, backshore and beyond. In *Handbook of Beach and Shoreface Morphodynamics,* edited by A. D. Short, pp.145–70. John Wiley, New York.

Hesp, P. A., and B. G. Thom. 1990. Geomorphology and evolution of active transgressive dunefields. In *Coastal Dunes: Form and Process,* edited by K. Nordstrom and N. Psuty, 253–288. John Wiley and Sons, New York.

Hinsdale, W. B. 1925. *Primitive Man in Michigan.* Michigan Handbook Series 1. University Museums, University of Michigan, Ann Arbor.

Hinsdale, W. B. 1931. *An Archaeological Atlas of Michigan.* Michigan Handbook Series 4. University Museums, University of Michigan, Ann Arbor.

Holman, M. B. 1978. *The Settlement System of the Mackinac Phase.* PhD dissertation, Department of Anthropology, Michigan State University, East Lansing. University Microfilms International, Ann Arbor.

Holman, M. B. 1984. Pine River Ware: Evidence for *in situ* development of the Late Woodland in the Straits of Mackinac region. *Wisconsin Archeologist* 65:32–48.

Holman M. B., and M. J. Hambacher. 1989. *An Archaeological Evaluation of the Porter Creek South Site (20MN100) and the Porter Creek Shovel Testing Program in the Manistee National Forest, Mason County, Michigan.* Archaeological Survey Report No. 101. Michigan State University Museum, Michigan State University, East Lansing.

Holman, M. B., and W. A. Lovis. 2008. Social and environmental constraints on mobility in the late prehistoric upper Great Lakes USA. In *The Archaeology of Mobility: Old*

World and New World Nomadism, edited by H. Barnard and W. Wendrich, 280–306. Cotsen Advanced Seminar 4. Cotsen Institute of Archaeology, University of California, Los Angeles.

Holmstadt, J. 2008. *Geomorphology and Geoarchaeology of a Small Foredune Complex Along the Eastern Shore of Lake Michigan.* MA thesis, Department of Geography, Michigan State University, East Lansing.

Hough, J. L. 1955. Lake Chippewa, a low stage of Lake Michigan indicated by bottom sediments. *Geological Society of America Bulletin* 66:957–968.

Hough, J. L. 1958. *Geology of the Great Lakes.* University of Illinois Press, Urbana.

Hough, J. L. 1963. The prehistoric Great Lakes of North America. *American Scientist* 51:84–109.

Hough, J. L. 1966. Correlation of glacial lake stages in the Huron-Erie and Michigan Basins. *Journal of Geology* 74:62–77.

Janzen, D. E. 1968. *The Naomikong Point Site and the Dimensions of Laurel in the Lake Superior Region.* Anthropological Papers No. 36. Museum of Anthropology, University of Michigan, Ann Arbor.

Jelgersma, S., J. de Jong, W. H. Zagwijn, and J. F. van Regteren Altena. 1970. The coastal dunes of the western Netherlands: Geology, vegetational history and archaeology. *Mededelingen Rijks. Geologische Dienst,* n.s. 21:93–167.

Johnson, T. C., R. D. Stieglitz, and A. M. Swain. 1990. Age and paleoclimatic significance of Lake Michigan beach ridges at Baileys Harbor, Wisconsin. In *Late Quaternary History of the Lake Michigan Basin,* edited by A. F. Schneider and G. S. Fraser, 67–74. Special Paper 251. Geological Society of America, Boulder, CO.

Karrow, P. F. 1980. The Nipissing transgression around southern Lake Huron. *Canadian Journal of Earth Sciences* 17:1271–1274.

Karrow, P. F., T. W. Anderson, A. H. Clarke, L. D. Delorme, and M. R. Sreenivasa. 1975. Stratigraphy, paleontology and age of Lake Algonquin sediments in southwestern Ontario. *Quaternary Research* 5:49–87.

Kaszicki, C. A. 1985. History of Lake Algonquin in the Haliburton region, south-central Ontario. In *Quaternary Evolution of the Great Lakes,* edited by P. F. Karrow and P. E. Calkin, 109–123. Special Paper 30. Geological Society of Canada, St. John's, Newfoundland.

Kinietz, W. V. 1929. *Survey Log: Oceana, Mecosta, and Newaygo Counties.* On file at the Museum of Anthropology, University of Michigan, Ann Arbor.

Lamb, H. H. 1995. *Climate History and the Modern World.* 2nd ed. Routledge, London.

Larsen, C. E. 1985a. *A Stratigraphic Study of Beach Features on the Southwestern Shore of Lake Michigan: New Evidence of Historical Lake Level Fluctuation.* Environmental Notes, Illinois State Geological Survey, Urbana.

Larsen, C. E. 1985b. Lake level, uplift and outlet incision: The Nipissing and Algoma Great Lakes. In *Quaternary Evolution of the Great Lakes,* edited by P. F. Karrow and P. E. Calkin, 63–77. Special Paper 30. Geological Society of Canada, St. John's, Newfoundland.

Larsen, C. E. 1987. *Chronological History of Glacial Lake Algonquin and the Upper Great Lakes.* Bulletin 180. United States Geological Survey, Washington, DC.

Larson, G. J., and R. J. Schaetzl. 2001. Origin and evolution of the Great Lakes. *Journal of Great Lakes Research* 27:518–546.

Leach, M. L. 1903. History of the Grand Traverse region. *Michigan Pioneer and Historical Collections* 32:14–175.

Lee, G. B., W. E. Janke, and A. J. Beaver. 1962. Particle-size analysis of Valders Drift in eastern Wisconsin. *Science* 138:154–155.

Leverett, F. 1897. *The Pleistocene Features and Deposits of the Chicago Area.* Bulletin 2. Geology and Natural History Survey, Chicago Academy of Sciences.

Leverett, F., and F. B. Taylor. 1915. *The Pleistocene of Indiana and Michigan and the History of the Great Lakes.* Monograph 53. U.S. Geological Survey, Washington, DC.

Lewis, C. F. M. 1969. Late Quaternary history of lake levels in the Huron and Erie basins. In *Proceedings 12th Conference on Great Lakes Research,* 250–270. International Associations for Great Lakes Research.

Lewis, C. F. M. 1970. Recent uplift of Manitoulin Island. *Canadian Journal of Earth Sciences* 7:665–675.

Lewis, C. F. M., and T. W. Anderson. 1989. Oscillations of levels and cool phases of the Laurentian Great Lakes caused by inflows for Glacial Lakes Agassiz and Barlow-Ojibway. *Journal of Paleolimnology* 2:99–146.

Lewis, C. F. M., and T. W. Anderson. 1992. Stable isotope (O and C) and pollen trends in eastern Lake Erie: Evidence for a locally-induced climatic reversal of Younger Dryas Age in the Great Lakes basins. *Climate Dynamics* 6:241–250.

Lichter J. 1995. Lake Michigan beach-ridge and dune development, lake level and variability in regional water balance. *Quaternary Research* 44:181–189.

Livingstone, I., and A. Warren. 1996. *Aeolian Geomorphology: An Introduction.* Addison Wesley Longman, Harlow, England.

Loope, W. L., and A. F. Arbogast. 2000. Dominance of an 150-year cycle of sand-supply change in Late Holocene dune-building along the eastern shore of Lake Michigan. *Quaternary Research* 54:414–422.

Lovis, W. A. 1973. *Late Woodland Cultural Dynamics in the Northern Lower Peninsula of Michigan.* PhD dissertation, Department of Anthropology, Michigan State University, East Lansing. University Microfilms International, Ann Arbor.

Lovis, W. A. (editor). 1976a. *Archaeological Investigations within Fisherman's Island State Park: 1976 Season.* Archaeological Survey Report Number 10. Michigan State University Museum, Michigan State University, East Lansing.

Lovis, W. A. 1976b. Quarter sections and forests: An example of probability sampling in the Northeastern Woodlands. *American Antiquity* 41:364–372.

Lovis, W. A. 1978a. A numerical taxonomic analysis of changing Woodland site location strategies on an interior lake chain. *Michigan Academician* 11(1):39–48.

Lovis, W. A. 1978b. A case study of construction impacts on archaeological sites in Michigan's Inland Waterway. *Journal of Field Archaeology* 5:357–360.

Lovis, W. A. 1990a. Site formation processes and the organization of space at the stratified Late Woodland O'Neil site. In *The Woodland Tradition in the Western Greet Lakes: Papers Presented to Elden Johnson,* edited by G. Gibbon, 195–211. Publications in Anthropology No. 4. University of Minnesota Press, Minneapolis.

Lovis, W. A. 1990b. Curatorial considerations for systematic research collections: AMS dating a curated ceramic assemblage. *American Antiquity* 55:382–387.

Lovis, W. A. 1990c. Accelerator dating the ceramic assemblage from the Fletcher Site: Implications of a pilot study for interpretation of the Wayne Period. *Midcontinental Journal of Archaeology* 15:37–50.

Lovis, W. A. 2009. Hunter-Gatherer adaptations and alternate perspectives on the Michigan Archaic: Research problems in context. In *Archaic Societies: Diversity and Complexity Across the Midcontinent,* edited by T. Emerson, D. McElrath, and A. Fortier, 725–754. State University of New York Press, Albany.

Lovis, W. A. N.d. Draft Wycamp Creek site analytic materials on file at the Division of

Anthropology, MSU Museum, Michigan State University, East Lansing.

Lovis, W. A., R. E. Donahue, and M. B. Holman. 2005. Long distance logistic mobility as an organizing principle among northern hunter-gatherers: A Great Lakes Middle Holocene settlement system. *American Antiquity* 70:669–693.

Lovis, W. A., and M. B. Holman. 1976. Subsistence strategies and population: A hypothetical model for the development of Late Woodland in the Straits of Mackinac/Sault Ste. Marie area. *Michigan Academician* 8:267–276.

Lovis, W. A., R. C. Mainfort, and V. E. Noble. 1976. *An Archaeological Inventory and Evaluation of the Sleeping Bear Dunes National Lakeshore.* Archaeological Survey Report No. 5. Michigan State University Museum, Michigan State University, East Lansing.

Lovis, W. A., G. Rajnovich, and A. Bartley. 1998. Exploratory cluster analysis, temporal change, and the Woodland ceramics of the Portage site at L'Arbre Croche. In *Papers in Honor of Ronald J. Mason,* guest edited by R. Birmingham and C. Cleland. *Wisconsin Archeologist* 79:89–112.

Lynott, M. J., F. Frost, H. Neff, J. W. Cogswell, and M. D. Glascock. 1998. Prehistoric occupation of the Calumet Dune Ridge, northwest Indiana. *Midcontinental Journal of Archaeology* 23:221–260.

Martin, S. R. 1985. *Models of Change in the Woodland Settlement of the Northern Great Lakes Region.* PhD dissertation, Department of Anthropology, Michigan State University, East Lansing. University Microfilms International, Ann Arbor.

Martin, S. R. 1989. A reconsideration of aboriginal fishing strategies in the northern Great Lakes region. *American Antiquity* 54:594–604.

Martin, T. J. 1980. Animal remains from the Winter site, A Middle Woodland occupation in Delta County, Michigan. *Wisconsin Archeologist* 61:91–99.

Martin, T. J. 1982. *Animal Remains from the Scott Point Site: Evidence for Changing Subsistence Strategies during the Late Woodland Period in Northern Michigan.* Paper presented at the 86th Annual Meeting of the Michigan Academy of Science, Arts, and Letters, Kalamazoo.

Mason, R. J. 1966. *Two Stratified Sites on the Door Peninsula of Wisconsin.* Anthropological Papers No. 26. Museum of Anthropology, University of Michigan, Ann Arbor.

Mason, R. J. 1967. The North Bay component at the Porte des Morts site, Door County, Wisconsin. *Wisconsin Archeologist* 48:267–345.

Mason, R. J. 1981. *Great Lakes Archaeology.* Academic Press, New York.

Mason, R. J. 1988. *Rock Island: Historical Indian Archaeology in the Northern Lake Michigan Basin.* Special publication, *Midcontinental Journal of Archaeology.* Kent State University Press, Kent, Ohio.

McPherron, A. 1967. *The Juntunen Site and the Late Woodland Prehistory of the Upper Great Lakes Area.* Anthropological Papers No. 30. Museum of Anthropology, University of Michigan, Ann Arbor.

Mejdahl, V., and H. H. Christiansen. 1994. Procedures used for luminescence dating of sediments. *Boreas* 13:403–406.

Mickelson, D. M., L. Clayton, D. S. Fullerton, and H. W. Borns. 1983. The late Wisconsin glacial record of the Laurentide ice sheet in the United States. In *Late-Quaternary Environments of the United States.* Vol. 1, *The Late Pleistocene,* edited by H. E. Wright Jr., 3–38. University of Minnesota Press, Minneapolis.

Monaghan, G. W. 2002. Geoarchaeology of the Marquette Viaduct Relocation Project sites 20BY28 and 20BY387, Bay City, Michigan. In *A Bridge to the Past: The Post-Nipissing Archaeology of the Marquette Viaduct Replacement Project Sites 20BY28 and 20BY386,*

Bay City, Michigan. edited by W. A. Lovis, 2.1–2.38. Michigan State University Museum and Department of Anthropology, Michigan State University, East Lansing.

Monaghan, G. W., and A. K. Hansel. 1990. Evidence for the Intra Glenwood (Mackinaw) Low Water Phase of Glacial Lake Chicago. *Canadian Journal of Earth Sciences* 27:1236–1241.

Monaghan, G. W., and D. H. Hayes. 1998. Millennial-scale patterns to Middle-to-Late Holocene alluviation in the Great Lakes and Mid Atlantic regions, USA. *Geological Society of America. Programs with Abstracts* 30:7.

Monaghan, G. W., and D. H. Hayes. 2001. Archaeological site burial: A model for site formation within middle-to-late Holocene alluvial settings of the Great Lakes and Mid-Atlantic regions, USA. Geological Society of America. *Programs with Abstracts* 33.

Monaghan, G. W., and G. J. Larson. 1986. Drift stratigraphy of the Saginaw Lobe, south-central Michigan. *Geological Society of America Bulletin* 97:329–334.

Monaghan, G. W., G. J. Larson, and G. D. Gephart. 1986. Late Wisconsinan drift stratigraphy of the Lake Michigan lobe in southwestern Michigan. *Geological Society of America Bulletin* 97:324–328.

Monaghan, G. W., and W. A. Lovis. 2002. *Middle and Late Holocene Lake Level and Environmental Stability: Implications for Archaic Settlement Potential in the Huron Basin.* Presented at the symposium *La Periode Pre-ceramique dans le Nord-Est,* 35th Annual Meeting of the Canadian Archaeological Association, Ottawa, Ontario.

Monaghan, G. W., and W. A. Lovis. 2005. *Modeling Archaeological Site Burial in Southern Michigan: A Geoarchaeological Synthesis.* Environmental Research Series No. 1. Michigan Department of Transportation, Michigan State University Press, East Lansing.

Monaghan, G. W., W. A. Lovis, and L. Fay. 1986. The Nipissing transgression in the Thumb area of Michigan. *Canadian Journal of Earth Sciences* 23:1851–1854.

Monaghan, G. W., W. A. Lovis, and K. C. Egan-Bruhy. 2006. The earliest *Cucurbita* from the Great Lakes Region, Northern USA. *Quaternary Research: An Interdisciplinary Journal* 65(2):216–222.

Morton, J. D., and H. Schwarcz. 2004. Paleodietary implications from stable isotopic analysis of residues on prehistoric Ontario ceramics. *Journal of Archaeological Science* 31:503–517.

Murray, A. S., and L. B. Clemmensen. 2001. Luminescence dating of Holocene aeolian sand movement, Thy, Denmark. *Quaternary Science Reviews* 20:751–754.

Murray, A. S., and A. G. Wintle. 2003. The single aliquot regenerative dose protocol: Potential for improvements in reliability. *Radiation Measurements* 37:377–381.

Olson, J. S. 1958a. Lake Michigan dune development: Wind-velocity profiles. *Journal of Geology* 66:254–263.

Olson, J. S. 1958b. Lake Michigan dune development: Plants as agents and tools in geomorphology. *Journal of Geology* 66:345–351.

Olson, J. S. 1958c. Lake Michigan dune development: Lake level, beach, and dune oscillations. *Journal of Geology* 66:473–483.

Peebles, C. S., and D. B. Black. 1976. *The Distribution and Abundance of Archaeological Sites in the Coastal Zone of Michigan.* Report to the Michigan History Division, Michigan Department of State, by the Division of the Great Lakes, Museum of Anthropology, University of Michigan, Ann Arbor.

Peltier, W. R. 1994. Ice age paleotopography. *Science* 265:195–201.

Peltier, W. R. 2002. Global glacial isostatic adjustment: Palaeogeodetic and space-geodetic tests of the ICE-4G (VM2) model. *Journal of Quaternary Science* 17:491–510.

Peltier, W. R. 2004. Global glacial isostasy and the surface of the ice-age earth: The ICE-5G

(VM2) model. *Annual Review of Earth and Planetary Science* 32:111–149.

Peske, G. R., and B. Kent. 1963. *Michigan Site Survey, Upper Peninsula, 1963*. On file at the Museum of Anthropology, University of Michigan, Ann Arbor.

Peterson, J. M., and E. Dersch. 1981. *A guide to sand dune and coastal ecosystem functional relationships*. Extension Bulletin E-1529. Michigan State University, East Lansing, MICHU-SG-81-501.

Prescott, J. R., and J. T. Hutton. 1994. Cosmic ray contributions to dose rates for luminescence and ESR dating: Large depths and long-term time variations. *Radiation Measurements* 23:497–500.

Quimby, G. I. 1966. The Dumaw Creek site: A seventeenth century prehistoric Indian village and cemetery in Oceana County, Michigan. *Fieldiana: Anthropology* 56(1):1–91.

Raviele, M. E. 2006. *Report of the Michigan State University Field School 2004: Excavations at the Antrim Creek Natural Area*. Report prepared for the Antrim Natural Area Commission, Bellaire, Michigan. Copy on file at the Division of Anthropology, MSU Museum, Michigan State University, East Lansing.

Reimer, P. J., M. G. L. Baillie, E. Bard, A. Bayliss, J. W. Beck, C. J. H. Bertrand, P. G. Blackwell, C. E. Buck, G. S. Burr, K. B. Cutler, P. E. Damon, R. L. Edwards, R. G. Fairbanks, M. Friedrich, T. P. Guilderson, A. G. Hogg, K. A. Hughen, B. Kromer, G. McCormac, S. Manning, C. B. Ramsey, R. W. Reimer, S. Remmele, J. R. Southon, M. Stuiver, S. Talamo, F. W. Taylor, J. van der Plicht and C. E. Weyhenmeyer. 2004. IntCal04 terrestrial radiocarbon age calibration, 0–26 cal kyr BP. *Radiocarbon* 46:1029–1058.

Richner, J. J. 1973. "Depositional History and Tool Industries at the Winter Site: A Lake Forest Middle Woodland Cultural Manifestation." MA thesis, Department of Anthropology, Western Michigan University, Kalamazoo.

Saarnisto, M. 1975. Stratigraphic studies on the shoreline displacement of Lake Superior. *Canadian Journal of Earth Sciences* 12:300–319.

Schaetzl, R. J., J. T. Darden, and D. Brandt (editors). 2009. *Michigan: Geography and Geology*. Pearson Custom Publishing, Boston.

Schoolcraft, H. R. 1851. *Personal Memoirs of a Residence of Thirty Years with the Indian Tribes on the American Frontiers*. Lippincott, Grambo and Co., Philadelphia.

Schulenburg, J. K. 2002. *The Point Peninsula to Owasco Transition in Central New York*. PhD dissertation, Department of Anthropology, Pennsylvania State University, University Park.

Scott, I. D. 1942. The dunes of Lake Michigan and correlated problems. *44th Annual Report, Michigan Academy of Science, Arts and Letters* 53–61.

Sella, G. F., S. Stein, T. H. Dixon, M. Craymer, T. S. James, S. Mazzotti, and R. K. Dokka. 2007. Observation of glacial isostatic adjustment in "stable" North America with GPS. *Geophysical Research Letters* 34, L02306:1–6.

Sherman, D. J., and B. O. Bauer. 1993. Dynamics of beach-dune systems. *Progress in Physical Geography* 17:413–447.

Short, A. D., and P. A. Hesp. 1982. Wave, beach and dune interactions in south eastern Australia. *Marine Geology* 48:259–284.

Sly, P. G., and C. F. M. Lewis. 1972. *The Great Lakes of Canada—Quaternary Geology and Limnology*. Guidebook for Field Excursions A43. 24th International Geological Congress, Montreal.

Smith, B. A. 1983. *Faunal Identifications for Seven Northern Michigan Archaeological Sites: Eckdahl-Goudreau Site, Portage Site, Halberg Site, Wood Site, Wood Extension Site, Mt. McSauba Site, Neff's A.H. Map Site*. Unpublished report on file at the Division of

Anthropology, MSU Museum, Michigan State University, East Lansing.

Smith, B. A. 1995. Summary of faunal findings from Porter Creek South site (20MN100). *Michigan Archaeologist* 41(2–3):101–118.

Smith, H. I. 1910. Preliminary list of the sites of aboriginal remains in Michigan. In *Publication 1, Biological Series 1*, 67–89. Michigan Geological and Biological Survey, Lansing.

Snyder, F. S. 1985. A *Spatial and Temporal Analysis of the Sleeping Bear Dunes Complex, Michigan (A contribution to the geomorphology of perched dunes in humid continental regions)*. PhD dissertation, Department of Geography, University of Pittsburgh, Pittsburgh.

Southwick, R. E. 1922. *Notes on Oceana County*. On file at the Museum of Anthropology, University of Michigan, Ann Arbor.

Spaulding, A. C. 1948. *Field notes on Charlevoix County*. On file at the Museum of Anthropology, University of Michigan, Ann Arbor.

Spencer, J. W. 1881. Deformation of the Algonquin beaches and the birth of Lake Huron. *American Journal of Science* 41:12–21.

Spencer, J. W. 1888. The St. Lawrence basin and the Great Lakes. *Science* 11:99–100.

Spencer, J. W. 1891. High level shores in the region of the Great Lakes, and their deformation. *American Journal of Science* 41:201–211.

Speth, J. D. 1972. Geology of the Schultz Site. In *The Schultz Site at Green Point: A Stratified Occupation Area in the Saginaw Valley of Michigan*, edited by J. E. Fitting, 53–75. Memoirs No. 4. Museum of Anthropology, University of Michigan, Ann Arbor.

Stanley, G. M. 1936. Lower Algonquin beaches of Penetanguishene Peninsula. *Geological Society of America Bulletin* 47:1933–1960.

Stanley, G. M. 1938. The submerged valley through Mackinac Straits. *Journal of Geology* 46:966–974.

Stevens, E. J. 1924. *Report on Van Buren's Archaeological Features by Townships*. On file at the Museum of Anthropology, University of Michigan, Ann Arbor.

Stuiver, M., and P. J. Reimer. 1993, Extended ^{14}C data base and revised CALIB 3.0 ^{14}C age calibration program, *Radiocarbon*, 35:215–230.

Stuiver, M., P. J. Reimer, and R. Reimer. 2006. *CALIB Radiocarbon Calibration v. 5.0.2*. Quaternary Isotope Laboratory, University of Washington, Seattle.

Sturdevant, J. T., and D. Bringelson. 2007a. *Archaeological Inventory and Testing at Sites Along the Dunes Creek Corridor 2002–2005, Indiana Dunes National Lakeshore, Porter County, Indiana*. Technical Report No. 101. Midwest Archaeological Center, National Park Service, Lincoln, NE.

Sturdevant, J. T., and D. Bringelson. 2007b. *An Archaeological Overview and Assessment of Indiana Dunes National Lakeshore, Porter County, Indiana*. Technical Report No. 97. Midwest Archaeological Center, National Park Service, Lincoln, NE.

Tague, G. C. 1946. *The Postglacial Geology of the Grand Marais Embayment, Berrien County, Michigan*. Publication 45, Geological Series No. 38. Michigan Geological Survey, Lansing.

Taylor, F. B. 1894. A reconnaissance of the abandoned shorelines of the south coast of Lake Superior. *American Geologist* 13:365–383.

Taylor, F. B. 1897. The Nipissing-Mattawa River, the outlet of the Nipissing Great Lakes. *American Geologist* 20:65–66.

Taylor, K. C., P. A. Mayewski, R. B. Alley, E. J. Brook, A. J. Gow, P. M. Grootes, D. A. Meese, E. S. Saltzman, J. P. Severinghaus, M. S. Twickler, J. W. C. White, S. Whitlow, and G. A. Zielinski. 1997. The Holocene–Younger Dryas transition recorded at Summit, Greenland. *Science* 278(5339):825–827.

Thompson, T. A. 1992. Beach-ridge development and lake-level variation in southern Lake Michigan. *Sedimentary Geology* 80:305–318.

Thompson, T. A., and S. J. Baedke. 1995. Beach-ridge development in Lake Michigan: Shoreline behavior in response to quasi-periodic lake-level events. *Marine Geology* 129:163–174.

Thompson, T. A., and S. J. Baedke. 1997. Strand-plain evidence for late Holocene lake-level variations in Lake Michigan. *Geological Society of America Bulletin* 109:666–682.

Thompson, T. A., and S. J. Baedke. 1999. Strandplain evidence for reconstructing late Holocene lake level in the Lake Michigan basin. In *Proceeding of the Great Lakes Paleo-Levels Workshop: The Last 4000 Years,* edited by C. E. Sellinger and F. H. Quinn, 30–34. Technical Memorandum ERL GLERL-113. NOAA, Ann Arbor, MI.

Thompson, T. A., and S. J. Baedke. 2000. A geologic perspective on Lake Michigan water levels. *Water Level Bulletin* (U.S. Army Corps of Engineers, Detroit District) 140:1–5.

Thompson, T. A., S. J. Baedke, and J. W. Johnston. 2004. Geomorphic expression of late Holocene lake levels and paleowinds in the upper Great Lakes. In *The Geology and Geomorphology of Lake Michigan's Coast,* edited by E. C. Hansen. *Michigan Academician* 35:355–371.

Timmons, E. A., T. G. Fisher, E. C. Hansen, E. Eisaman, T. Daly, and M. Kashgarian. 2007. Elucidating aeolian dune history from lacustrine sand records in the Lake Michigan coastal zone, USA. *Holocene* 17:789–801.

United States Department of Agriculture. 1962. *Soil Survey Manual.* Handbook No. 18. 3rd ed. Agricultural Research Administration, United States Department of Agriculture, Washington, DC.

Van Dijk, D. 2004. Contemporary geomorphic process and change on Lake Michigan coastal dunes: An example from Hoffmaster State Park, Michigan. *Michigan Academician* 35:425–53.

Van Oort, M., A. F. Arbogast, E. C. Hansen, and B. Hansen. 2001. Geomorphological history of massive parabolic dunes, Van Buren State Park, Van Buren County, Michigan. *Michigan Academician* 33:175–88.

Vreeland, F. 1924. *Resume of Trip to Northern Michigan, Report from Log Book.* On file at the Museum of Anthropology, University of Michigan, Ann Arbor.

Walcott, R. I. 1970. Isostatic response to loading of the crust in Canada. *Canadian Journal of Earth Sciences* 7:716–727.

Wilson, S. E. 2001. *Michigan's Sand Dunes: An Michigan's Study of Coastal Sand Dune Mining in Michigan.* Pamplet 7. Geological Survey Division, Michigan Department of Environmental Quality, Lansing. http://www.michigan.gov/documents/deq/PA07_304669_7.pdf.

Winchell, A. 1866. *A Report on the Geological and Industrial Resources in the Counties of Antrim, Grand Traverse, Benzie and Leelanau in the Lower Peninsula of Michigan.* Dr. Chase's Steam Printing House, Ann Arbor.

Winslow, D. A. 1876. Early history of Berrien County. *Michigan Pioneer and Historical Collections* 1:120–125.

Wintle, A. G., and A. S. Murray. 2006. A review of quartz optically stimulated luminescence characteristics and their relevance in single-aliquot regeneration dating protocols. *Radiation Measurements* 41:369–391.

Wright, G. F. 1918. Explanation of the abandoned beaches about the south end of Lake Michigan. *Geological Society of America Bulletin* 29:235–244.

Wright, J. V. 1967. *The Laurel Tradition and the Middle Woodland Period.* Bulletin 217. National Museum of Canada, Ottawa.

Index

A

aboriginal sites, 19

absolute ages. *See* AMS; OSL; radiocarbon dating

Accelerator Mass Spectrometer. *See* AMS

accretion, 52, *53*, 73, *131*, *132*

alluviation records, 77

Alma College, 20

AMS (Accelerator Mass Spectrometer): dating, 10, 11, 14, 35, 46; dune growth and, 46, 83, 108; Mt. McSauba site, 98; Scott Point site, 93, 94; Winter site, 87. *See also* radiocarbon dating

Anderton, J. B., 45

Antrim County, 20, 26, 27, 28, 29

Antrim Creek Natural Area site: archaeological site, *18*, 29–30, 130, 142, *143*, 144; dunes, 53, 113; eolian deposits, 29; importance of, 35; occupation, 29; OSL dates, 130, 138, 144; transect, 138

Arbogast, Alan F., 7, 8, 91, 96, 103, 105–6, 114, 119; Lake Michigan shore dunes and, 45–50

Arcadia area, 53, *112*, 113, 114, *118*

archaeological sites: abandonment of, 127, 128; Archaic, 33, *121*, 125, 130, 138, 143, 146; artifact location, 33; buried, 5, 125, 127–28, 133, *134*; climate and, 122, 127; coastal zone, 10, 15–16, 26–27, 29, 33–34, 130, 133; in coastal eolian dunes, *18*, *112*; cyclical burial, 5, 127; dating augmentation, 35; deflated, 25, 32, 105; deflated surfaces in, 27, 31, 33, 92, 107, 128, 142; discoverable, 128, 133; dune activation and, 35, 127; dune-based, 9–12, 21, 33–34, *129*, 133, *134*; dune formation and, 122–28; dune growth hiatuses and, 50, 56, 114, 120, 122–26; dune swales and, 34, 133; eolian deposition and, 23, 88, 94, 133; ephemeral occupation at, 98, 106; erosion, 133; field techniques and, 128; foredune complex, 3; formation and destruction of, 9; geological dunes and, 12; hinge line and, 34, 130; ISTEA buried site taphonomy/dune activation and cycling project, 83; Lake Michigan glacier, 16; middens, 85, *85–86*; occupation, cultural, 108, 109–10, 122; occupation, human, 98, 106, 126, 127; Paleo-Indian period, 19, 23, 44; precontact, 16; preservation, 6, 17, 28, 33, 35, 126–28, 133; preservation, artifacts, 127; Saginaw Bay area, 38, 106; settlement patterns and burial of, 126; shoreline processes create specific types of, *129*; soil development and, 122; survey histories, 19–20; taphonomic processes and, 35, 126, 127, 128; taphonomy, 35, 83, 110, 128, 133; topography and preservation of, 133; Wisconsin Period, 8; Woodland periods, 29, 61, 87, 108–9, 130. See also *individual archaeological sites by name*

Archaic Period, *19, 132*; archaeological sites, 20, 30, 33, *121*, 130, 138, 143, 146; artifact location, 33; buried occupation sites, 125; cultural period, 61; dune formation, 33, 127; lake levels, 67; phases, 29, 61, *61*. See also Antrim Creek site; Eastport site; North Manitou Island site; Solomon Seal site; Whisky Creek site

Arkona Stage, Saginaw basin, *61*

artifacts, 28, 83; aceramic, 26; copper, 21, 88; hearth, *3, 35*, 98, *104, 106*, 106–7, 133; lithics, 21, 24, 30, 128; points, 88; preservation of, 127; projectiles, 88. See also ceramics; sherd, samples

B

Baedke, S. J., 46, 49, *67*, 80

Bagnold, R. A., 39

Barnes Township Park site, *18*, 26, *112*, *118*, 142

Bayport chert, 106, *106*. See also chert

beach: early, 4, 46, 69; eolian, 26, 49, 120; erosion, 63, 64, 69, 73, *129*; foredunes, 17, 41, *129*; formation, 69–71, 74, 77, 116, 120; fossil, 31, 69, 70; gravel, *5*, 51, 52, 98, *99*, *105, 107*; high-water stages and, 63; hinge-line origins theory and, 71; Holocene, 69, 72, 73, 74, *79, 89, 129*; hydrographs, 74, *75;* isostatic rebound and, 69; lake level and, 73, 74, *129;* littoral sand and, 117, *118;* Nipissing-Algoma, *75;* post-Nipissing, 24, *132;* preservation, 70–71, 74, 76; raised, 42, 60, *75*, 82, 116, *129;* rebound influences on, 69, 71, *75;* recessional, *93;* regressional, 71, 73, *75*, 76, *79*, 82, 116, *132;* regressive dunefields and, 44; as sand source, 42, 43, 48, 62, 63, *131, 132;* shoreline processes create specific types of, 63, *129;* stages, 41, 71; terraces, 26, 28, *133;* transgressional, 71, *79*, 82, 116, *132;* uplift and, *75*, 76, *79*, 86, 109, 117, *129*

Bells Bay State Park Campground, 100

Benzie County, 31

Bianchi, Thomas, 24, 84, 85, 87

Big Bay de Noc, 22, 23, 84, 85

Binford, L. R., 20, 30, 89, 92

bioturbation, 33

Black, Deborah B., 10, 17, 20

Black, Glenn A., 13

Blumer, Brad, *40*, 105, 114

Bogdan-Lovis, Elizabeth, 91, 93

Bois Blanc Island. See Juntunen site

Braidwood, Robert, 21

Brashler, Janet G., 28, 29, 100

Brevort area, dunes, *53*, 74, *75*, 76, *76*, 112

Bringelson, D., 33

Brose, David S., 23, 32

Brouwer, Marieka, 105

Buckler, W. R., 43, 56, 73, 82

Buckmaster, Marla, 10, 22, 25, 92, 93, 94

C

^{14}C. See radiocarbon dating

Camp Miniwanca site, *3;* Arbogast at, 103, 105; archaeological site, *18*, 32, 107; artifact assemblage, 106, *106;* basal deposit, 107; buried hearth, *104*, 106–7, *106*, 133; ^{14}C dating at, *105*, 107; ceramics, 105, 106, 107; chert, 106, *106;* dunes, 51, *53*, 103, 105, 107; eolian sand, 102; flora, 106; location and description, 102–3, *103, 104*, 106; Lovis at, 103, 105; occupation, 105, 106, 107; OSA files and, 103; OSL dating, 107; paleosols, 103, 106, 107; stratigraphy, 103, 105, *105*, 106, 107; tools, 106; UMMA files and, 103; Woodland, 3, 105–6, 133

carbon-14 dating (^{14}C). See radiocarbon dating

Case Western Reserve University, 32

ceramics: children's, 88; coastal dune, 34; cord-impressed, 31, 105; cross-dating, 34, 85; Dane Incised, 88; dating of, 21, 24, 25–26, 34, 35; Door Peninsula,

88; Drag and Jab, 94, 95; Havana-like Middle Woodland, 88; introduction of, *19;* Juntunen, 91, 93–95; Mackinac Ware, 94, 95; NMU and, 92–93; Oneota, 94; prehistoric, 96; preservation of, 127; shell-tempered, 94, 95; sherd residue, 10; Spring Creek ware, 106; as Woodland period marker, 31, 34, 88, 91, *97,* 105–6, 109. *See also* Barnes Township Park site; Mt. McSauba site; North Bay site, 24, 87; Porter Creek site, 109; Scott Point site; South Manitou site, 31; Winter site
chalcedony, 88
Charlevoix County, 26–29, 31, 34–35, 95–102, 106, *139*
Cheboygan, 27, 62
chert, 30, 88, 101, 106, 144
Chippewa low phase, Lake Michigan, 45
Chippewa-Stanley phase lakes, *19, 61,* 65, 66, *67, 140*
Clark, J. A., 74
Cleland, Charles, 26, 92
climate: archaeological site burial and, 122, 127; drought, 7, 122; flooding, *19,* 64, *121;* Holocene "climate controlled" lake levels, 76, 77, 78; required for dune formation, 9; soil formation and, 122–25; weathering, *97,* 127; wind and dune formation, 17, 34, *39,* 40, 48, 62, 78, 114
Cordoba-Lepczyk, X., 49
coring, difficulties, 35. *See also* GeoProbe; Dual-tube; OSL; Winter site
Cowles, H. C., 41–42
Crane, H. R., 91
Cross Village survey, 20

D

Darling, Birt, 103
dating, site. *See* AMS; GeoProbe; OSL; radiocarbon dating; soil
Delta County, 84, *84. See also* Winter site
Door Peninsula, Wisconsin, *18;* archaeological sites, 22, 85, 87, 88, 125; bluff recession, 73; ceramics, 88; high water levels, 22; middens, 85; Middle Woodland deposits, 125; Middle Woodland high-water stage, 85; North Bay projectiles, 88; occupations, 87; radiocarbon dating, 87; research at, 22, 23; uplift, 4
Dorr, J. A., 43
Dow, K. W., 12, 42, 45–46, 47
downcutting, outlet channel, 68
drought, and dune growth, 7, 122
Dumaw Creek site, 32
dune activation, 1, 23, 51; age ranges for sample locales, *53;* cyclic, 5, 7, 8, 25, 33, 49, 54, 59, 63; cyclic lake level and, *54,* 76, 80; ISTEA buried site taphonomy/dune activation and cycling project, 83; models, 4, 5, 6–9, 11, 14, 45; research, 9–11, 35
dune formation, 1, 5; archaeological site burial cycles and chronology in, 125–28; archaeological site formation and northeastern coastal, 122–25; Archaic, 33, 127; backslope, 40; chronology, 111–14; crest, 40; drought and, 7, 122; earliest, 113; embayments in, 120; eolian deposition and archaeological sites, 23, 88, 94; eolian processes, 30, 37, 39, 41, 114, 117, 119–20; erosion and, 120, 122; factors required for simultaneous coastal, 9, 62, 78, 114–15; growth hiatuses and, 50, 56, 114, 120, 122–26; hinge line and, 117; isostatic uplift rate and, 115–17; lake-level cycles and, 42, 63, 115–16, 117, 119, 122; Holland Paleosol and, 120; lag hypothesis, 120, 122, *126;* late Holocene, 115–16, 119, 122; Late Woodland, 33, 34, 127, 128; middle Holocene, 13, 51, 74, 111, 113, 115, *131–32;* occupied surface stabilization and burial in, 123–25; OSL dating and reconstruction of, 111, 114, 119; rebound and, 115, 116; reconstructions, 55–56, 111, 114, 119; regressions and, 115; sampling areas, *112;* sand movement and, 37, *39, 39,* 40–43, *40–41,* 117; sand supplies, 42, 43, 63, 78; sand supplies, high, 114–15, 117, 119–20, 122; shoreline processes create specific types of, 129; size theories, 46–47, 120; slip face, 40; spatial-temporal model for coastal,

114, 115, 117–22; stabilization and, 111; storm deposit effect on, 22, 63, 85, 122; swales, 28, 34, 109; transgressions and, 115, 122; wind and, 17, 34, *39,* 40, 48, 62, 78, 114

dune sites: geological, 3, 12; mapped dunefields, 2; rare, 119

dunefield: activation, 1, 5, 6–9, 23, 51, *53;* activation models, 4, 5, 6–9, 11, 14, 45; activation research, 9–11, 35; bathymetric map of, *118;* coastal, 13; eolian, 29; evolution, 6–9; geomorphic evolution, 45; Grand Sable, 45; isostatic rebound in, 13; locations, *2, 38,* 45–46, *53,* 109, 113, 133; Moran, 51, *53,* 114, 117; Nodaway, 46; perched, *44,* 45, 57; regressive, *44;* renewed, 50; space-time relationship within, 117; transgressive, *44;* variations, locational, 56. *See also* embayment; foredunes; sand

dunes: ages of, 31, 34, 42, 45, 46, 113, 137; archaeological sites with, 9–12, 21, 33–34, *129, 133, 134;* backdunes, 47, 48, 55, 56; barrier, 43, 47, 56; basswood and, 41; bathymetric map of, *118;* changes over time, 4–5; coastal, *2,* 4, 8, 32, 37, *38,* 137; Cowles' five classifications for, 41; cultural material, 33; dating, 43, 48; dynamics, long-term, 1; embayments, 34, 52, 56, 117, *118,* 120; embryonic, 41; erosion of, 33; established, 41; evolution, 59; field data collection, 12–13; geological deposits and, 83; geological events and, *19, 61,* 62, 73; geological sites and, 11, 12, 40, 64, *141, 143, 145, 147;* geological studies of, 3, 4, 5, 13, 15; growth, 43–45, 48, 54–56; growth, size theories based on lake levels, 46–47; hinge line and shape of, 42; interior, 6, 7, 8, 24, 32, *38;* isostatic rebound and growth of, 13, 16, 51, 108–10, 122; ISTEA buried site taphonomy/dune activation and cycling project, 83; lag, 5, 7; lag deposits, 29, 105, 120, 122, *126,* 127, 138, 142; lag pavement, 27, 31, 33, 34, 127, 133; Lake Michigan, history of, 47–48; Lake Michigan coastal zone, 137; largest, 13, 43, 120, 140; management, 1; marram grass and, 41, 43; model, perched, 45; morphology, 41, 43, *44, 47,* 48, 52, 144; morphology, parabolic, 49–50, 56, 98, 103, 105, 107; Native American uses of, 1; oldest Michigan, 34, 42, 45; OSL dating of, 46, 111, 114, 119, 125, 138; perched, 31, 32, 42, 43, *44,* 45, 50, 57; precontact, 1; primary, Scott's, 42; rapid uplift and, 133, *134;* rare Nipissing, 119; regressive, *44;* research, early, 40–43; research, modern, 32, 37, 43, 45, 49–57, 63; sand sources, 41, 42; Saugatuck, 38; sensitivity to human changes and, 1, 4; stabilization, 1, 17, 33, 46, 51; swales, 28, 34, 109; transgressive, *44,* 50; uses, 1–2; wandering, 41; Wisconsin geology and, 64. *See also* dunes, morphologic characteristics; foredunes

dunes, morphologic characteristics: blowout, 29, 42, 43, 48, 56, 95–99, 103, 107; domal, 43; Lake Michigan shoreline, *44,* 137; linear, 49; onlap, 49; overlapping, 43, 56; parabolic, 35, 42, 43, 49, 56, 103; ridges, 43, 56, 57, 117, 120; shadow, 43. *See also* foredunes

E

Early Woodland Period, 29, 61. *See also* Late Woodland Period; Middle Woodland Period

Eastport site: age, 125; archaeological site, *18,* 30; Archaic, 125, 140, *141,* 142; chert, 30; coring problems, 35; dunes, 30, 113, 140, *126;* Greenman survey of, 20; growth hiatus, 125–26; importance of, 35; as a lag pavement, 34; lithic assemblage, 30; oldest dunes, 113; OSL dating, 125; UMMA excavations, 30

Ekdahl-Goodreau site (Seul Choix), 12; Arbogast at, 91; archaeological site, *18;* basal sands, 91; ^{14}C dating, 24, 34, 89, *90,* 91; ceramics, 34, 91; destruction at, 89, 91; discovery of, 89; dating, 30; dunes, *53;* eolian activity, 25; GeoProbe core sampling, 13, 52, *90,* 91; importance of, 22, 35; location and

description of, 89, *89,* 91; Lovis at, 91; Monaghan at, 91; occupations, 23, 24, 35, 125; OSL at, 91; paleosols, 91; rebound effects at, 25; at Seul Choix point, 24; stratigraphy, 25, *90,* 91; UMMA excavation of, 23, 24, 89, 91; UMMA survey of, 22; unpublished reports, 24, 83, 89, 91; Woodland period, 24, 89, 91, 108, 125
embayment, 34, 56, 117, *118,* 120; Algoma, 74, 75; Grand Marais, 42; Holocene, 51, 52
Emmet County, 19, 28
Entisols, 47
eolian processes, 37, 39, *40;* activation, 7, 29, 108; archaeological sites and, *18,* 31, 33, 34; features, human use, 6; lacustrine deposits and, 6, 49, 50, *85, 86, 87, 126;* Lake Michigan, 49–51, 56; landforms, 17, 146; sand deposits, 8, 48, 105, 146; supply variations, 117. *See also* sand; stratigraphy
Erie. *See* Huron-Erie Lobe
erosion, 29; beach, 73–74; bluff, 4, 8, 43, 45, 63–64, 73, 119, *131–32;* coastal, 122; gravel and, 5; archaeological site, 15, 33; dune, 40–41, *41,* 42, 48–49; lag deposits, 142; lake evolution from, 66, 68–69; measurement tools, 48; outlets and, 66, 78, 81, 115; sand supplied by, 119, 120, *129;* shoreline, 32, 73, 76, 82, 116; topography dictated by, 62; transgressions and, 64; uplift, 76, 77, *79;* vertical dune face, 25; wave, 120
Eschman, D. F., 43, *67*

F
Farrell, J. P., 45
fauna: deer, 88, 92, 98; Winter site, 84–85, 88
Fenelon Falls outlet, 71
Fisher, T. G., 49, 50
Fisher Lake site, *18,* 32
Fisherman's Island Campground, 100, 138, 144, *147*
Fisherman's Island State Park site: age, 51; archaeological sites, 26, *100,* 144; campground, 100, 138, 144, *147;* dunes, 51, *53, 100,* 113, 117, 138; eolian landforms, 146; GeoProbe sampling, 13, 26, *100;* Holocene embayments, 52; MSUM survey of, 28–29, *139;* OSL dating, 138, 146; Solomon Seal stratigraphy, 12, 29, 100–102, *102,* 130; transect, *100,* 113, 117, 146, *147;* Young State Park and, 13, 26, 138. *See also* Whisky Creek site
Fitting, James, 10, 22
flint, 92, 105, 128
flooding, *19,* 64, *121*
floodplain, 121, *121*
food, cultigens, *19*
foredunes, *3, 89, 93, 99,* 105, *129,* 138; evolution, 17, 34, 37, 41, 42, 43, 48; lake level cycles and formation of, 43; models, 43, 45–48, 51–52, 82, 116–17, 120, 124, *131–32, 134;* as primary dune form, 42; sand sources, 42; stratigraphy, 45; swales, 34, 109, 110, 140, 144; troughs with archaeological artifacts and, 27, 28, 29, 34, 142
Fort Hill site, 19
Fort Wayne outlet channel, *60, 61*
Frazer, G. S., 45
Fritz Trail site, 144

G
Garden Peninsula, 21, 57, 108, 109, 110; archaeological sites, 15, 22, 23, 88. *See also* Spider Shelter site; Winter site
geochronological modeling, 30, 45
geomorphology: coring and, 30; Cowles' five dune divisions, 41; data models, 3, 14, 45, 49, 72, 73; dune evolution and, 6, 41–49, *44,* 51–53, 59, 144; gaps, 13; Holocene, 51, 59; lake evolution and, 64, 65; Lake Michigan, 16, 52, 56, 57, 73, 105; Lake Michigan basin, 62, 64, 65; Lake Michigan coastline, 5, 12, 30, 57, 59–60, 62, 69, *143;* Lake Michigan vegetation, 41; Lake Michigan-Huron, 7, *75;* lake terrace, 8, 42, 46–47, 49; linear dunes, 49, 109, 142; map units, 49; Mt. McSauba site, 28; onlap dunes, 49; paleosols and, 48; parabolic dunes, 49–50, 56, 98, 103, 105, 107; processes,

73; rebound and, 69; research, modern dune, 32, 37, 43, 45, 49–50, 57, 63; research variance, 15; shadow dunes, 49; variables affecting foredunes, 48

GeoProbe: core sampling, 11; difficulties with, 31; dual-tube, 52, 154, 155; OSL and, 52, 101; stratigraphy and, 52, 86, 101. *See also* Ekdahl-Goudreau site; Fisherman's Island State Park site; Solomon Seal site; Torch Bay Nature Preserve; Winter site

Getz, James R., 89

Glacial Grand River, *60*

Glenn A. Black Laboratory of Archaeology, 13

Goldthwait, J. W., 71, 81

Grand Marais Embayment, 42

Grand Sable, dunefield, 45, 46

Grand Traverse Bay, 19, 26, 27, *118*

Grand Traverse County, 26

Grand Valley State University, 20

Grassmere-Elkton, *61*

Great Lakes Basin: channels, 140; "climate-controlled," 67; deglaciation, 69; hinge line, 71–72; Holocene adjustment, 16; Lake Nipissing, 30; lake stages, *19*, 64, 66–67, *67*; modern configuration of, 81, 116; "outlet controlled" framework, 65, 67, 72, 78; outlets, 60–61, 66, 72; post-glacial, 4, 6, 7; rebound, 68; as a single lake, 81; stabilization, 14; transgression, 72, 81; uplift, 12, 68, 69, 137; water levels, 60, 62, 65, 66–67, *67*, 72. *See also* shoreline, Michigan

Greatlakean advance, *19*

Greatlakean till, 85

Greenman, Emerson F., 19–20, 29

Griffin, J. B., 30, 91

groundwater, 4, 7

H

Hamlin Lake site, 32

Hansen, Edward C., 47–48, 50, 103

Heins Creek site, 22

hinge line, 12, *60*, *70*, *79*, 109, 110, 117, 138; Algonquin and Nipissing, *112*; archeological sites, 34, *112*, 130; defined, 71–72; dune shape and, 42; Goldthwait concept of, 71–73, 81–82, 116; origin theory, 71; uplift and rebound affected by, 23, 42

Hinsdale, Wilbur B., 18, 20, 31

Hoffmaster, P. J., 50

Hoffmaster State Park, 48

Holland dunes, 8, *38*, 47–48, 113

Holland Paleosol, 48, 50, 56, 57–58, 125, *131*; dune formation and, 114, 120

Holmstadt, Jennifer, 96, 144

Holocene: conditions, 16; embayments, 51, 52; lake level variations, 80; lake hydraulics, 67; Lake Michigan evolution in, 51, 65, 78, *86*, 116, *131–32*, 137; lake sequences, 60, 64–67, *67*, 77; shoreline development, *131*, *134*; time stratigraphy, *61*

Holocene, early, 8

Holocene, late: beach evolution, 72, 79, *129*; "climate controlled" lake levels in, 76, 77, 78; dune formation, 115, 117, *131–32*; isostatic rebound, 78; lake level changes, 77, 78, 80, 115; lake sequences, 51, 64, 65, 77; outlet controls, lake, 78; recessional beaches, 89, 93; regional climate change, 59; regressions, 76, 81; sand deposits, 119; shoreline erosion, 4, 78; static lakes, 77; subsidence, 116; transgressions, 72, 76, 81, *86*; uplift, 116

Holocene, middle, 77; beaches, *79*; cultural horizons, 86; dune formation, 13, 51, 74, 111, 113, 115, *131–32*; lake evolution, 58; lake level changes, 80, 81, 115; Nipissing Transgression and, 52, 78, 81, 115; shoreline erosion, 4, 82; subsidence, 116; uplift, 57, 79, 116

Hough, J. L., 30, 71

Hughes, J. D., 45

Huron-Erie Lobe, 60, 64

Huron-Manistee National Forest, 10

Huron lake basin, 7, 14, 65–66, 71

Huron occupation, 22

hydrographs, 45, 74, *75, 75*, 76, *76*, 121

Hypsithermal warming, 8

I

Indiana Dunes National Lakeshore, 33, 34,

48, 57
Indiana Dunes National Recreation Area, 20
Indiana University, 13
Initial Woodland Period, 87. *See also* Middle Woodland Period, Laurel
Inland Waterway Project (MSUM), 27
Interstade, *61*
Intra-Glenwood low (Lake Michigan), 61
Inwood Creek site, *18,* 26, 29, 146
isostatic depression, 68, 69, 115
isostatic rebound, *59;* defined, 4, 69; beach growth and, 69, 71; Brevort-St. Ignace area, *75;* Chicago, 69, *79,* 81; dune growth and, 13, 16, 42, 51, 108–10, 122, 130; Holocene, 69, 78, *79;* lake levels and, 4, 9, 12, 16, 66, 68, *70,* 71; Lake Michigan, 69, 72, *75–76,* 76–77, 108, 116; Lake Michigan-Huron, *75;* locales, 72; North Bay outlet, post-glacial, 68; radiocarbon dates and, 107; rate variability, 23, 59, *60,* 69, 71; Sault Ste. Marie, 81, 115; Scott Point, 25, 109; Summer Island, 23
isostatic uplift, 12, 65, 66, 71, 108; deglaciation and, 69; lake levels and, 69, 77; measurements of, 68–69, *70;* modern, 68; post-Nipissing stability and, *132;* rates, Lake Michigan, 23, 69, 115, 116; ridge formation and, 117; sand deposition and shore, 109; sand supply affected by, 110; Scott's, 42; stability of, 69; Summer Island, 23; Upper Great Lakes, 69–70, *70*
ISTEA buried site taphonomy/dune activation and cycling project, 83. *See also* Antrim Creek Natural Area site; Eastport site; Fisherman's Island State Park site; Ludington State Park; Manistique Sand Pits; Moran dunefield; O'Neil site; Portage site; Porter Creek site; Summer Island site; Torch Bay Nature Preserve site; Wilderness State Park dunefield; Wycamp Creek site

J
Jackson Quarry site, 8
Juntunen site, Bois Blanc Island, 22; archaeological site, *18,* 21; burials at, 21; ceramics, 91, 93–95; Late Woodland, 25

K
Kalamazoo morainal system, 62
Kalkaska County, 26
Kent, B., 25
Kinietz, Vernon, 20
Kirkfield low phase (Lake Algonquin), *61,* 71
Kirkfield outlet, 71

L
lacustrine deposits (littoral), 51, *90,* 113–14; eolian sand mobilization and, 6, 49, 50, *85, 86,* 87, *126;* glacial, *44, 86*
Lake Algoma: Algoma II, 81; formation of, 66; Holocene, *67, 74, 75, 79;* Lake Michigan dune levels and, 74; levels, *80,* 81; receding, 74; short duration of, 66; stages, *19,* 29, 46, *61,* 115, *121;* uplift, *79*
Lake Algonquin: deposits, 66; discharge outlets, 66; low levels, 7, 66; phases, *19, 61,* 70, 71
Lake Arkona, *61*
Lake Border morainal system, 62
Lake Charlevoix, 26
Lake Chicago: Calumet phase, 61, 68; discharge outlets of, *60,* 64, 66, 68; Glenwood phases, *61;* Intra-Glenwood low, *61;* Two Creeks low phase, 61
Lake Chippewa, 66, *67*
Lake Chippewa-Stanley, *19, 61,* 65, 66, *67,* 140
Lake Erie basin, 64
Lake Grassmere, 7, *61*
Lake Hough, 66
Lake Hough-Stanley, *61*
Lake Huron: discharge outlets of, 66; Early Algonquin, *19, 61;* formation, 78; phases, 81; rivers draining into, *60;* uplift changes to, *79,* 115; water levels, 65, *67,* 70, 81, 115
lake levels: climate-controlled model for, *67,* 81; climatic effects on, 68; cyclical fluctuations in, 9, 48, *60,* 62, 63, 119;

deglaciation uplift and, 69; factors producing changes in, 68, 80–81; Holocene, 81, 115–16; ice-margin fluctuation and, 62, 64–65, 68, 69, *70, 71, 72*; isostatic rebound and, 68; modern, 81, 116; outlet channels and, 68; "outlet controlled" framework, 65, 67, 72, 78; outlet erosion and, 68; sand supply and high, 117; transgressive-regressive events, 81, 116; water levels, 65, 115; water volume changes and, 68. *See also* Lake Huron; Lake Michigan; Lake Nipissing; Lake Superior

Lake Michigan: beach transgression, 82, 116; Chippewa low phase, 45; climate controls on water levels of, 81, 116; discharge outlets, 59; dune ages, 52–57, 137; dune formation, 62–64, 81, 82, 116–17, 137; dune history, 47, 55–58, 74, 117, 137; dune models, 45; dune types, 56–57; dunefields, 133; Early Algonquin phase, *19*; eastern shore, 32, 45–47; eolian deposits and, 138; erosion, 82, 116, 138; geomorphological evolution of, 16, 52, 56, 57, 73, 105; hinge line, 137; Holocene evolution of, 51, 65, 78, *86*, 116, *131–32*, 137; ice lobes, *60*, 62, 64; isostatic rebound, 69, 72, *75–76*, 76–77, 108, 116; isostatic stability of, 69, 81, 82; Nipissing transgression, *131–32*; phases of, 81; radiocarbon dating, 57; regression and, 116; rivers running into, 142, 144; shoreline, 22, 56, 82, *131*; shoreline, northeast, 26–32, 51–52, 111–14; shoreline subsidence in, 72–77, 81, 82, 116, *131*; southeastern areas, 32–33, 50–51, 56, 57; stabilized, *13, 132*; stratigraphy, 51, *53, 57*; uplift, 57, 74, 78, *79*, 81, 82, 116; water levels, 62, 65, *67*, 74, 81, 115–16. *See also* Lake Chippewa-Stanley

Lake Michigan basin, 16, 19, 21, 57–58, 59, 65

Lake Michigan-Huron basin, 7, 59, *76*; hydrograph, 75, *75*; outlets, 60, 62, 65, 81; rebound, *75*; separation, 81

Lake Michigan Lobe, *60*, 62, 64

Lake Nipissing: beach altitude, 113; beach gravels, 52; causes of transgression, 78; discharge outlets of, 66, 78; dune formation correlates with, 113; dunes, 42, 45, 51, 117, 119; dune formation during, 46, 52, 55, *131–32*; eolian formation, 43, 45, 117, 119–20; formation of, 66, 78; high-water phase, 7, 8; high stand of, 46, 47, 49, 50, 55–56, 57; lake levels of, 66, 78; location, 78; Nipissing I phase, *61*; Nipissing II phase, 48, 49, *61*, 81, 115; Nipissing II high stand, 45, 46, 47, 48, 49, 50, 56, 81; Nipissing II transgression, 49, *67*, 70, 71, 115; phases, 70, 101; post-glacial, 30; rare dunes, 119; regression of, 120; shoreline, *93*; shoreline erosion, 120; timing of, 78; transgression, lake level changes and, 63, 66, 78, 115; transgression, Lake Michigan basin, *19*, 49, *61*, 65; transgression and coastal erosion, 120, *129, 132*; uplift, effects on, 66, 68, 75

Lake Ontario, 65, 78

Lake Saginaw, 7, 30, 64. *See also* Saginaw Bay area

Lake Stanley: formation of, 7, 65–66; Lake Chippewa-Stanley, 19, *61*, 65, 66, *67*, 140; as low-level lake, *61*, 66, 140; Nipissing transgression and, *67*, 70, 71; stability, 66

Lake Superior, 2, *38, 53*; basin, 60; climate controlled, 67; discharge outlets, 60, 62, 65, 81; formation of, 78; Holocene, 60, 66, 78, 115; hydrologic properties, 78, 115; north shore Upper Sable dunefield, 45; phases, 60, 65, 66, 78, *80*, 81, 115–16; separation, 81, 115; southern shore dune growth, 46, 119; water levels, 65

Lake Warren, *61*

Lake Whittlesey, *61*

land use practices, 130, 138

landforms, 17, 34, *75, 132*, 146

L'Arbre Croche. *See* Portage site

Larsen, C. E., 74, 77

Late Woodland Period: archaeological sites, 3, 24–25, 32, *97*, 130, 142; ceramics, 31, 34, 91, *97*, 105–6, 109; dune formation, 127, 128; hearth, buried, 3, 133; Juntunen Phase, 25, 91; key cultural events of, *19;* key geological events of, *61;* Mackinac phase, 25, 28; Nipissing age features, 142; occupation, 24, 34, 89, *89,* 91, 105–6, *134;* OSL sampling, 91, *97;* Petoskey area, 109; settlement patterns, 121. *See also* Camp Miniwanca site; North Manitou Island site; Mt. McSauba site; O'Neil site; Portage site; Solomon Seal site; Wycamp Creek site. *See also* Early Woodland Period; Middle Woodland Period

Laurentide Ice Sheet, 16, 17, 59; basins free of, *19;* crustal subsidence and, 72, 73, 82; glacial sediment and, 42; lobes, *60, 62, 64, 68;* mantle material, 72; margin fluctuation, 62, 64–65, 68, 69, *70,* 71, 72. *See also* isostatic rebound

leaching, 95, 123

Leelanau County, 27; Site 20LU2, 31; Site 20LU24, 31

Leelanau Peninsula, *18,* 21, 73

lithic assemblage, 21, 30, 88, 89, 101, 106; coastal scatters, 128; Woodland, 24

Little Ice Age, *19,* 77, 81, 116, 122

Little Point Sable, dunes, 57

Little Traverse Bay, 27, 117, *118*

littoral deposits, 6, 56, 63, 117, *118. See also* lacustrine deposits

lobes (glaciers), 16, *60, 62, 64, 68*

Loope, W. L., 7, 8, 45–46, 49, 50, 119

Lovis, William A., 7, 31, 65, 74, 77, 80, 91, 93; at Camp Miniwanca site, 103, 107; at Mt. McSauba site, 96

Ludington area, 16, 32, 58, 66, 113; dunes, *38,* 56; sand pits, *3, 5*

Ludington State Park, *18,* 32, 51–52, *53,* 113–14

M

Mackinac. *See* Straits of Mackinac

Mackinac County. *See* Moran dunefield; Scott Point site

Mackinac Ware, 94, 95

Manistee-Ludington dunes, 8, *38*

Manistee morainal system, 62

Manistee Moraine, 42, 43, 62

Manistique dunes, 113, 117

Manistique Sand Pits, 51, *53,* 54, *76,* 112, *167*

Manistique sand quarry, *112,* 166

Manitoulin Island, *79*

Martin, S. R., 98

Martin, Terrance, 84, 85, 88

Mason, Ronald J., 10, 22, 23, 85

Mason County, 83

McGeach Creek, 146

McPherron, Alan, 21

Medieval Warm Period, 8, *19,* 77, 81, 116

Mero site, 22, 87

Michigan: archaeological sites, 15–16; coastline evolution, 17; cultural periods, *19;* geological events, *19;* Lower Peninsula, 16, 27, 65; prehistory, time periods, *19;* Upper Peninsula, 16, 27, 45, 74, 88; Upper Peninsula dunes, 113, 114, 138

Michigan Department of Transportation (MDOT), 3

Michigan/Huron basin, 6, 7, 8, *19, 61*

Michigan State Parks, 29

Michigan State University (MSU), 21; Department of Anthropology, 28, 29

Michigan State University Museum (MSUM), 10; excavations with ^{14}C dating, 30; Inland Waterway Project, 27; research sites recorded by, 26, 27–29, 30–31, 92, *93*

Michigan Technological University, 20

Mid-Holocene age (Nipissing): archaeology, 129; dunes, 113, *141;* largest dune sizes, 13; shoreline, *84, 93*

midden sites, 85, *85–86,* 87, *121*

Middle Woodland Period, , 87; archaeological excavations, 24, 108–9; ceramics, 88; Door Peninsula deposits, 125; high-water stage, 85; Hopewell, 130; Initial, 87; Laurel, 23, 24, 26, 120, 125, 130; occupation, coastal, 127, *134;* occupation dates, 22, 24, *121;* stratified or buried component locations, 108–9, 125; Wisconsin, 125; Wycamp Creek site, 108, 109, 125. *See also* Ekdahl-Goudreau site; Fisher Lake

229

site; Middle Woodland Period, Laurel; Rock Island site; Summer Island site; Winter Site. *See also* Early Woodland Period; Late Woodland Period
Milner, Claire McHale, 92
Monaghan, G. William, 7, 65, 74, 77, 80, 91, 96
Morainal systems, 42, 43, 62, 64, 142; arcuate, *118*
Moran dunefield, 51, *53*, 114, 117
morphology, dune, 41, 43, *44*, 47, 48, 52, 144
mound burial, *19*
MSU. *See* Michigan State University
MSUM. *See* Michigan State University Museum
Mt. McSauba City Park, 27
Mt. McSauba site, *53*, 117; AMS used at, 98; Arbogast at, 96; archaeological site, *18*, 27; artifacts, 98; buried soils, 98; ^{14}C dating, *99*; ceramics, 28, 96, 98; coring problems, 35; dune blowout, 95, 98; dune stratigraphy, 51, *97*, 98–100, *99*; eolian sands, 99; fauna, 98; geomorphology, 28; hearths, 98; Holmstadt at, 96; importance of, 35; location and description, 95–97, *96*, 98; Lovis at, 96; MSU work at, 10; MSUM survey of, 27–28, 98; occupation, 98; OSL dating, 35, 96, *97*, 98–99; paleosols, 96, *97*, 98, *99*; reconstructed history, 99; sherds, 96, *97*; site visibility and wind conditions at, 98; stratigraphy, 98, 99, *99*; unpublished report on, 83; Woodland, 27, *97*
Muskegon: coastal areas, 4, 13, 21; dune-based sites near, 15, 32, 46; Holocene dunes, 113; isostatic areas near, 16

N

National Environmental Policy Act, 20
National Forest system, 10, 20
National Historic Preservation Act, 20
National Park Service, 20, 31
National Park Service Midwest Archaeological Center, 10, 20
National Register of Historic Places, 92, 93
Native Americans, 1, 19

near-surface sites, 52, 128, 133
near-surface wind velocity gradient, *39*
Neff's A. H. Map site, 26
Neville Public Museum, 21, 22
New Toilet Area site, 26
Niagara River, outlet channel, *60*
Nipissing age: archaeological sites, 133, 140; coastline features, 28, 140; dating, 138; dune features, 28, 74; dune growth, 56, 137; largest dune sizes, 13, 43, 120, 140
Nipissing-Algoma, dune complex, *75*
Nipissing-Algoma high-water phase, *19*
NMU. *See* Northern Michigan University
Nodoway dunefield, 46
Nordhouse Dunes, 8, *18*, 32
North Bay, Ontario, 65
North Bay outlet, 65, 66, 68, 71, 78, *79*, 88
North Bay site: ^{14}C dating, 66; ceramics, 24, 87; Plain sherds, 87; projectiles, 88; Punctate vessels, 87
North Manitou Island site, *18*, 30–31, 34, *112*
North Manitou 3 Site, *18*, 31
Northern Michigan University (NMU), 20, 22, 23, 25, *93*, 93–94; ceramics identification, 92
Nugent Quarry, 8, 30, 88, 101, 106, 144

O

obsidian, 88
Oceana County, 20, 83, 102, 106
Office of the State Archaeologist (OSA) (Michigan), 10, 24, 92, 93, 103
Olson, J. S., 42–43, 46, 47, 48, 57, 120
O'Neil site, 11, 25, 27, *100*, *112*, *118*, *147*; archaeological site, *18*, 35; ^{14}C dating of, 28, 34, 109, 138; ceramics, 34; eolian dunes, 26; hydrograph, *76*; occupation, 28, 35; strata layers, 35, 108, 123, 124, 138; swales, 28, 34, 109; uses for, 125; Woodland, 27, 28, 34
Oneota occupation, 25, 94, 95
Optically Stimulated Luminescence. *See* OSL
OSA. *See* Office of the State Archaeologist
O'Shea, John, 10
OSL (Optically Stimulated Luminescence):

accurate age estimates with, 111; Antrim Creek area and, 130, 138, 144; archaeological site dating with, 10, 111, 125; dating, 7, 8, 10, 11, 13, 24, 46, 74–75, 83; dune history reconstruction using, 46, 111, 114, 119, 125, 138; Eastport area dating by, *126*; eolian sand deposit dating by, 117; Garden Peninsula and, 57; GeoProbe and, 52, 101; Holland backdunes, 47–48; Lake Michigan basin, 52–55, *53–54*; Nodaway dunefield, 46; Nipissing beach sand dating, 75, *76*; Petoskey State Park, 49; Probability Density Diagram and, 120, *121*; processing, 13; sampling, 14, 35; Sleeping Bear Dune area, 31; techniques, 35; Torch Bay area and, *126*, 138, *141*; U.S. 31 area, *143*. *See also* Camp Miniwanca site; Ekdahl-Goudreau site; Fisherman's Island State Park site; Mt. McSauba site; Solomon Seal site; Winter site, *86–87*

Ottawa National Forest, 10

outlet, *60, 64, 66, 68*

P

Paleo-Indian period: archaeological sites, 19, 23, 144; chert, 144; cultural events, *19, 33, 61*; earliest, *19*; geological events, *71*; lake stages, *67*; in Michigan, 19; subsistence patterns in, 5, 21, 36–34, 64, 126

paleosol, 83; archaeological features and strata, *52*, 86, 105–8, 120–21, 123, *129, 134*; buried within dunes, 51; dune stabilization and, 111; Early-Middle Woodland sites with, 134; ephemeral, *134*; Holland Paleosol, 48, 50, 56, 57–58, 114, 120, 125, *131*; Indiana Dunes area lacking, 33; Ludington area, 5, 56; modern Lake Michigan bluff, *131*; Muskegon area, 35; Nipissing dune, *131–32, 133*; over-thickened, 48; poorly developed, 50; Sable Creek Soil and Sleeping Bear Dunes site, 45; uplift samples, *129*; well-developed, 91. *See also* Antrim Creek site; Camp Miniwanca site; Eastport site; Ekdahl-Goudreau site; Mt. McSauba site; Porter Creek site; Solomon Seal site; Torch Bay site; Van Buren State Park site; Winter site

Papworth. M. L., 20, 30

PDD. *See* Probability Density Diagram

Peebles, Christopher S., 10, 17, 20

Pine River Channel site, 26

peninsula. *See* Door Peninsula; Garden Peninsula; Leelanau Peninsula; Michigan

Peske, G. R., 25

Petoskey, 27; archaeological sites, 109–10; buried strata, 110; dating, 108, 109; hydrograph, *76*; lake levels and, 74; Portage site, *18*, 109; site preservation, 110

Petoskey State Park: cyclic dune building at, 49, 114, 117; dating, 49, 113; dune evolution, 49, 124; dune forms, 117; dune swale, 34, 109; dunefield, 27, 109, 114, 117, *118*; eolian sand, 124; geomorphic map units at, 49; research, 12, 49, 51; sand supply reduction, 114. *See also* Portage site

Pi-wan-go-ning (Pe-Wan-Go-Ning) Quarry site, *18*, 144

Pigeon River State Forest, 144

Pine River Channel site, 26

Point Arcadia site, *53, 112*, 113, 114, *118*

Point Patterson. *See* Scott Point site

Point Scott. *See* Scott Point site

points, Wisconsin stemmed, 88

Port Huron, 59, *61*; drift underlies, 68; hydrograph, 75, 76, *76, 121*; modern lake levels at, 66–67, 78, 81, 115, *132*; outlet, 66, 68, 74; uplift, 74, *79*, 120, *121*

Port Huron advance, *19*

Port Huron outlet channel, 59, *60*, 66–69, 71, 81; erosion, 78, 81, 115, *132*; lake levels controlled by, 74, 76, 115; sill, 78

Port Huron Stade, *61*

Portage site: archaeological site, *18*; in a dune swale, 109; L'Arbre Croche area, 11, 27, 34, 108–9, 123, 125; Petoskey State Park area, 109, *112, 118*; radiocarbon dates, 109; Woodland, 11, *18*, 27, 34, 108–9, 123

Porter Creek South: archaeological site, *18*, 32, 108, 109; ceramics, 109; paleosol, 108; South bank site, 12, *112*, 123, 125; stratified dune deposits, 32
post-Algoma, *19*, 26
post-Nipissing: dune growth history, 137, 140, 146; Eastport dune site, 30, 140; fluctuations, *132*; landforms, 34, *75*, *132*; recession, 140, 142; regressional phase, 99, 115, 127, *132*; Solomon Seal land forms during, 29, 146; stability, *132*, 146; U.S. 31 case studies and, 140; Whisky Creek, 26, 29; Winter site beach, 24
Prahl, Earl, 24, 91
"premodern" flood phase, *19*
Probability Density Distribution (PDD), 54, *54*, 55, 120, *121*
projectiles, 88

Q

quarry sites, 8, 110, *112*, *144*. See also *quarries by name*
quartzite, 88
Quimby, George I., 23, 89, 92

R

radiocarbon dating, 3, 6, 10, 12–14, 24, 34, 35, 108; buried soils, 119; calibration, 11; coastal dune, 21, 43, 45–52, 57; cultural phases, *19*; dune formation, 111, 114, 119; dune, 45, 74; erroneous, 111; Holocene lake levels, *67*, 81; lithic assemblage, 106; MSU excavation, 28; MSUM excavation, 30; published data, 83; stratified site compilation of, 11; Woodland hearth, 3. See also Camp Miniwanca site; ceramics; Ekdahl-Goudreau site; Grand Sable; Mt. McSauba; North Bay; O'Neil site; Petoskey; Portage site; Saginaw Bay area; Scott Point site; Sleeping Bear Dunes; Snyder's dune; Winter site. See also AMS, dating
resources, plant and animal, 16, 64, 127
Richner, Jeffrey, 84, 85, *86*, 87
Rock Island site, 22, 109, 125
Rosy Mound site, 8

S

Sable Creek Soil, 45
Saginaw Bay area: archaeological sites, 38, 106; chert, 106; dunes, *38*; dune dynamics in, 7; floodplain activity in, *121*; Holocene beach uplift in, 79; hydrograph, *75*, *76*; lake levels, *61*, 64, *75*, *76*, *77*, *121*; Lake Saginaw, 7, 30, 64; lowland, *60*; Saginaw River basin, 106
St. Clair River, 59–60, *60*, 66, 68
St. Ignace area, 19, *75*, *75*, *76*
St. Mary's River, 60, *60*, 66
sampling: archaeological site, 14, *112*; Camp Miniwanca, 105; dune strategies, 50–52; equipment, 13; Holocene dune, 111; U.S. 31 area, 144; vertical techniques, 119. See also AMS; GeoProbe; OSL
sand, 83; creep, 39, *40*; dating, 43, 45; eolian, 3, 7, 8, 17, 39–41, *40*, 47; Lake Michigan ice lobe drift and, 62; saltation, 39, *40–41*; supply sources for dune formation, 9, 39, 41, 42, 43, 62, 63, 78; supply variations and high water, 117, *118*, 119; transport of, 40
sand sheets, 63, 88, 109, 110, 138
Saugatuck dunes, *38*
Sault Ste. Marie, 46, 67; isostatic rebound, 81, 115; rapids, 60, 66, 78, *78*, *80*, 81, 115
Schoolcraft County. See Ekdahl-Goudreau site; Manistique Sand Pits
Scott, I. D., 42, 117
Scott Point site (Point Patterson, Point Scott): AMS dating at, 93, 94; archaeological site, *18*, 25, 35; artifacts, 92; beaches and dunes, *93*; Buckmaster at, 22, 25, 92, 93, 94; ^{14}C dating, 25, 34, *94*, *94*; ceramics, 25–26, 34, 35, 92–94, *95*; Cleland at, 92; collection displays, 92; coring problems, 35, 93; eolian sand, 93, 94–95; excavations, 92, 93; fauna, 92; hearths, 35; importance of, 23, 25, 35; location and description, 26, 92, *93*; MSUM work at, 92; on National Register of Historic Places, 92, 93; NMU work at, 10, 23, 25, 92, 93; OSA and, 92, 93; paleosols, 93; rebound effects at, 25; sherds, 82; stratigraphy, 25,

93–94, *94;* surface debris, 92; UMMA survey of, 22, 25, 92
sediment: glacial, 42; transport, 40
Sella, G. F., 74
settlement, human, 17, 21, 23; climate, dune growth, and, 64, 77, 126–27; dune growth hiatus and, 126
Seul Choix. *See* Ekdahl-Goudreau site
Sheffield Centre for International Drylands Research, 13, 96, 98
sherd, samples: carbonized, 11; ceramic, 10, 35, 93; Juntunen Drag and Jab, 94, *95;* Late Woodland, *97;* Mackinac Ware, 94–95; North Bay Plain, 87; North Bay Punctate, 87; rimsherds, 87, 93–96, *97, 106,* 107. *See also* Camp Miniwanca site; Mt. McSauba site; Scott Point site
shoreline, Michigan, 5, 9; Algonquin, *100,* 142, *143, 145, 147;* Antrim Creek, 142, *143;* archaeology, 34–35, 74, 83, 98, 111; bluffs, 62, 64, 116, *131;* Charlevoix area, 34–35, *147;* complexes, *75;* configurations, *129;* dune-dominated, 73, 82; dune formation requirements and, 63, 114–15, 120, *129;* dune morphological types, *44;* dune strategies for, 50–52; dunes, northern basin, 55; eastern, 83, 109; embayments, 56, 120; environmental effects on, 4; eolian deposits, 88; erosion of, 32, 62, 64, 73–76, 82, 116, 130–*131,* 142; Holland Paleosol, 56; Holocene, 4, 78, 82, *93,* 98, 115, *129,* 130, 133; human settlement along, 126; isostatic uplift influences, 68–69; lake-level variation and, 129; Nipissing, *93,* 98, 126, *131, 141*–*43, 147;* northeast, 21; northern coastal, 138; northwest, 22; Petoskey area, 109, 114; post-Nipissing, *132;* preservation, 81, 130; processes create specific sites, 63, *129;* regression, 63, 73, *131;* sand sheets, 109–10; southeast, 11; southern, 32, 72, 83; subsidence, 72, 73, 82; Torch Bay area, *126, 141;* transgression, 63, 82, 116, *131;* uplift, *75,* 76, 81, *129,* 133, *134;* Upper Peninsula, 138; Whisky Creek area, 144, *145. See also* Scott Point site; Solomon Seal site; Winter site
Silver Lake State Park, 49, 50
Skegemog phase, 28
Sleeping Bear Dunes, 17, 21, 31, *38, 112, 118;* archaeological site, *18;* growth model, 45; National Lakeshore, 12, 20, 46, 51, 113; North Manitou Island sites in, 31; perched dunes, 43; Snyder at, 43, 45, 57; survey, 12, 28, 31, 57
Smith, Beverley, 98
Smith, Harlan I., 19
Snyder, F. S., 43, 45, 57
soil: alleviation and, 77; analysis, dating, 46; buried, 51–52, 98, 101, 122–25; climate effects on, 122; cyclical, 45; components, 123; dune formation dynamics and, 123–25; dune surface, 8; formation, 108, 123; occupation-organic site, 123–25
Solomon Seal site: age of, 29, 34; Archaic, 29, 34, 100, 130, 146; archaeological site, *18;* artifacts, 28, 29, 100, 101; bone, 101; chert, 101; coring, 13, 100, 101; dunes, *38,* 113, 117; eolian sand, 29, 102; excavations, 100, 101; first reported, 100; GeoProbe sampling, 13, 101, *102;* lithic assemblage, 101; location and description, 99, *100,* 101; MSUM excavations at, 29; Nipissing age features, 28, 34, 100, 101, 146; occupation, 100–101; OSL and, 100, 101–2; stratigraphy, 12, 29, 100–102, *102,* 130; Woodland, 29. *See also* Fisherman's Island State Park site
South Manitou Island: archaeological sites, *18,* 30, 31, 35, 109, *112,* 123, 125; ceramics, 31, 34; eolian dunes, 31; GeoProbe sampling, 11, 31; lighthouse, 1, 109, 123, 125; radiocarbon dating, 34–35; Sleeping Bear Dunes National Lakeshore survey and, 31
South Manitou 1 Site (20LU24), 31, 35
South Manitou Light Site 1, 109, 123, 125
South Point, Michigan, 138
Spencer, J. W., 71
Spider Shelter site, 22, 88
Spodosol, 8, 45, 47, 48, *102,* 114
State of Michigan Archaeological Site

Files, 91
storm deposits, 22, 63, 85, 122
Stony Lake, 102
Straits of Mackinac, 21, 26, 57, 59, *61*, 66, 69, 83; on Bathymetric map, *118*; outlet channel, *60*; survey, 19; uplift at, 57, 73, 79, 81, 116
stratigraphy: Antrim Creek area, 12, 130; archaeological site, 125, *129*, 130; dating and, *53*, 85–86, *86*, *91*; GeoProbe and, 52, 86, 101; Holland dunes, 47; Lake Michigan, 51, *53*, *57*; perched dune, 45; salient aspects of, 11–13, 125; shoreline uplift and, *129*; time and events, 61, 85, *101*, 130; uses for, 14, 51–52. *See also* Camp Miniwanca site; Eastport site; Ekdahl-Goudreau site; Mt. McSauba site; Scott Point site; Solomon Seal site; Summer Island site; Torch Bay site; Winter site; Wycamp Creek site
Sturdevant, J. T., 33
Sturgeon Bay Point, 51, *52*, 72, 114, *118*
subsidence, 72, 73, 82
subsistence patterns, 5, 21, 36–34, 64, 126
Summer Island site: age, 34; archaeological site, *18*; ceramics, 34; eolian dunes, 22–23; importance of, 22; location, 23; occupation, 23, 35; stratigraphy, 23, 109; UMMA survey and excavations at, 22; uplift and rebound rates, 23, 69, 71, 115, 116; Woodland, 108, 109, 125

T
Tague, G. C., 42, 48
taphonomy: archaeological site formation processes, 14, 15, 17, 21, 28, 110, *129*; archaeological site project, 35, 83, 110, 128, 133; archaeological site taphonomic processes and, 35, 126, 127, 128; coastal sites, 23, 110; dune models, 6; ISTEA buried site taphonomy/dune activation and cycling project, 83; site preservation and, 126, 127, 128
Telewski, Frank, 106
Thompson, T. A., 46, 49, *67*, 80
Timmons, E. A., 49–50
Torch Bay, 117

Torch Bay Nature Preserve site, 51, 126, 142; dunes, *53*, 56, *76*, *126*; GeoProbe sampling, 13, 52, 138; post-glacial channel, 140; transect, *126*
Torch Lake, Michigan, 138, 140
transgressions, 49, *67*, 70, 71, 115
transgressive dune systems, *44*, 50. *See also* Dunes
Trent River, *60*
Two Creeks low, *61*

U
UMMA. *See* University of Michigan Museum of Anthropology
University of Illinois–Chicago, 13, 101
University of Michigan, 19, 20, 21, 22–23, 25
University of Michigan Museum of Anthropology (UMMA), 9, 10, 20; coastal surveys, 22–23; Eastport excavation, 30; Ekdahl-Goudreau excavation, 23, 24, 91; Scott Point work, 22, 25, 92; Spider Shelter site work, 22
uplift: beach, 75, 76, 79, 86, 109, 117, *129*; dune and shoreline development in rapid, 133, *134*; erosion and, 76, 77, 79; Great Lakes basin, 12, 68, 69, 137; hinge line and, 23, 42; Holocene, 57, 79, 116; Lake Huron, *79*, 115; Lake Michigan, 74, 78, 79, 81, 82, 116; Nipissing transgression, 66, 68, *75*; Port Huron, 74, *79*, 120, *121*; shoreline, *75*, 76, 81, *129*, 133, *134*; Straits of Mackinac, 57, 73, *79*, 81, 116; Winter site, 74; Wisconsin, 4. *See also* dunes
Upper Mississippian, *19*. *See also* Oneota occupation
Upper Sable dunefield, 45
U.S. 31, case study: Segment 1-Torch Lake to T32N, 140, *141*, 142; Segment 2-T32N to North of Antrim Creek, 142, *143*, 144; Segment 3-North of Antrim Creek to North of Whisky Creek, 144, *145*; Segment 4-North of Whisky Creek to South Point, 144, 146, *147*
U.S. 31, shoreline area: Algonquin

shoreline, 142, 144; archaeological sites, 138, 140, 144, 146; Archaic age, 144, 146; bedrock exposed near, 144, 146; buried sites, 142, 144, 146; case study locations, *139, 145*; ceramics, 144; dating, 138, 142, 146; disturbance activities, 146; drumlin field, 142; dunes, 140, 144, 146; Eastport area, 140, *141*, 142; eolian features, 140, 142, 144, 146; follows the Algonquin terrace, 142, 146; follows the Lake Nipissing terrace, 140, 142; gravels, 146; highway corridor, 130, 142, 146; Late Woodland, 142; location and description, 138, *139, 143*; Nipissing dune near, 138, 140, 146; Nipissing shoreline, 140, 144; Norwood chert in, 144; OSL dates, 138, 140, 146; paleosols, 142, 144; quarry sites along, 144; right-of-way, 138; sampling, 138, 144; State of Michigan ownership, 144; transects, 138, 146; wetlands, 142

USDA Forest Service, 20

V

Van Buren State Park, dunes, 47, 48, 50, 113, 133

Van Dijk, D., 48

Van Oort, M. A., 47

Vreeland, F., 20

W

Warren dunes, *38*

Warren Dunes State Park, *38*, 50, *53, 112,* 133

water budget, and lake levels, 68

wave processes, 17

Wayne State University, 20

weather: drought and dune formation, 7, 122; weathering, 97, 127; wind and dune formation, 17, 34, *39*, 40, 48, 62, 78, 114

Western Michigan University (WMU), 20, 22, 23–25

Whisky Creek site: aceramic artifact deposit, 26; Archaic deposits, 29; archaeological site, *18*, 26; MSUM excavations at, 29, 33

Wilderness State Park dunefield, *53,* 56, *112,* 113–14, 117, *118*

wind: dune formation and, 17, 34, *39,* 40, 48, 62, 78, 114; velocity, 37, 39, 40

Winter site, 12; AMS dating, 87; archaeological site, *18*; architecture, 88; basal sands, 87; ^{14}C dating of, *53,* 85–86, *86,* 88, 108; ceramics, 24, 87, 84, 88; chert, 88; coring, 86–87; dunes, *53;* excavation by Richner, 84, 85, *86,* 87; fauna, 84–85, 88; GeoProbe sampling, 13, 52, *86;* hydrograph, *76;* importance of, 35; lithic assemblage, 88, 89; location and description of, *18,* 23, 83, 84, 84–85, 88, *112;* middens, 85, *85, 86,* 87; occupation, 35, 87–88; OSL dating, 87; paleosols, 88; projectiles, 88; stratigraphy, 28, 35, 85, *86,* 86–87, 88, 108; surface burial at, 108, 123, 124, 125; uplift and, 74; wind direction and, 34; Winter I and Winter II assemblages, 87; WMU excavations at, 22, 23–24, 25; Woodland, 84–85, 87, 88, 108

Winters, H. A., 73, 82

Wisconsin: archaeological sites, 10, 23, 73, 22; Door Peninsula sites, *18,* 22, 85, 87, 125; dune formation, 8; dune site research, 21; Rock Island site, 22, 109, 125; stemmed points, 88; uplift, 4

Wisconsin Period, 16, *19, 61,* 62, 65, 68, 77, *86, 100, 131*

Woodland Period. *See* Early Woodland Period; Initial Woodland Period; Late Woodland Period; Middle Woodland Period; Oneota occupation

Wycamp Creek site: archaeological site, *18;* coring problems, 35; geoarchaeology, 3, 6, 15, 26, 30, 108; Greenman survey of, 19–20; importance of, 35; MSU work at, 10, 20, 26; Woodland, 26, 109

Y

Yansa, Catherine, 106

Young State Park, 13, 26, 138

Younger Dryas climatic interval, 7